Evolving Pathways
Key Themes in Evolutionary Developmental Biology

Evolutionary developmental biology, or 'evo-devo', is the study of the relationship between evolution and development. Dealing specifically with the generative mechanisms of organismal form, evo-devo goes straight to the core of the developmental origin of variation, the raw material on which natural selection (and random drift) can work. *Evolving Pathways* responds to the growing volume of data in this field, with its potential to answer fundamental questions in biology, by fuelling debate through contributions that represent a diversity of approaches. Topics range from developmental genetics to comparative morphology of animals and plants alike, including palaeontology. Researchers and graduate students will find this book a valuable overview of current research as we begin to fill a major gap in our perception of evolutionary change.

ALESSANDRO MINELLI is currently Professor of Zoology at the University of Padova, Italy. An honorary fellow of the Royal Entomological Society, he was a founding member and Vice-President of the European Society for Evolutionary Biology. He has served as President of the International Commission on Zoological Nomenclature, and is on the editorial board of multiple learned journals, including *Evolution & Development*. He is the author of *The Development of Animal Form* (2003).

GIUSEPPE FUSCO is Assistant Professor of Zoology at the University of Padova, Italy, where he teaches evolutionary biology. His main research work is in the morphological evolution and post-embryonic development of arthropods, with particular reference to the role of segmentation in the Chilopoda.

Evolving Pathways
Key Themes in Evolutionary Developmental Biology

Edited by
ALESSANDRO MINELLI
GIUSEPPE FUSCO
University of Padova, Italy

CAMBRIDGE
UNIVERSITY PRESS

CAMBRIDGE UNIVERSITY PRESS
Cambridge, New York, Melbourne, Madrid, Cape Town,
Singapore, São Paulo, Delhi, Mexico City

Cambridge University Press
The Edinburgh Building, Cambridge CB2 8RU, UK

Published in the United States of America by Cambridge University Press, New York

www.cambridge.org
Information on this title: www.cambridge.org/9781107405455

© Cambridge University Press 2008

This publication is in copyright. Subject to statutory exception
and to the provisions of relevant collective licensing agreements,
no reproduction of any part may take place without the written
permission of Cambridge University Press.

First published 2008
First paperback edition 2011

A catalogue record for this publication is available from the British Library

ISBN 978-0-521-87500-4 Hardback
ISBN 978-1-107-40545-5 Paperback

Cambridge University Press has no responsibility for the persistence or
accuracy of URLs for external or third-party internet websites referred to in
this publication, and does not guarantee that any content on such websites is,
or will remain, accurate or appropriate.

Contents

Part IV   **317**
Evolving body features

Contributors

Claudio R. Alonso
University of Cambridge, Department of Zoology, Cambridge, United Kingdom

Wallace Arthur
National University of Ireland at Galway, Department of Zoology, Galway, Ireland

Jaume Baguñà
Universitat de Barcelona, Facultat de Biologia, Departament de Genètica, Barcelona, Spain

Paul M. Brakefield
Leiden University, Institute of Biology, Leiden, The Netherlands

Ariel D. Chipman
The Hebrew University of Jerusalem, Department of Evolution Systematics and Ecology, Jerusalem, Israel

Wim G. M. Damen
Universität zu Köln, Institut für Genetik, Köln, Germany

Jean S. Deutsch
Université Pierre et Marie Curie (Paris 6), Biologie du Développement and CNRS, Paris, France

Cassandra G. M. Extavour
Harvard University, Department of Organismic and Evolutionary Biology, Cambridge, Massachusetts, United States of America

David E. K. Ferrier
University of St Andrews, The Gatty Marine Laboratory,
St Andrews, Scotland, United Kingdom

Valentin Grob
Universität Zürich, Institut für Systematische Botanik und Botanischer Garten, Zürich, Switzerland

C. Jill Harrison
University of Oxford, Department of Plant Sciences, Oxford,
United Kingdom

Joachim T. Haug
Universität Ulm, Sektion Biosystematische Dokumentation, Ulm,
Germany

Nigel C. Hughes
University of California Riverside, Department of Earth Sciences,
Riverside, California, United States of America

Ronald A. Jenner
University of Bath, Department of Biology and Biochemistry, Bath,
United Kingdom

Jane A. Langdale
University of Oxford, Department of Plant Sciences, Oxford,
United Kingdom

Philippe Lopez
Université Pierre et Marie Curie (Paris 6), Systématique Adaptation
Évolution and CNRS, Paris, France

Pere Martinez
Universitat de Barcelona, Facultat de Biologia, Departament de
Genètica, Barcelona, Spain

Michel C. Milinkovitch
Université Libre de Bruxelles, Institute for Molecular Biology &
Medicine, Laboratory of Evolutionary Genetics, Gosselies, Belgium

Gerd B. Müller
University of Vienna, Department of Theoretical Biology, Vienna,
Austria, and Konrad Lorenz Institute for Evolution and Cognition
Research, Altenberg, Austria

Claus Nielsen
University of Copenhagen, The Natural History Museum of
Denmark, Zoological Museum, Copenhagen, Denmark

Jordi Paps
Universitat de Barcelona, Facultat de Biologia, Departament de
Genètica, Barcelona, Spain

Evelin Pfeifer
Universität Zürich, Institut für Systematische Botanik und
Botanischer Garten, Zürich, Switzerland

Nikola-Michael Prpic
Georg-August-Universität, Johann-Friedrich-Blumenbach-Institut
für Zoologie und Anthropologie, Abteilung für Entwicklungsbio-
logie, Göttingen, Germany

Marta Riutort
Universitat de Barcelona, Facultat de Biologia, Departament de
Genètica, Barcelona, Spain

Rolf Rutishauser
Universität Zürich, Institut für Systematische Botanik und
Botanischer Garten, Zürich, Switzerland

Isaac Salazar-Ciudad
University of Helsinki, Institute of Biotechnology, Developmental
Biology Program, Helsinki, Finland

Einhard Schierenberg
Universität zu Köln, Institut für Zoologie, Köln, Germany

Gerhard Scholtz
Humboldt-Universität zu Berlin, Institut für Biologie/
Vergleichende Zoologie, Berlin, Germany

Jens Schulze
Universität zu Köln, Institut für Zoologie, Köln, Germany

Ralf J. Sommer
Max-Planck Institute for Developmental Biology, Department for
Evolutionary Biology, Tübingen, Germany

Angelika Stollewerk
Queen Mary, University of London, School of Biological and
Chemical Sciences, United Kingdom

Athanasia C. Tzika
Université Libre de Bruxelles, Institute for Molecular Biology &
Medicine, Laboratory of Evolutionary Genetics, Gosselies, Belgium

Dieter Waloszek
Universität Ulm, Sektion Biosystematische Dokumentation, Ulm,
Germany

Hans Zauner
Max-Planck Institute for Developmental Biology, Department for
Evolutionary Biology, Tübingen, Germany

Preface

Two important events marked the year 2006 in the still short history of evolutionary developmental biology. The first European Workshop on Evolutionary Developmental Biology, held in Venice on 5–6 May 2006, offered some 30 researchers from most of the European teams active in this field a timely perspective on key issues in the discipline, and opened a lively discussion on where to move next, in terms of problems, model organisms, and levels of investigation. The second event was the Founding Congress of the European Society of Evolutionary Developmental Biology, held in Prague on 16–19 August 2006, which was attended by more than 300 biologists from all over the world.

This book is based on a selection from the papers contributed to the Venice workshop, plus five additional essays expressly written for this work.

The Venice workshop was generously sponsored by the Istituto Veneto di Scienze Lettere ed Arti and hosted in the wonderful Palazzo Cavalli-Franchetti. We are very grateful to Leopoldo Mazzarolli, the President of the Istituto, for sympathetically offering this academy's spaces for the first evo-devo event at European scale; our sincere thanks are also extended to Alessandro Franchini, Antonio Metrangolo, and to the whole technical staff of the Istituto for steadily helping in the organization of the meeting.

The book has benefited from the enormous help provided by numerous colleagues in reviewing more or less advanced drafts of the chapters. For this, we thank Ron Amundson, Peter Barlow, Richard Bateman, Geoff Boxshall, Carlo Brena, Paolo Burighel, Leo Buss, Fernando Casares, Eric Davidson, Claude Desplan, Frank Ferrari, Jordi Garcia-Fernàndez, Brian Hall, Steffen Harzsch, Peter Holland, Christian Klingenberg, Tim Littlewood, Kenneth McNamara, Stuart Newman, Paolo Piazza, Günter Purschke, Michael Richardson, Frederick Schram,

Miltos Tsiantis and Paul Whitington. Obviously, the reviewers are entirely absolved from any responsibility for the final contents of this book. We also thank a number of authors who, in addition to contributing a chapter, also served as reviewers, Leandro Drago, Claudio Friso and Diego Maruzzo who helped us to edit the manuscript, and our copy editor Lindsay Nightingale.

Last but not least, our warmest thanks to our Cambridge University Press editors Katrina Halliday and Dawn Preston, who enthusiastically endorsed our proposal for this book and helped us to translate the initial project into what we hope will be a useful contribution to the growth and visibility of evolutionary developmental biology.

Introduction: Pathways of change

> The molecular mechanisms that bring about biological form in modern-day embryos . . . should not be confused with the causes that led to the appearance of these forms in the first place . . . selection can only work on what already exists.
>
> G. B. Müller and S. A. Newman 2003: 3

> The evolution of form is . . . descent with modification (of development).
>
> S. B. Carroll 2005: 294–295

Combining words into new formulas is an all too easy exercise. But in our case it could turn into a dangerous trick, if evolutionary developmental biology (evo-devo) does not prove to be a fruitful new adventure in science. The question is increasingly acute, as a rapidly rising number of researchers are lured by the new flag, more and more resources are put into experimental and theoretical efforts under this banner, and evo-devo is finally getting public acknowledgment in the form of dedicated university chairs, specialised journals, workshops, and the launch of new professional societies.

Over the past few years, the nature, or the identity, of evo-devo has been passionately debated. However, it would be unwise to attempt to crystallise this discipline's content in a brief formula. Taking a historical perspective, hardly any facet of contemporary science would recognise itself in what in the past would have seemed an adequate definition of a research field under the same name. Scientific disciplines change along the years. In biological parlance, one could say that scientific disciplines develop, or that they evolve. So it would be ironic to attempt to fix the meaning of a discipline whose field is growing at the frontier between the two traditional fields of investigation of change – ontogeny and phylogeny – in biological systems.

Exactly at this frontier is the place where we want to focus here. What is really going on at this cutting edge? Is there only a peaceful coexistence of two distinct research agendas, or is there, on the contrary, evidence of a symbiotic relationship growing? In other words, is evo-devo, with respect to its parental disciplines – developmental biology and evolutionary biology – just a *multidisciplinary* collective enterprise, or is it instead an effective *interdisciplinary* venture?

We are convinced that a cross-fertilisation is really occurring between the two parent disciplines, and that the hybrid is already showing its higher fitness in a disciplinary environment where either developmental biology or evolutionary biology alone would not fare too well.

This inspection of the dynamics of evo-devo's growing tip will probably suggest that little of what is currently advertised under the new discipline's name really belongs unequivocally to it. That is, quite a lot of fashionable evo-devo is in fact the product of repeated backcross within either one or the other of the parent disciplines, which is developmental biology more often than evolutionary biology. This backcross progeny has its established place and, with its undeniable success within the province of one of the two parent disciplines, is steadily providing evo-devo with wonderful tools to be used in pursuing its more specific aims on which we actually want to focus attention here.

The first need for cross-fertilisation between formerly separate fields is the development of a common language. Difficulties to be overcome are not so much in the existence of terms specific to either field, and completely ignored in the other, but rather in the different meaning that the same word may take in either field. Are we sure we are asking the same question, in evolutionary biology and in developmental biology, when we ask what is gastrulation, or a carpel? Again, is a larva, or a metamorphosis, in any reasonable sense the same thing in both fields?

Clearly, recognising this problem and addressing it in a search for common ground is a basic, constructive way towards the identification of shared, and possibly overlooked questions. This is actually what is already going on.

To be sure, this is not necessarily done for free, as it may require re-thinking many of our cherished terms and concepts. This does not simply affect those concepts or categories that perhaps belong more to philosophy than to everyday science, such as cause, or change. Sooner or later, a cultural revolution will also affect more technical concepts such as developmental stage, segment or gene. This happens because all these things are not 'given', as we quite often take for granted in many traditional biological disciplines.

Taking an evolutionary perspective, we are forced to acknowledge that all these interesting things we are speaking about are the products of evolutionary history. As soon as we acknowledge the historically contingent nature of all these objects, be they body axes, or meristems, we find ourselves right within the field of evolutionary developmental biology.

We cannot systematically address questions in developmental biology as if our study objects were something independent of time, independent of evolutionary change. Everything we deal with is the product of history. This nearly trivial truth casts a deep shadow over the traditional, uncritical use of 'model systems' as satisfactory ways to discover 'the rules' of developmental pathways.

By refining or fine-tuning our historical sensibility, we discover that most of our questions in developmental biology have been framed until now in terms of end products (the adult, the organ, the gross trait of body structure) to be eventually obtained, and offered to the action of natural selection, rather than in terms of existing kinds of organisation, from which the system can move towards alternative states within a range only defined by the starting conditions and by the rules of change.

This is one side of the coin, that is, the effect of injecting an evolutionary dimension, or perspective, into developmental biology. But there is also the other side of the coin, that is, the effect of the awareness that all traits offered to selection are the products of developmental processes.

The weakest point in the standard theory of evolution is, indeed, understanding the origin of variation. We can hardly content ourselves with explaining it only in terms of mutation and sexuality. We may well dispute whether the long neck of the giraffe evolved under the selective pressure of critically important food items only available in the canopy of the acacias during the dry season in the savannah or, as some researchers suggest, as an effect of sexual selection. But what neither of these Darwinian scenarios will eventually offer is an explanation of why the giraffe's extra-long neck is supported by no more than seven cervical vertebrae, exactly the same number we find nearly universally in mammals. This is exactly the point where developmental biology can reciprocate in shedding light on an evolutionary problem by revealing what developmental processes can, or cannot, make available to selection. By understanding the rules of development, we may come closer to understanding the origins of variation; and so, in the end, closer to understanding evolution.

Thus, following a few years of lively and yet somewhat chaotic cross-fertilization, it would be very difficult to maintain that evolutionary biology and developmental biology are still the same as they were before the beginning of the dialogue. This ongoing intercourse has probably blurred many traditional distinctions between research agendas, but this seems unavoidable, in a science of change as biology largely is. The title of this book, *Evolving Pathways*, alludes to this cross-fertilising dialogue, as it can be read either as 'pathways that evolve', or as 'to evolve pathways', moving from development or evolution, respectively.

Scientists looking for general principles should become aware of the historically determined nature of the *kinds of* systems they investigate. There is probably no universal recipe for disentangling the search for 'laws', or general principles, from the study of historically contingent and unique events.

REFERENCES

Carroll, S. B. 2005. *Endless Forms Most Beautiful: The New Science of Evo Devo and the Making of the Animal Kingdom*. New York, W. W. Norton.
Müller, G. B. & Newman, S. A. 2003. *Origination of Organismal Form: Beyond the Gene in Development and Evolution*. Cambridge, MIT Press.

Part I Thinking about evolution by taking development on board

What is *evo-devo*? Undoubtedly this is a shorthand for *evolutionary developmental biology*. There, however, agreement stops. Evo-devo has been regarded as either a new discipline within evolutionary biology or simply a new perspective upon it, a lively interdisciplinary field of studies, or even necessary complement to the standard (neo-Darwinian) theory of evolution, which is an obligate step towards an expanded New Synthesis. Whatever the exact nature of evo-devo, its core is a view of the process of evolution in which evolutionary change is the transformation of (developmental) processes rather than (genetic or phenotypic) patterns. Thus our original question could be more profitably rephrased as: What is evo-devo for? This section contributes many-faceted insights into the identity and scope of evo-devo.

According to Gerd Müller (Chapter 1), evo-devo is a discipline in its own right, because it asks a specific set of questions, solves biological problems that could not be solved by other approaches, and affects our understanding of evolutionary theory. After a short reflection on evo-devo history, the chapter examines in detail a set of evo-devo big questions. All these have at their core two interrelated components, namely how evolution affects development, and how the properties of developmental systems affect the course of evolution. Finally the author considers current evo-devo research programs, and discusses the impact of evo-devo on the theory of evolution.

Isaac Salazar-Ciudad (Chapter 2) critically reviews advantages and disadvantages of three 'schools of thought' in evolutionary biology that differ with respect to their views on the origin of variation: neo-Darwinism, the developmental constraints school and the developmental genetics school. He then presents a new set of concepts and studies that try to avoid the drawbacks of the three schools and argues that some aspects of the evolution of morphology and development are predictable if information is available about development and about the selective pressures that were operative in previous generations.

Wallace Arthur (Chapter 3) questions whether mega-evolution is more than just a result of the accumulation of micro/macro-evolutionary events, or, alternatively, if evolution is effectively a 'scale-independent' process. This question is approached by comparing magnitude, type and developmental timing of changes involved in high- and low-level divergence of lineages. He discusses three competing hypotheses: that mega-evolutionary changes are something quite apart from everyday changes; that mega-evolutionary divergences are statistically different from their lower-level counterparts; and that all levels of evolution are the same in both the absolute and the statistical sense.

Why do species show the patterns of diversity and disparity they do? Combining an exploration of how phenotypic variation is produced at each generation with an analysis of how this variation is influenced by natural selection and other extrinsic processes can provide the means for a comprehensive understanding of evolutionary patterns. Paul Brakefield (Chapter 4) presents a well-documented case study that illustrates an integrative approach linking the evolution of developmental mechanisms with the role of selection in the evolution of wing eyespots and other traits in *Bicyclus* butterflies.

Evo-devo aims to provide a mechanistic explanation of how developmental mechanisms have changed during evolution, and how these modifications are reflected in changes of organismal form. Thus, in contrast with studies on natural selection, which aim to explain the 'survival of the fittest', the main target of evo-devo is to determine the mechanisms behind the 'arrival of the fittest'. At the most basic level, the mechanistic question about the arrival of the fittest involves changes in the function of genes controlling developmental programs. Thus it is important to reflect on the nature of the elements and systems underlying inheritable developmental modification using an updated molecular background. Claudio Alonso dedicates a chapter (Chapter 5) to precisely this task.

In the search for evo-devo identity, Ronald Jenner (Chapter 6) starts from the perspective of an important, but neglected, epistemological dualism in a science like biology, that is, idiographics vs. nomothetics. Idiographics pertains to the description of unique and historically contingent particulars, while nomothetics pertains to the search for law-like regularities or generalities. Thus, idiographically, evo-devo aims to document the unique effects of changes in evolutionary developmental mechanisms on the origin of novelties and the evolution of body plans. Nomothetically, it attempts to establish the general effects of evolutionary developmental mechanisms on determining the overall direction of phenotypic evolution. Recognising the dualism is not only conceptually important, but has also practical consequences, for example in the choice of model organisms.

Evo-devo as a discipline

GERD B. MÜLLER

Since its inception in the early 1980s, evo-devo has evolved into a mature discipline. This is manifest in the naming of research groups, scientific journals and books, professional meetings and societies. Despite such formal attributes of a scientific discipline it is often unclear what constitutes its conceptual distinctiveness. Does evo-devo have its own set of specific questions and research methods? Does it solve biological problems that cannot be solved by other approaches? And does it represent a significant change in the theoretical understanding of development and evolution? That is, in which way do the goals, the empirical programs and the theories of evo-devo research differ from those of neighbouring disciplines such as developmental biology or evolutionary biology? The present chapter provides a concise overview of the current status of evo-devo as a discipline. This requires a short reflection on its history.

CONCEPTUAL FOUNDATIONS

The parallels between embryonic stages and the 'scale of beings' had already been contemplated in pre-Darwinian times, and the foundation of a scientific theory of evolution was significantly influenced by embryological arguments. Darwin called embryology 'by far the strongest single class of facts in favour of a change of form', and his first sketches of a phylogenetic tree seem to have been inspired by tree-like renderings of embryological differences between species (Richards 1992). Much of the early work in evolutionary biology focused on the uses of embryonic characters for taxonomical purposes. Francis Balfour, William Brooks, Karl Gegenbaur, Fritz Müller and many others applied the comparative

Evolving Pathways: Key Themes in Evolutionary Developmental Biology, ed. Alessandro Minelli and Giuseppe Fusco. Published by Cambridge University Press.
© Cambridge University Press 2008.

method to embryology and could thus discern hitherto unknown phylogenetic relationships. Alexandre Kowalevsky's (1866) discovery, for instance, that larval traits such as a notochord, gill slits and neural folds relate the ascidians to the vertebrates was one of the great successes of this method.

These comparative endeavours were soon followed by more mechanistically oriented and theoretically grounded programs. One sprang from the joining of the concept of recapitulation with a mechanism for effecting developmental change. Recapitulation, a widespread notion in late eighteenth-century Naturphilosophie, was elaborated by Ernst Haeckel into a mechanistic concept of morphological evolution (Haeckel 1866) by uniting it with developmental timing as a key mechanism for embryonic change. Under Haeckel's patronage this approach assumed programmatic and even ideological status. Recapitulation remained the only thinkable way by which ontogeny and phylogeny could be tied together until well into the twentieth century. The rise of experimental embryology on the one hand, and that of genetics on the other, stifled the – by then often exaggerated – recapitulationist claims. Eventually, the new paradigm of genetic variation and differential inheritance eclipsed recapitulation as a general explanatory principle for the progression of organic life. In the subsequent disregard for recapitulation theory it was often overlooked that it had contained a mechanism for evolutionary change, namely the modification of development through heterochrony, a point notably resurrected in the late 1970s (Gould 1977).

The study of environmental influences on embryogenesis, and the maintenance in subsequent generations of the effects thus induced, was another major movement that related ontogeny to phylogeny during the first half of the twentieth century. Most of these endeavours were carried out in a neo-Lamarckian vein, testing the possibility of an inheritance of acquired characters. An extensive amount of data was generated by ingenious modifications of external parameters in the development of insects (Jollos 1934), amphibians (Kammerer 1923), and other taxa (Kammerer 1925, Hämmerling 1929). Entire institutions, such as the Vivarium Institute in Vienna (1902–1945), devoted their efforts to the study of the environment–development–evolution interaction. The conclusiveness of the results was debated heatedly (e.g. E. W. McBride vs. opponents in *Nature* during the 1920s). Eventually the neo-Lamarckian interpretations lost credibility. But these early attempts to combine environmental modification with breeding experiments represent a body of evidence that merits attention independently from their

Lamarckian interpretations. Recently the importance of 'enduring modification' and 'epigenetic inheritance' has been reconsidered (Rubin 1990, Jablonka and Lamb 1995), and plasticity research actively readdresses the issue of environmental influences on development and evolution (Gilbert and Bolker 2003).

Another conceptual root of evo-devo arose with early attempts to include the genetics of development into evolutionary theory, based on theoretical considerations and experimental quantitative genetics. Among these concepts ranged reaction norms (Wolterek 1909), rate genes (Goldschmidt 1940), assimilation (Waddington 1956) and the whole field of epigenetics (in the Waddingtonian sense). These initiatives took place before the rise of DNA genetics and in the absence of molecular tools for genetic analysis. But the calls for a more prominent role of these mechanisms in evolutionary theory, such as expressed by Goldschmidt and Schmalhausen, and later by Waddington, went largely unheard. Attention concentrated on transmission genetics and quantitative genetics, whereas developmental genetics, and developmental biology for that matter, were left aside.

These initiatives all addressed facets of the ontogeny–phylogeny or development–evolution interface and thus kept the connections between the fields alive even during prolonged periods of their largely separate study in the twentieth century. Except for certain conceptual traces not much of these traditions has survived in modern evo-devo, and none of them can be considered its immediate forerunner. Two developments were more directly responsible. One stimulus was the increasing awareness of explanatory deficits in the prevailing paradigm of evolutionary theory. Neo-Darwinism worked well for the population genetic phenomena it concentrated on, but in the late 1970s and early 1980s concern accumulated about its difficulty to account for many characteristics of phenotypic evolution. Such phenomena included biased variation (Alberch 1982, Maynard Smith *et al.* 1985), rapid changes of form (Eldredge and Gould 1972), the occurrence of non-adaptive traits (Gould and Lewontin 1979), and the origination of higher-level phenotypic organisation such as homology and body plans (Riedl 1978). Most of the criticisms attributed the explanatory deficits of neo-Darwinism to its neglect of the generative processes that relate genotype to phenotype and to the exclusion of developmental theory from the evolutionary synthesis (Hamburger 1980, Reid 2007).

The rising interest in these topics during the early 1980s was reflected in scientific meetings (such as those in Dahlem 1981, Sussex

1982, Plzen 1984, Columbia 1985 or Woods Hole 1985) and books (Bonner 1982, Goodwin *et al.* 1983, Raff and Kaufman 1983), which began to concentrate on the intersections between development and evolution. Empirical research took up the theme (e.g. Katz *et al.* 1981, Alberch and Gale 1983, 1985, Raff *et al.* 1984, Müller 1986), using classical techniques of comparative and experimental embryology at first, and later, increasingly, the methodologies of molecular biology. This new agenda, which aimed at defining the role of developmental processes in organismal evolution, was initially called 'ontophyletics' (Katz *et al.* 1981, Katz 1983) or 'evolutionary embryology' (Müller 1991), until 'evolutionary developmental biology' (Hall 1992, Wake 1996) became the generally accepted label. Besides heterochrony, developmental constraints were a central topic in this early period of evo-devo (Alberch 1982, Maynard Smith *et al.* 1985).

In the mid 1980s a second major boost for modern evo-devo came from the rise of molecular developmental genetics, which brought the cloning of regulatory genes and the techniques for the visualisation of their activation in the embryo. This created a completely new approach to comparing the development of different taxa and led to the discovery of unexpected similarities in gene regulation among distantly related species (McGinnis *et al.* 1984). During the following years these similarities were found to extend to the spatial and temporal sequences of early gene expression in anatomically very different embryos such as insects and mammals (Duboule and Dollé 1989, Graham *et al.* 1989). In contrast to earlier notions that took the diverse ways in which animals develop to be the result of an equally diverse genetic apparatus, it became increasingly clear that relatively few genetic regulators are implicated in the embryonic foundations of all animal body plans (Akam 1989, Holland 1992, Holland *et al.* 1996). The search for commonalities and differences in gene expression patterns and gene regulation gained rapid momentum and led to a much improved understanding of the molecular underpinnings of development (Carroll *et al.* 2005, Davidson 2006). Today, the evolution of the developmental genome and of gene regulatory networks has become the most popular theme in empirical evo-devo research. High-throughput genomics is adding another methodological level to this comparative developmental genetics.

THE QUESTIONS OF EVO-DEVO

Evo-devo starts from the postulate that a causal-mechanistic interaction must exist between the processes of individual development and the processes of evolutionary change. Understanding these interactions and their consequences for organismal evolution represents the central research goal. Hence, the core question of evo-devo has two interrelated components: evolution's influence on development and development's influence on evolution. This reciprocal interrelationship constitutes a genuinely dialectical and systemic research agenda. The following will be a brief characterisation of the major research questions that arise from this general agenda.

How did development originate?

This question relates to the origins of multicellularity and the evolution of life cycles. John Bonner, a major influence in triggering the evo-devo revolution, early on reflected on the relations between organism size, internal complexity, reproductive success and life-cycle selection (Bonner 1965, 1988). Most of these ideas were based on the study of extant colonial or aggregating unicellular organisms such as cellular slime moulds. In early multicellular aggregates competition among cells to become the ones that propagate the next generation was possibly an important factor. The transition between the cell as the unit of selection and the multicellular individual as the unit of selection would have been the key evolutionary event at the origin of development (Buss 1987).

A different approach targets the physical properties of cells and tissues. Single-cell organisms that existed before the emergence of multicellularity possessed liquid-like viscoelasticity, adhesiveness and chemical excitability. Consequently, protometazoan cell aggregates must have had an inherent capacity to self-organise spatial patterns. Development would have arisen at the point when certain cells achieved organisational control over other cells, e.g. by releasing diffusible chemical substances, and this capacity would have resulted in cell aggregates consisting of non-uniformly distributed cell states. In conjunction with differential adhesion (Steinberg 1963) and other generic physical mechanisms (Newman 1994) such simple systems can produce an array of 'generic forms', whose shapes and sizes are much determined by the physico–chemical conditions of the environment in which they form (Newman *et al.* 2006). Because of this strong environmental influence, it is assumed that in early forms of development the close correlation between genotype and phenotype observed in modern organisms would not have existed yet. Rather the genotype–phenotype relation might have been one-to-many during what has been called a 'pre-Mendelian phase' of evolution (Newman and Müller 2000). Subsequent selectional fixation and genetic routinisation would have resulted in the robust forms of development and the faithful Mendelian kind of inheritance seen in extant organisms.

How did the developmental repertoire evolve?

This question is predominantly approached at the genetic level, e. g. through the study of gene duplications, especially of the regulatory genes (McGinnis and Krumlauf 1992, Holland 1999), and the evolution of gene regulatory networks (Davidson *et al.* 1995, Wray and Lowe 2000). The genetic redundancy generated by such mechanisms can be exploited through the acquisition of new functions for these genes, a process referred to as recruitment (Keys *et al.* 1999) or cooption (True and Carroll 2002). Present summaries of the evolution of developmental pathways rely almost exclusively on genetics (Wilkins 2002, Carroll *et al.* 2005), but the epigenetic mechanisms controlling gene activation also evolve, including the processes of cell and tissue interaction and embryonic induction, which had been considered in earlier treatments of the evolutionary roles of epigenesis (Løvtrup 1974, Hall 1983, Edelman 1988).

Modularity constitutes a principle connecting the genetic and epigenetic facets of evolving developmental repertoires in recognising that developmental systems are decomposable into components that operate according to their own intrinsically determined principles (Schlosser and Wagner 2004, Callebaut and Rasskin-Gutman 2005). Such modules can be characterised as integrated structural and process units that depend on input from other components and, in turn, influence other components by their outputs, represented, for example, by gene signalling pathways or inductive interaction networks. The evolutionary function of developmental modules would be their phenotypic selectability. A selectable developmental module can consist of a set of genes, their products and their developmental interactions, including the resulting character complex and the functional effect of that complex. The genes affecting the modular character complex would be characterised by a high degree of internal integration and a low degree of external connectivity: that is, pleiotropic connections would be largely within-module. Modularity could thus become one of the most productive approaches to the evolving genotype–phenotype relationship (von Dassow and Munro 1999).

How are established processes of development modified through evolution?

The empirical study of changes in developmental gene regulation occupies much of the present research effort (see below and contributions in this volume). A broader concept is heterochrony, i.e. evolutionary changes in the relative timing and rates of developmental processes. This classical idea has been revived by Gould (1977) and Raff and Kaufman (1983) and has since been elaborated into a powerful explanatory framework (McKinney and McNamara 1991, Parichy *et al.* 1992, McNamara 1997). Different forms and mechanisms of heterochrony are associated with different life-history strategies and produce different phenotypic results (Hall 1984, Raff and Wray 1989). Heterochrony has been documented in most groups of organisms, and its study is now taken to molecular and genetic levels (Parks *et al.* 1988, Wray and McClay 1989, Kim *et al.* 2000). Mutations that directly affect developmental timing have been demonstrated in animals (Ruvkun and Giusto 1989) and plants (Dudley and Poethig 1991). A number of genetic mechanisms affecting developmental timing have been tested experimentally (Dollé *et al.* 1993, Zákány *et al.* 1997). Without doubt heterochrony based on gene regulatory changes represents a powerful mode for altering morphological characters and body plans (Duboule 1994). But it remains difficult to distinguish between heterochronic phenomena that are simply a consequence of any change to development and those cases in which heterochrony of a particular process represents the causal mechanism for the evolutionary modification of a trait.

Does development play a role in phenotypic variation?

The extent to which the properties of developmental systems influence the variational and directional dynamics of phenotypic evolution is a question primarily addressed by the concept of developmental constraint. This was one of the themes that triggered evo-devo (Alberch 1982, Maynard Smith *et al.* 1985), and it is still relevant today. The empirical evidence for constraints is extensive, including data from comparative morphology (e.g. Wake 1982, Bell 1987, Vogl and Rienesel 1991, Caldwell 1994), comparative and experimental embryology (e.g. Alberch and Gale 1983, 1985, Müller 1989, Webb 1989, Streicher and Müller 1992), plant biology (e.g. Donoghue and Ree 2000) and quantitative genetics (e.g. Cheverud 1984, Rasmussen 1987, Wagner 1988). Whereas early conceptualisations of constraint concentrated on the limitations of phenotypic variation, later treatments emphasised also the heightened potential for change in particular aspects of the phenotypic character space (Arthur 2001). A taxon's capacity to generate heritable phenotypic variation or innovation will depend on mechanisms that reduce or overcome constraints, a controversial issue (Eberhard 2001, Wagner and Müller 2002). Much of the present work on plasticity and evolvability discussed below equally relates to the issue of constraint.

What is the contribution of development to the origin of phenotypic novelty?

Innovation and phenotypic novelty is one of the areas of evolutionary biology to which evo-devo could make a genuine contribution (Wagner 2000, Love 2003, Müller and Newman 2005b). While earlier conceptions concerning innovation were based on function shift (Mayr 1960), macro-mutation (Goldschmidt 1940) and symbiosis (Margulis and Fester 1991), evo-devo approaches concentrate on the role of development. One specific proposal is epigenetic causation, the idea that developmental systems do not merely transform genetic change into phenotypic change but also represent a generative component in phenotypic evolution (Müller 1990, Newman and Müller 2000). The starting point is the distinction between general selectional trends and the specificity of phenotypic response conferred by the developmental system. Selection acting on overall organismal features, such as shape, proportion, function or behaviour, can elicit epigenetic by-products that arise from the generic properties of developing cell and tissue systems, e.g. following changes in blastema size or mechanical load. New structural elements, skeletal parts for instance, can arise through this mode, without having been selected for, as a side-effect of the evolutionary modification of general developmental parameters (Müller 1990, Newman and Müller 2005). Thus, epigenetic mechanisms could have had a significant role in the origination of body parts and organismal form (Müller and Newman 2003, Love and Raff 2005).

A related approach is the origination of innovation through environmental induction (West-Eberhard 2003). This approach relies less on the physical properties of developmental systems, concentrating more on phenotypic plasticity and reaction norms as discussed below.

Does development affect the organisation of the phenotype?

The origin of higher-level organisational phenomena (homology, body plans) is one of the central questions of evo-devo (Raff 1996, Minelli

2003). Many of the new ideas on these topics were triggered by the discovery of the surprisingly high conservation of the gene regulatory apparatus in very diverse organisms. This has led to gene-based definitions of homology (Holland *et al.* 1996, Abouheif 1997), whereas others have pointed out the shortcomings of such reasoning (Bolker and Raff 1996, Minelli 1998). While the most notoriously conserved developmental control genes, the homeobox genes, exhibit non-homologous expression domains in vertebrate and invertebrate embryos, the reverse also applies: homologous structures can be specified by non-homologous genes (Wray 1999). New developmental concepts of homology concentrate on commonalities of developmental pathways (Wagner 1989, 1996) or on the modularity of developmental systems (Minelli 1998, Gilbert and Bolker 2001). Other positions emphasise that the establishment of homology goes through different stages in which development has an important generative role, but eventually achieves independence from the underlying generative mechanisms (Müller 2003, Love and Raff 2005). Here the evolution of homology and body plans is viewed as a consequence of phenotypic integration that maintains the identity of building elements despite variation in their molecular, developmental and genetic makeup.

How does the environment interact with development and evolution?

Once thought of as crucial for understanding evolution, this question had been marginalised for several decades because of its seemingly Lamarckian connotations. But new data on the genetic and environmental aspects of developmental phenomena such as phenocopy, polyphenism and plasticity have revitalised the interest in their evolutionary roles and led to proposals of an enlarged scope of evo-devo research (Gilbert 2001, Hall *et al.* 2003). The foundational concept in this domain is phenotypic plasticity (Pigliucci 2001, West-Eberhard 2003). It provides a unifying theoretical framework for the interpretation of quantitative genetic, developmental and morphological responses to environmental influences in evolving populations. The concept of plasticity is tightly interconnected with that of reaction norm (Schlichting and Pigliucci 1998, Sarkar 2003), i.e. the range of variation and phenotypes that can result from a single genotype as a response to different environmental conditions. Developmental plasticity, the mechanistic realisation of this responsiveness, is thought to represent in itself an adaptive trait of a taxon. But plasticity can also refer to evolutionary modifications of development that do not have a significant effect on the phenotypic outcome, a phenomenon often observed in species-level comparisons (Chipman 2002).

The maturation of evo-devo has led to a diversification of empirical research that does not coincide so much with the conceptual questions discussed above but rather represents different methodological emphases. At least four major 'programs' can be distinguished in modern evo-devo, although the individual scientists and research groups engaging in these studies do not necessarily define their particular approach in such a way, and many overlaps exist between these programs.

The comparative embryology program

Comparative embryology is still required in modern evo-devo, even though much of it involves molecular tools. However, major contributions in this domain do not come from extant organisms but, increasingly, from palaeontology. The palaeontological data include direct embryological evidence, such as information from fossilised dinosaur eggs (Carpenter *et al.* 1996) and from early stages of invertebrate development (Bengtson and Zhao 1997). Palaeontology also provides indirect information about the contribution of development to phenotypic evolution by documenting heterochrony, constraint and innovation (McNamara 1997, Vrba 2003). This is especially true of characters that represent 'frozen stages' of development such as teeth or hard shells. The fossil record yields morphospaces against which the developmental capacities of extant taxa can be tested (McGhee 2007). The relationships between variational data from fossils and the developmental processes responsible for their generation are increasingly addressed in explanations of phenotypic character evolution (Shubin *et al.* 1997). At the same time the comparative approach generates the data for what could be called the systematics program of evo-devo. This work provides robust phylogenetic reference systems for the interpretation of the evolution of developmental mechanisms, an essential prerequisite for evo-devo (Mabee 2000). Many more applications of the comparative approach are summarised in Minelli (2003).

The epigenetic and experimental program

The aim of this approach is to probe developmental systems with regard to their intrinsic capacities to generate evolutionarily relevant phenotypes. Perturbations of cell number, cell cycle, developmental timing or inductive interactions, in the traditions of classical experimental embryology, have been shown to produce both ancestral character states as well as phenocopies of derived states (e.g. Alberch and Gale 1983, 1985, Müller 1989). Such experiments address the developmental and, hence, the evolutionary dissociabilities of temporal, spatial and functional interactions and highlight the roles of physical properties and tissue geometries in developmental self-organisation. They also demonstrate the existence of thresholds and constraints in developmental processes which contribute to the evolvability of a lineage, influencing phenotypic evolution and providing new heritable variation and novel character states. In particular, the experimental approach exposes mechanisms through which quantitative selectional trends may be transformed into qualitative phenotypic change.

Today, the classical perturbation method is expanded by genetic and molecular tools, such as gain of function and loss of function experiments, and the attempt to redesign phenotypes with well-chosen mutations. This designer approach (Dworkin *et al.* 2001, Larsen 2003) tests the range of possible morphologies that can be achieved by a developmental system through small changes in cell behaviour. While it does not automatically follow that new structures actually arose through similar mutations, such experiments indicate that even highly conserved phenotypes are not necessarily strongly constrained. Using mutations to compare the relative stability of characters, this approach further elucidates the nature of developmental constraints and will assist in revealing the genetic backgrounds that are required for stabilising phenotypic innovations. Combining experimental studies of what forms can be generated by a developmental system with theoretical morphospace concepts (see below) could lead to further evo-devo insights.

The epigenetic and experimental program intersects with the study of developmental plasticity, especially as it relates to environmental influences in what has recently been termed ecological developmental biology or eco-devo (Gilbert 2001, Hall *et al.* 2003). The essence here is that the regulating factors of developmental processes (and their evolutionary modifiability) do not all reside within the embryo but depend substantially on the ecological context. Although these effects eventually feed into developmental-genetic pathways, the causality resides in an interplay between internal and external factors including diet, pH, humidity, temperature, photoperiod, seasonality, population density, predator presence and many more. Particular attention is paid to physiological and metabolic processes that mediate interactions between the environment and development, such as endocrine and hormone activity (Davey 2003, Rose 2003). Other relevant data come from predator-induced polyphenisms (Tollrian and Harvell 1999), changing nutrient regimes (Newlon *et al.* 2003), environmental regulation (Nijhout 1999) and other fields (see Schlichting and Pigliucci 1998, Gilbert 2001, Hall *et al.* 2003). Although these kinds of study have a long tradition there is a new awareness that developmental plasticity and environmental induction have an important function in the origination of evolutionary novelty (West-Eberhard 2003), an opinion pioneered by Ryuichi Matsuda (Matsuda 1987) whose work is receiving renewed attention (Hall *et al.* 2003). The importance of epigenetic parameters is also increasingly recognised in hominid evolution (Lovejoy *et al.* 1999).

The evolutionary developmental genetics program

This research program, often called evo-devo as such, is 'focused on the developmental genetic machinery that lies behind embryological phenotypes' (Arthur 2002). The rapid cloning of an increasing number of regulatory genes, and the development of techniques that enable their expression in the embryo to be visualised, has made this the most active area of empirical evo-devo. Two overlapping subprograms can be distinguished. One is the elucidation of molecular body plans, aiming for an understanding of the role of developmental control genes in the patterning of phylogenetically and anatomically diverse organisms, such as arthropods (Akam 1994), vertebrates (Holland 1992) and other taxa (Carroll *et al.* 2005). This program now extends beyond the usual model organisms and reveals interesting expression patterns that are associated with body plan novelty (Lee *et al.* 2003). An impressive amount of information on the commonalities and differences in the deployment of regulatory genes has been accumulated, including evidence that the evolutionary modification of major body regions is associated with the axial shifts of *Hox* gene expression boundaries (Burke *et al.* 1995), or that the evolution of mesopodial limb elements is associated with shifts in *Hoxa-11/a-13* expression regions (Wagner and Chiu 2001). In insects, the differences in bristle patterns in different species of *Drosophila* are associated with a variation of *Ubx* expression (Stern 1998) and the evolution of butterfly eyespot patterns involves recruitment of a *hedgehog* regulatory circuit (Keys *et al.* 1999). A future goal must be to ascertain that the observed shifts and changes of gene expression were actually causal for the derived phenotypic condition (Wagner 2001).

A second subfield of evolutionary developmental genetics could be called the developmental regulation program. Here the research concentrates on changes in the evolving architecture of the regulatory circuitry. Tremendous progress has been achieved in the understanding of gene regulatory pathways and networks (Wilkins 2002, Davidson 2006). Increasingly complex gene interaction networks are unravelled, and a kind of regulatory cladistics is emerging. The conclusions following from this program posit that the evolution of organismal form is much less a direct consequence of mutational genetic innovation, as believed earlier, but rather depends on continuing shifts, recruitments and re-wiring of regulatory interactions in development. Evolution seems to favour the generation of alternative genetic circuits which are subsequently coopted into new regulatory functions. The intricate details of these molecular networks are going to keep the program active for a long time to come, especially as it is moving into genomics and proteomics.

The theoretical biology program

Theoretical biology, as the science concerned with the formulation of general rules and mathematical abstractions of biological processes, but also with theory analysis, modelling and simulation, has recently taken a heightened interest in evo-devo. This is because, on the one hand, genotype–phenotype mapping, plasticity, modularity and other evo-devo events require formalisation for their integration into the theoretical framework of evolution. On the other hand, the complexities of developmental processes benefit greatly from the new computational tools of visualisation, quantification and simulation. Several strategies are followed in the quest for a theoretical biology of evo-devo.

One is the computation of morphogenesis. Since developmental evolution resides in the modification of the dynamics of gene, cell and tissue interactions, the precise topology, timing and quantity of gene activity as related to actual changes of cell behaviour and tissue properties becomes a target issue of evo-devo. These requirements have led to the development of computational tools for the three-dimensional reconstruction and quantification of gene expression in developing embryos (Jernvall *et al.* 2000, Streicher *et al.* 2000, Sharpe *et al.* 2002, Weninger *et al.* 2006), and new bioinformatic techniques for the analysis of such data are explored (Costa *et al.* 2005). The aim is to understand the topological evolution of gene expression patterns in a given developmental system in order to determine the spatio-temporal modifications of gene activity that are associated with phenotypic variation.

Advanced morphometrics and statistics further refine the quantitative approach, in particular as landmark-based tools begin to be widely used (Bookstein 1991, Eble 2002), including ontogenetic shape trajectories (Prossinger and Bookstein 2003, Cobb and O'Higgins 2004, Mitteroecker *et al.* 2004). Multivariate shape analyses not only help quantify evolutionary modifications but are essential for defining their ontogenetic locations and, hence, assist the identification of the developmental pathways that are responsible for effecting these changes. In addition the multivariate approach provides a means to link evo-devo with quantitative genetics and the study of morphological integration (Cheverud 1982).

The theoretical biology program also includes another aspect of modelling evo-devo, an area with much future potential (Collins *et al.* 2007). On the one hand the quantitative developmental data can be used for the biomorphic modelling of concrete developmental systems, such as tooth development (Jernvall 2000, Jernvall *et al.* 2000) or limb development (Hentschel *et al.* 2004), illustrating how the differential activation of genes and gene products can affect morphogenesis and evolutionary variation or innovation. On the other hand, modelling can identify important general properties of evolving developmental networks and regulatory circuits, demonstrating, for instance, that evolution has a tendency to substitute emergent networks by hierarchical networks (Salazar-Ciudad *et al.* 2001). This indicates that the routinised and genetically entrenched ontogenies of extant species, from which our knowledge of development is derived, constitute a highly evolved and stabilised condition, whereas greater flexibility and innovative potential may have existed in primitive systems (Newman and Müller 2000). Such scenarios, emerging in part from a modelling approach, call for a re-evaluation of the earlier concepts of canalisation and assimilation (Waddington 1942).

The joining of bottom-up modelling of the interaction between genes, cell behaviour and tissue organisation with the concepts of generative morphospaces provides a framework in which a set of given rules produces a range of possible patterns that can be compared with forms that did or did not appear in natural systems (Thomas and Reif 1993, Rasskin-Gutman 2003, McGhee 2007). These models can be used to characterise the generative capacities of developmental systems and they permit predictions about potential phenotypic variation and innovation. Morphospace modelling indicates that only a limited number of phenotypic solutions can be obtained from a given developmental system, even in the presence of ample genetic variation. But these effects are not only limitational. Certain morphological solutions, for example, are more likely to arise than others, independent from the molecular and genetic circuitry associated with their generation, pointing to inherent properties of the developmental systems involved. These considerations include the interactions of environmental parameters with evolving developmental systems (Collins *et al.* 2007).

With regard to conceptual advancement, maybe the greatest potential of evo-devo lies in the theoretical biology program. A host of new tools for visualisation, quantification and mathematical analysis are being adapted for embryonic studies and can be used in a comparative way. Computational modelling and simulation will be able to address many kinds of evo-devo questions and will assist the formal characterisation of the epigenetic rules in genotype–phenotype relations. They will also help identify new biological questions for empirical study. The multiple modelling strategies that emerge in evo-devo represent important heuristic, conceptual, explanatory and integrative tools. The detailed analysis and subsequent integration of these different modelling strategies will promote the theoretical integration of evo-devo (Laubichler and Müller 2007).

THE IMPACT OF EVO-DEVO ON EVOLUTIONARY THEORY

The conceptual framework of evo-devo arose as a response to the incompleteness of the neo-Darwinian synthesis (Callebaut *et al.* 2007). The paradigm of the synthesis was based on the correlation of phenotypic character variation with statistical changes of gene frequencies in populations. Adaptive change as a population genetic event was the explanandum. The paradigm of evo-devo, by contrast, represents a causal-mechanistic approach towards the explanation of phenotypic change in evolution. Here the evolutionary alteration of developmental parameters (gene, cell and tissue properties, and their interactions) and their effects on phenotypic evolution are the explananda, whether adaptive or not. This means that the explanatory reach of evolutionary theory is significantly expanded by evo-devo, because it not only includes adaptive variation but extends also to other features of phenotypic evolution, such as the generation of new structural elements (novelty), the establishment of standardised building units (modularity, homology), the arrangement of such units in lineage specific combinations (body plans), and the repeated generation of similar forms in independent taxa (homoplasy). In addition, evo-devo aims at explaining how development itself evolves and how the control of developmental processes is effected by the interplay between genetic, epigenetic and environmental factors.

Evo-devo has prompted the formulation and improvement of a host of concepts and models regarding the evolution–development interface, such as recapitulation, heterochrony, constraint, modularity, plasticity and many others (Arthur 2002, Müller 2007). These new conceptual approaches of evo-devo have enormously stimulated biological research, both empirically and theoretically. Developmental models have become evolutionary, and evolutionary models have become developmental, reflecting a more pluralistic approach to phenotypic evolution (Fusco 2001). The major expansions that evo-devo represents with regard to the classical Evolutionary Synthesis can be characterised by three terms: evolvability, emergence and inherency.

Evolvability, the intrinsic potential of a given lineage to produce phenotypic variation, has long been studied primarily from a genetic point of view, focusing on mutator genotypes or the changes in genotype to phenotype mapping (Wagner and Altenberg 1996, Wagner 2005). Evo-devo adds the generative potentialities of development to this concept. These are determined by the continued efficacy of gene regulatory and epigenetic interdependencies that have been acquired during evolution of a lineage, including the dynamics of developmental interactions and the non-programmed physical properties of the involved tissues (Kirschner and Gerhart 1998, Deacon 2006, Newman *et al.* 2006). Evolvability can now be analysed in terms of the developmental plasticity of specific organ systems and their responses to changing environmental conditions such as in plants (Sultan 2003), butterflies (Beldade *et al.* 2005) or beetles (Emlen 2000).

Evolvability has the potential to integrate the classical population genetic approach with the results from plasticity research (Schlichting and Pigliucci 1998, Pigliucci 2001). The possibility of correlating data from ecology with physiological parameters, developmental reaction norms and gene regulatory pathways in a quantitative way facilitates various new modelling strategies in evo-devo (Collins *et al.* 2007). Although such models have not yet reached the same status as, for instance, the 'adaptive landscapes' in the Modern Synthesis, significant heuristical effects emanate from these new perspectives on the role of evolvability in organismal evolution.

Emergence refers to the fact that through the integration of such concepts as modularity, plasticity and innovation, evolutionary theory becomes explanatory not only about what is being varied and maintained in organismal evolution but also about what could possibly arise. The gene-centric position of neo-Darwinism glossed over this problem by tacitly assuming that genes are directly responsible for structure in an additive fashion. Hence it was sufficient to focus on the dynamics of alleles in populations, assuming the prior existence of the phenotypic entities to which they correspond. No feedback between genes, gene products, the material properties of developmental systems and their environments was accounted for. Yet the capacities for emergence lie in these systems interactions, and evo-devo addresses precisely this aspect.

A theory of emergence complements the theory of adaptation in that it introduces the non-deterministic developmental and environmental factors that are responsible for the origins of novelty. An important starting point for this conceptual change is the recognition that novelties represent a distinct class of phenotypic change, not based on character variation and not a direct consequence of natural selection. Selection cannot set in until there are entities to select. Empirical evo-devo research has begun to concentrate on these issues (Müller and Newman 2005a), and the role of emergence in evolutionary theory is critically evaluated (Reid 2007). Natural selection as the unique guiding force of evolution is challenged by evo-devo.

Inherency refers to the fact that through the inclusion of evo-devo into evolutionary theory there is a shift of focus from the external and contingent to the internal and generic. Whereas historical contingency is a significant element in the standard evolutionary framework, accounting for the lawful dependence on conditions that involve a large component of chance, inherency is something that will always happen because the potentiality is immanent to the system and can actually only be inhibited (Newman and Müller 2006, Newman *et al.* 2006). This position not only permits one to account for convergence and parallel evolution, provided for by the traditional adaptive explanation, but also includes the frequent instances of similar forms (homoplasy) that arise in phylogenetically distant species.

If phenotypic evolution is predictable to a certain degree from the material properties and generative rules of the constituent developmental components (see Vermeij 2006) it implies that a new principle, inherency, is added to the external selection paradigm of evolutionary theory. It posits that the causal basis for phenotypic evolution resides not merely in population genetic events or, for that matter, in gene regulatory evolution, but in the inherent features of evolving developmental systems themselves. This defies the program notions that abound in present-day genetics (Moss 2003, Neumann-Held and Rehmann-Sutter 2006) in the sense that although genes specify cellular properties to a certain extent, the specific tissue structures and biological forms that result from cell interactions are not encoded in the genome in any deterministic fashion. Evo-devo aims at defining rules for these mechanistic genotype–phenotype relationships in evolution.

Through its inclusion of evolvability, emergence and inherency evo-devo takes evolutionary theory beyond the standard scope of the Modern Synthesis. This does not invalidate its neo-Darwinian core as a theory that accounts for the quantitative variation and fixation of traits in populations, but it shifts the attention to the qualitative phenomena of higher order phenotypic organisation and its mechanistic causes. The shifting emphases in evolutionary biology initiated by evo-devo are recognised by a growing number of metatheoretical, historical and philosophical accounts (Robert 2004, Amundson 2005, Laubichler and Maienschein 2007, Sansom and Brandon 2007). Without doubt the concepts introduced by evo-devo will represent important components of an extended evolutionary synthesis.

ACKNOWLEDGEMENTS

I dedicate this chapter to the memory of Pere Alberch, whose groundbreaking experimental and theoretical work in the early 1980s was a major inspiration of what we call evo-devo today. This is a reworked and updated version of an earlier essay in the *Handbook of Evolution* (Wuketits and Ayala 2005). I thank Alessandro Minelli and Giuseppe Fusco for their invitation to participate in the Venice meeting and this volume.

REFERENCES

Abouheif, E. 1997. Developmental genetics and homology: a hierarchical approach. *Trends in Ecology and Evolution* **12**, 405–408.

Akam, M. 1989. Hox and HOM: homologous gene clusters in insects and vertebrates. *Cell* **57**, 347–349.

Akam, M. 1994. The evolving role of Hox genes in arthropods. *Development 1994 Supplement*, 209–215.

Alberch, P. 1982. Developmental constraints in evolutionary processes. In J. T. Bonner (ed.) *Evolution and Development*. Berlin: Springer, pp. 313–332.

Alberch, P. & Gale, E. A. 1983. Size dependence during the development of the amphibian foot. Colchicine-induced digital loss and reduction. *Journal of Embryology and Experimental Morphology* **76**, 177–197.

Alberch, P. & Gale, E. A. 1985. A developmental analysis of an evolutionary trend: digital reduction in amphibians. *Evolution* **39**, 8–23.

Amundson, R. 2005. *The Changing Role of the Embryo in Evolutionary Thought*. Cambridge: Cambridge University Press.

Arthur, W. 2001. Developmental drive: an important determinant of the direction of phenotypic evolution. *Evolution & Development* **3**, 271–278.

Arthur, W. 2002. The emerging conceptual framework of evolutionary developmental biology. *Nature* **415**, 757–764.

Beldade, P., Brakefield, P. & Long, A. D. 2005. Generating phenotypic variation: prospects from "evo-devo" research on *Bicyclus anynana* wing patterns. *Evolution & Development* **7**, 101–107.

Bell, M. A. 1987. Interacting evolutionary constraints in pelvic reduction of three-spine sticklebacks, *Gasterosteus aculeatus* (Pisces, Gasterosteidae). *Biological Journal of the Linnean Society* **31**, 347–382.

Bengtson, S. & Zhao, Y. 1997. Fossilized metazoan embryos from the earliest Cambrian. *Science* **277**, 1645–1648.

Bolker, J. A. & Raff, R. A. 1996. Developmental genetics and traditional homology. *BioEssays* **18**, 489–494.

Bonner, J. T. 1965. *Size and Cycle*. Princeton: Princeton University Press.

Bonner, J. T. (ed.) 1982. *Evolution and Development*. Berlin: Springer.

Bonner, J. T. 1988. *The Evolution of Complexity by Means of Natural Selection*. Princeton: Princeton University Press.

Bookstein, F. L. 1991. *Morphometric Tools for Landmark Data: Geometry and Biology*. Cambridge: Cambridge University Press.

Burke, A. C., Nelson, C. E., Morgan, B. A. & Tabin, C. 1995. *Hox* genes and the evolution of vertebrate axial morphology. *Development* **121**, 333–346.

Buss, L. W. 1987. *The Evolution of Individuality*. Princeton: Princeton University Press.

Caldwell, M. W. 1994. Developmental constraints and limb evolution in Permian and extant lepidosauromorph diapsids. *Journal of Vertebrate Paleontology* **14**, 459–471.

Callebaut, W., Müller, G. B. & Newman, S. A. 2007. The organismic systems approach: EvoDevo and the streamlining of the naturalistic agenda. In R. Sansom & B. Brandon (eds.) *Integrating Evolution and Development: From Theory to Practice*, in press. Cambridge, MA: MIT Press.

Callebaut, W. & Rasskin-Gutman, D. (eds.) 2005. *Modularity: Understanding the Development and Evolution of Complex Natural Systems*. Cambridge, MA: MIT Press.

Carpenter, K., Hirsch, K. F. & Horner, J. R. 1996. *Dinosaur Eggs and Babies*. Cambridge: Cambridge University Press.

Carroll, S. B., Grenier, J. K. & Weatherbee, S. D. 2005. *From DNA to Diversity*. Malden: Blackwell Science.

Cheverud, J. M. 1982. Phenotypic, genetic, and environmental morphological integration in the cranium. *Evolution* **36**, 499–516.

Cheverud, J. M. 1984. Quantitative genetics and developmental constraints on evolution by selection. *Journal of Theoretical Biology* **110**, 155–171.

Chipman, A. D. 2002. Variation, plasticity, and modularity in anuran development. *Zoology* **105**, 97–104.

Cobb, S. N. & O'Higgins, P. 2004. Hominins do not share a common postnatal facial ontogenetic shape trajectory. *Journal of Experimental Zoology B (Molecular and Developmental Evolution)* **302**, 302–321.

Collins, J. P., Gilbert, S., Laubichler, M. & Müller, G. B. 2007. Modeling in Evo-Devo: how to integrate development, evolution, and ecology. In M. Laubichler, & G. B. Müller (eds.) *Modeling Biology: Structures, Behaviors, Evolution*. Cambridge, MA: MIT Press, pp. 355–378.

Costa, L. D. F., Travençolo, B. A. N., Azeredo, A. *et al.* 2005. A field approach to three-dimensional gene expression pattern characterization. *Applied Physics Letters* **86**, 143901–143903.

Davey, K. G. 2003. Evolutionary aspects of thyroid hormone effects in invertebrates. In B. K. Hall, B. J. Pearson & G. B. Müller (eds.) *Environment, Development, and Evolution.* Cambridge, MA: MIT Press, pp. 279–295.

Davidson, E. H. 2006. *The Regulatory Genome: Gene Regulatory Networks in Development and Evolution.* San Diego: Academic Press.

Davidson, E. H., Peterson, K. & Cameron, R. A. 1995. Origin of the adult bilaterian body plans: evolution of developmental regulatory mechanisms. *Science* **270**, 1319–1325.

Deacon, T. W. 2006. Reciprocal linkage between self-organized processes is sufficient for self-reproduction and evolvability. *Biological Theory* **1**, 136–149.

Dollé, P., Dierich, A., LeMeur, M. *et al.* 1993. Disruption of the *Hoxd-13* gene induces localized heterochrony leading to mice with neotenic limbs. *Cell* **75**, 431–441.

Donoghue, M. J. & Ree, R. H. 2000. Homoplasy and developmental constraint: a model and an example from plants. *American Zoologist* **40**, 759–769.

Duboule, D. 1994. Temporal colinearity and the phylotypic progression: a basis for the stability of a vertebrate bauplan and the evolution of morphologies through heterochrony. *Development 1994 supplement*, 135–142.

Duboule, D. & Dollé, P. 1989. The structural and functional organisation of the mouse *Hox* gene family resembles that of *Drosophila* homeotic genes. *EMBO Journal* **8**, 1497–1505.

Dudley, M. & Poethig, R. S. 1991. The effect of a heterochronic mutation, Teopod2, on the cell lineage of the maize shoot. *Development* **111**, 733–739.

Dworkin, I. M., Tanda, S. & Larsen, E. 2001. Are entrenched characters developmentally constrained? Creating biramous limbs in an insect. *Evolution & Development* **3**, 424–431.

Eberhard, W. G. 2001. Multiple origins of a major novelty: moveable abdominal lobes in male sepsid flies (Diptera: Sepsidae), and the question of developmental constraints. *Evolution & Development* **3**, 206–222.

Eble, G. J. 2002. Multivariate approaches to development and evolution. In N. Minugh-Purvis & K. J. McNamara (eds.) *Human Evolution Through Developmental Change.* Baltimore: Johns Hopkins University Press, pp. 51–78.

Edelman, G. M. 1988. *Topobiology.* New York: Basic Books.

Eldredge, N. & Gould, S. J. 1972. Punctuated equilibria: an alternative to phyletic gradualism. In T. J. M. Schopf (ed.) *Models in Paleobiology.* San Francisco: Freeman, Cooper, pp. 82–115.

Emlen, D. J. 2000. Integrating development with evolution: A case study with beetle horns. *BioScience* **50**, 403–418.

Fusco, G. 2001. How many processes are responsible for phenotypic evolution? *Evolution & Development* **3**, 279–286.

Gilbert, S. F. 2001. Ecological developmental biology: developmental biology meets the real world. *Developmental Biology* **233**, 1–32.

Gilbert, S. F. & Bolker, J. A. 2001. Homologies of process and modular elements of embryonic construction. *Journal of Experimental Zoology B (Molecular and Developmental Evolution)* **291**, 1–12.

Gilbert, S. F. & Bolker, J. A. 2003. Ecological developmental biology: preface to the symposium. *Evolution & Development* **5**, 3–8.

Goldschmidt, R. 1940. *The Material Basis of Evolution.* New Haven: Yale University Press.

Goodwin, B. C., Holder, N. & Wylie, C. C. (eds.) 1983. *Development and Evolution*. Cambridge: Cambridge University Press.

Gould, S. J. 1977. *Ontogeny and Phylogeny*. Cambridge, MA: Harvard University Press.

Gould, S. J. & Lewontin, R. C. 1979. The spandrels of San Marco and the Panglossian paradigm: a critique of the adaptationist programme. *Proceedings of the Royal Society of London B* **205**, 581–598.

Graham, A., Papalopulu, N. & Krumlauf, R. 1989. The murine and *Drosophila* homeobox gene complex have common features of organisation and expression. *Cell* **57**, 367–378.

Haeckel, E. 1866. *Generelle Morphologie der Organismen*. Berlin: Reimer.

Hall, B. K. 1983. Epigenetic control in development and evolution. In B. C. Goodwin, N. Holder & C. G. Wylie (eds.) *Development and Evolution*. Cambridge: Cambridge University Press, pp. 353–379.

Hall, B. K. 1984. Developmental processes underlying heterochrony as an evolutionary mechanism. *Canadian Journal of Zoology* **62**, 1–7.

Hall, B. K. 1992. *Evolutionary Developmental Biology*. London: Chapman & Hall.

Hall, B. K., Pearson, B. J. & Müller, G. B. (eds.) 2003. *Environment, Development, and Evolution*. Cambridge, MA: MIT Press.

Hamburger, V. 1980. Embryology and the modern synthesis in evolutionary theory. In E. Mayr & W. B. Provine (eds.) *The Evolutionary Synthesis: Perspectives on the Unification of Biology*. Cambridge: Harvard University Press, pp. 96–112.

Hentschel, H. G., Glimm, T., Glazier, J. A. & Newman, S. A. 2004. Dynamical mechanisms for skeletal pattern formation in the vertebrate limb. *Proceedings of the Royal Society of London B* **271**, 1713–1722.

Holland, L. Z., Holland, P. W. & Holland, N. D. 1996. Revealing homologies between body parts of distantly related animals by *in situ* hybridization to developmental genes: amphioxus versus vertebrates. In J. D. Ferraris & S. R. Palumbi (eds.) *Molecular Zoology*. New York: Wiley-Liss, pp. 267–295.

Holland, P. 1992. Homeobox genes in vertebrate evolution. *BioEssays* **14**, 267–273.

Holland, P. W. 1999. Gene duplication: past, present, and future. *Seminars in Cell and Developmental Biology* **10**, 541–547.

Hämmerling, J. 1929. *Dauermodifikationen*. Berlin: Borntraeger.

Jablonka, E. & Lamb, M. J. 1995. *Epigenetic Inheritance and Evolution*. Oxford: Oxford University Press.

Jernvall, J. 2000. Linking development with generation of novelty in mammalian teeth. *Proceedings of the National Academy of Sciences of the USA* **97**, 2641–2645.

Jernvall, J., Keranen, S. V. & Thesleff, I. 2000. Evolutionary modification of development in mammalian teeth: quantifying gene expression patterns and topography. *Proceedings of the National Academy of Sciences of the USA* **97**, 14444–14448.

Jollos, V. 1934. Inherited changes produced by heat treatment in *Drosophila melanogaster*. *Genetica* **16**, 476–494.

Kammerer, P. 1923. Breeding experiments on the inheritance of acquired characters. *Nature* **111**, 637–640.

Kammerer, P. 1925. *Neuvererbung oder Vererbung erworbener Eigenschaften*. Stuttgart: Walter Seifert Verlag.

Katz, M. J. 1983. Ontophyletics: studying evolution beyond the genome. *Perspectives in Biology and Medicine* **26**, 323–333.

Katz, M. J., Lasek, R. J. & Kaiserman-Abramof, I. R. 1981. Ontophyletics of the nervous system: eyeless mutants illustrate how ontogenetic buffer mechanisms channel evolution. *Proceedings of the National Academy of Sciences of the USA* **78**, 397–401.

Keys, D. N., Lewis, D. L., Selegue, J. E. *et al.* 1999. Recruitment of a hedgehog regulatory circuit in butterfly eyespot evolution. *Science* **283**, 532–534.

Kim, J., Kerr, J. Q. & Min, G. S. 2000. Molecular heterochrony in the early development of *Drosophila*. *Proceedings of the National Academy of Sciences of the USA* **97**, 212–216.

Kirschner, M. & Gerhart, J. 1998. Evolvability. *Proceedings of the National Academy of Sciences of the USA* **95**, 8420–8427.

Kowalevsky, A. 1866. Entwicklungsgeschichte der einfachen Ascidien. *Mémoires de l'Académie Impériale des Sciences de St.-Pétersbourg* (7) **10**, 11–30.

Larsen, E. 2003. Genes, cell behavior, and the evolution of form. In G. B. Müller & S. A. Newman (eds.) *Origination of Organismal Form*. Cambridge, MA: MIT Press, pp. 119–131.

Laubichler, M. D. & Maienschein, J. (eds.) 2007. *From Embryology to Evo-Devo: A History of Developmental Evolution*. Cambridge, MA: MIT Press.

Laubichler, M. D. & Müller, G. B. (eds.) 2007. *Modeling Biology: Structures, Behaviors, Evolution*. Cambridge, MA: MIT Press.

Lee, P. N., Callaerts, P., De Couet, H. G. & Martindale, M. Q. 2003. Cephalopod Hox genes and the origin of morphological novelties. *Nature* **424**, 1061–1065.

Love, A. C. 2003. Evolutionary morphology, innovation, and the synthesis ofevolutionary and developmental biology. *Biology and Philosophy* **18**, 309–345.

Love, A. C. & Raff, R. 2005. Larval ectoderm, organizational homology, and the origins of evolutionary novelty. *Journal of Experimental Zoology B (Molecular and Developmental Evolution)* **306**, 18–34.

Lovejoy, C. O., Cohn, M. J. & White, T. D. 1999. Morphological analysis of the mammalian postcranium: a developmental perspective. *Proceedings of the National Academy of Sciences of the USA* **96**, 13247–13252.

Løvtrup, S. 1974. *Epigenetics*. London: Wiley.

Mabee, P. M. 2000. Developmental data and phylogenetic systematics: evolution of the vertebrate limb. *American Zoologist* **40**, 789–800.

Margulis, L. & Fester, R. (eds.) 1991. *Symbiosis as a Source of Evolutionary Innovation*. Cambridge: MIT Press.

Matsuda, R. 1987. *Animal Evolution in Changing Environments with Special Reference to Abnormal Metamorphosis*. New York: Wiley.

Maynard Smith, J., Burian, R., Kauffman, S. *et al.* 1985. Developmental constraints and evolution. *The Quarterly Review of Biology* **60**, 265–287.

Mayr, E. 1960. The emergence of evolutionary novelties. In S. Tax (ed.) *Evolution After Darwin*. Chicago: University of Chicago Press, pp. 349–380.

McGhee, G. R. 2007. *The Geometry of Evolution*. Cambridge: Cambridge University Press.

McGinnis, W., Garber, R. L., Wirz, J., Kuroiwa, A. & Gehring, W. J. 1984. A homologous protein-coding sequence in *Drosophila* homeotic genes and its conservation in other metazoans. *Cell* **37**, 403–408.

McGinnis, W. & Krumlauf, R. 1992. Homeobox genes and axial patterning. *Cell* **68**, 283–302.

McKinney, M. L. & McNamara, K. J. 1991. *Heterochrony*. New York: Plenum Press.

McNamara, K. J. 1997. *Shapes of Time: The Evolution of Growth and Development*. Baltimore: Johns Hopkins University Press.

Minelli, A. 1998. Molecules, developmental modules, and phenotypes: a combinatorial approach to homology. *Molecular Phylogenetics and Evolution* **9**, 340–347.

Minelli, A. 2003. *The Development of Animal Form: Ontogeny, Morphology, and Evolution*. Cambridge: Cambridge University Press.

Mitteroecker, P., Gunz, P., Bernhard, M., Schaefer, K. & Bookstein, F. 2004. Comparison of cranial ontogenetic trajectories among hominoids. *Journal of Human Evolution* **46**, 679–697.

Moss, L. 2003. *What Genes Can't Do*. Cambridge, MA: MIT Press.

Müller, G. B. 1986. Effects of skeletal change on muscle pattern formation. *Bibliotheca Anatomica* **29**, 91–108.

Müller, G. B. 1989. Ancestral patterns in bird limb development: a new look at Hampé's experiment. *Journal of Evolutionary Biology* **2**, 31–47.

Müller, G. B. 1990. Developmental mechanisms at the origin of morphological novelty: a side-effect hypothesis. In M. H. Nitecki (ed.) *Evolutionary Innovations*. Chicago: Chicago Press, pp. 99–130.

Müller, G. B. 1991. Experimental strategies in evolutionary embryology. *American Zoologist* **31**, 605–615.

Müller, G. B. 2003. Homology: the evolution of morphological organization. In G. B. Müller & S. A. Newman (eds.) *Origination of Organismal Form*. Cambridge, MA: MIT Press, pp. 51–69.

Müller, G. B. 2007. Six memos for EvoDevo. In M. D. Laubichler & J. Maienschein (eds.) *From Embryology to Evo-Devo: A History of Embryology in the 20th Century*, in press. Cambridge, MA: MIT Press.

Müller, G. B. & Newman, S. A. (eds.) 2003. *Origination of Organismal Form: Beyond the Gene in Development and Evolution*. Cambridge, MA: MIT Press.

Müller, G. B. & Newman, S. A. (eds.) 2005a. Evolutionary innovation and morphological novelty. *Special Issue of the Journal of Experimental Zoology B (Molecular and Developmental Evolution)* **304** (6).

Müller, G. B. & Newman, S. A. 2005b. The innovation triad: an EvoDevo agenda. *Journal of Experimental Zoology B (Molecular and Developmental Evolution)* **304**, 487–503.

Neumann-Held, E. & Rehmann-Sutter, C. 2006. *Genes in Development: Rereading the Molecular Paradigm*. Durham, NC: Duke University Press.

Newlon, A. W., III, Yund, P. O. & Stewart-Savage, J. 2003. Phenotypic plasticity of reproductive effort in a colonial ascidian, *Botryllus schlosseri*. *Journal of Experimental Zoology A (Comparative and Experimental Biology)* **297**, 180–188.

Newman, S. A. 1994. Generic physical mechanisms of tissue morphogenesis: a common basis for development and evolution. *Journal of Evolutionary Biology* **7**, 167–188.

Newman, S. A., Forgacs, G. & Müller, G. B. 2006. Before programs: the physical origination of multicellular forms. *International Journal of Developmental Biology* **50**, 289–299.

Newman, S. A. & Müller, G. B. 2000. Epigenetic mechanisms of character origination. *Journal of Experimental Zoology B (Molecular and Developmental Evolution)* **288**, 304–317.

Newman, S. A. & Müller, G. B. 2005. Origination and innovation in the vertebrate limb skeleton: an epigenetic perspective. *Journal of Experimental Zoology B (Molecular and Developmental Evolution)* **304**, 593–609.

Newman, S. A. & Müller, G. B. 2006. Genes and form: inherency in the evolution of developmental mechanisms. In E. Neumann-Held & C. Rehmann-Sutter (eds.) *Genes in Development: Rereading the Molecular Paradigm*. Durham, NC: Duke University Press, pp. 38–73.

Nijhout, H. F. 1999. Control mechanisms of polyphenic development in insects. *BioScience* **49**, 181–192.

Parichy, D. M., Shaffer, H. B. & Mangel, M. 1992. Heterochrony as a unifying theme in evolution and development. *Evolution* **46**, 1252–1254.

Parks, A. L., Parr, B. A., Chin, J.-E., Leaf, D. S. & Raff, R. A. 1988. Molecular analysis of heterochronic changes in the evolution of direct developing sea urchins. *Journal of Evolutionary Biology* **1**, 27–44.

Pigliucci, M. 2001. *Phenotypic Plasticity: Beyond Nature and Nurture*. Baltimore: Johns Hopkins University Press.

Prossinger, H. & Bookstein, F. 2003. Statistical estimators of frontal sinus cross section ontogeny from very noisy data. *Journal of Morphology* **257**, 1–8.

Raff, R. 1996. *The Shape of Life*. Chicago: Chicago University Press.

Raff, R. A., Anstrom, J. A., Huffman, C. J. *et al.* 1984. Origin of a gene regulatory mechanism in the evolution of echinoderms. *Nature* **310**, 312–314.

Raff, R. A. & Kaufman, T. C. 1983. *Embryos, Genes, and Evolution*. New York: Macmillan.

Raff, R. A. & Wray, G. A. 1989. Heterochrony: developmental mechanisms and evolutionary results. *Journal of Evolutionary Biology* **2**, 409–434.

Rasmussen, N. 1987. A new model of developmental constraints as applied to the *Drosophila* system. *Journal of Theoretical Biology* **127**, 271–299.

Rasskin-Gutman, D. 2003. Boundary constraints for the emergence of form. In G. B. Müller & S. A. Newman (eds.) *Origination of Organismal Form*. Boston, MA: MIT Press, pp. 305–322.

Reid, R. G. B. 2007. *Biological Emergences: Evolution by Natural Experiment*. Cambridge, MA: MIT Press.

Richards, R. J. 1992. *The Meaning of Evolution*. Chicago: University of Chicago Press.

Riedl, R. 1978. *Order in Living Organisms*. Chichester: John Wiley & Sons.

Robert, J. S. 2004. *Embryology, Epigenesis, and Evolution: Taking Development Seriously*. Cambridge: Cambridge University Press.

Rose, C. S. 2003. Thyroid hormone-mediated development in vertebrates: what makes frogs unique. In B. K. Hall, B. J. Pearson & G. B. Müller (eds.) *Environment, Development, and Evolution*. Cambridge, MA: MIT Press, pp. 197–237.

Rubin, H. 1990. On the nature of enduring modifications induced in cells and organisms. *American Journal of Physiology* **258**, 19–24.

Ruvkun, G. & Giusto, J. 1989. The *Caenorhabditis elegans* heterochronic gene *lin-14* encodes a nuclear protein that forms a temporal developmental switch. *Nature* **338**, 313–319.

Salazar-Ciudad, I., Newman, S. A. & Solé, R. V. 2001. Phenotypic and dynamical transitions in model genetic networks. I. Emergence of patterns and genotype-phenotype relationships. *Evolution & Development* **3**, 84–94.

Sansom, R. & Brandon, B. (eds.) 2007. *Integrating Evolution and Development: From Theory to Practice*. Cambridge, MA: MIT Press.

Sarkar, S. 2003. Generalized norms of reaction for ecological developmental biology. *Evolution & Development* **5**, 106–115.

Schlichting, C. & Pigliucci, M. 1998. *Phenotypic Evolution: A Reaction Norm Perspective*. Sunderland: Sinauer.

Schlosser, G. & Wagner, G. P. (eds.) 2004. *Modularity in Development and Evolution*. Chicago: University of Chicago Press.

Sharpe, J., Ahlgren, U., Perry, P. *et al.* 2002. Optical projection tomography as a tool for 3D microscopy and gene expression studies. *Science* **296**, 541–545.

Shubin, N., Tabin, C. & Carroll, S. 1997. Fossils, genes and the evolution of animal limbs. *Nature* **388**, 639–648.

Steinberg, M. S. 1963. Reconstruction of tissues by dissociated cells. *Science* **141**, 401–408.

Stern, D. L. 1998. A role of Ultrabithorax in morphological differences between *Drosophila* species. *Nature* **396**, 463–466.

Streicher, J., Donat, M. A., Strauss, B. *et al.* 2000. Computer based three-dimensional visualization of developmental gene expression. *Nature Genetics* **25**, 147–152.

Streicher, J. & Müller, G. B. 1992. Natural and experimental reduction of the avian fibula: developmental thresholds and evolutionary constraint. *Journal of Morphology* **214**, 269–285.

Sultan, S. E. 2003. Phenotypic plasticity in plants: a case study in ecological development. *Evolution & Development* **5**, 25–33.

Thomas, R. D. K. & Reif, W.-E. 1993. The skeleton space: a finite set of organic designs. *Evolution* **47**, 341–360.

Tollrian, R. & Harvell, C. D. (eds.) 1999. *The Ecology and Evolution of Inducible Defenses.* Princeton: Princeton University Press.

True, J. R. & Carroll, S. B. 2002. Gene co-option in physiological and morphological evolution. *Annual Review of Cell and Developmental Biology* **18**, 53–80.

Vermeij, G. J. 2006. Historical contingency and the purported uniqueness of evolutionary innovations. *Proceedings of the National Academy of Sciences of the USA* **103**, 1804–1809.

Vogl, C. & Rienesel, J. 1991. Testing for developmental constraints: carpal fusions in urodeles. *Evolution* **45**, 1516–1519.

von Dassow, G. & Munro, E. 1999. Modularity in animal development and evolution: elements of a conceptual framework for EvoDevo. *Journal of Experimental Zoology B (Molecular and Developmental Evolution)* **285**, 307–325.

Vrba, E. S. 2003. Ecology, development, and evolution: perspectives from the fossil record. In B. K. Hall, B. J. Pearson & G. B. Müller (eds.) *Environment, Development, and Evolution.* Cambridge, MA: MIT Press, pp. 85–105.

Waddington, C. H. 1942. Canalization of development and the inheritance of acquired characters. *Nature* **150**, 563–565.

Waddington, C. H. 1956. Genetic assimilation. *Advances in Genetics* **10**, 257–290.

Wagner, A. 2005. *Robustness and Evolvability in Living Systems.* Princeton: Princeton University Press.

Wagner, G. P. 1988. The influence of variation and of developmental constraints on the rate of multivariate phenotypic evolution. *Journal of Evolutionary Biology* **1**, 45–66.

Wagner, G. P. 1989. The biological homology concept. *Annual Review of Ecology and Systematics* **20**, 51–69.

Wagner, G. P. 1996. Homologues, natural kinds, and the evolution of modularity. *American Zoologist* **36**, 36–43.

Wagner, G. P. 2000. What is the promise of developmental evolution? Part I: Why is developmental biology necessary to explain evolutionary innovations? *Journal of Experimental Zoology B (Molecular and Developmental Evolution)* **288**, 95–98.

Wagner, G. P. 2001. What is the promise of developmental evolution? Part II: A causal explanation of evolutionary innovations may be impossible. *Journal of Experimental Zoology B (Molecular and Developmental Evolution)* **291**, 305–309.

Wagner, G. P. & Altenberg, L. 1996. Complex adaptations and the evolution of evolvability. *Evolution* **50**, 967–976.

Wagner, G. P. & Chiu, C. H. 2001. The tetrapod limb: a hypothesis on its origin. *Journal of Experimental Zoology B (Molecular and Developmental Evolution)* **291**, 226–240.

Wagner, G. P. & Müller, G. B. 2002. Evolutionary innovations overcome ancestral constraints: a re-examination of character evolution in male sepsid flies (Diptera: Sepsidae). *Evolution & Development* **4**, 1–6.

Wake, D. B. 1982. Functional and developmental constraints and opportunities in the evolution of feeding systems in urodeles. In D. Mossakowski & G. Roth (eds.) *Environmental Adaptation and Evolution.* Stuttgart: Gustav Fischer, pp. 51–66.

Wake, D. B. 1996. Evolutionary developmental biology – prospects for an evolutionary synthesis at the developmental level. *Memoirs of the California Academy of Science* **20**, 97–107.

Webb, J. F. 1989. Developmental constraints and evolution of the lateral line system in teleost fishes. In S. Coombs, P. Gäner & H. Münz (eds.) *The Mechanosensory Lateral Line: Neurobiology and Evolution*. New York: Springer-Verlag, pp. 79–97.

Weninger, W. J., Geyer, S. H., Mohun, T. J. *et al.* 2006. High resolution episcopic microscopy: rapid 3D-analysis of gene expression and tissue architecture. *Anatomy and Embryology* **211**, 213–221.

West-Eberhard, M. J. 2003. *Developmental Plasticity and Evolution*. Oxford: Oxford University Press.

Wilkins, A. 2002. *The Evolution of Developmental Pathways*. Sunderland: Sinauer Associates.

Wolterek, R. 1909. Weitere experimentelle Untersuchungen über Artveränderung, speziell über das Wesen quantitativer Artunterschiede bei Daphniden. *Verhandlungen der Deutschen Zoologischen Gesellschaft* 1909, 110–172.

Wray, G. A. 1999. Evolutionary dissociations between homologous genes and homologous structures. In G. R. Bock & G. Cardew (eds.) *Homology*. Chichester: Wiley, pp. 189–203.

Wray, G. A. & Lowe, C. J. 2000. Developmental regulatory genes and echinoderm evolution. *Systematic Biology* **49**, 151–174.

Wray, G. A. & McClay, D. R. 1989. Molecular heterochronies and heterotopies in early echinoid development. *Evolution* **43**, 803–813.

Wuketits, F. M. & Ayala, F. J. (eds.) 2005. *Handbook of Evolution*, Vol. 2. Weinheim: Wiley-VCH.

Zákány, J., Gerard, M., Favier, B. & Duboule, D. 1997. Deletion of a *HoxD* enhancer induces transcriptional heterochrony leading to transposition of the sacrum. *EMBO Journal* **15**, 4393–4402.

2

Making evolutionary predictions about the structure of development and morphology: beyond the neo-Darwinian and constraints paradigms

ISAAC SALAZAR-CIUDAD

According to Darwinism, evolution occurs because, in populations, there is heritable phenotypic variation, and then ecological factors differentially affect the contribution of each of those variants to the next generation. Thus, to understand the way in which phenotypes in a population change over generations (this is the direction of evolutionary change) two questions need to be addressed: (1) which phenotypic variants arise in each generation, and (2) which of these variants are filtered out by ecological factors in each generation. In each generation, and assuming no dramatic genomic rearrangements, developmental dynamics determine which morphological variation arises from genetic and environmental variation. Developmental dynamics are currently not very well understood and thus the question of which phenotypic variants arise in each generation is not well understood either. A different emphasis is given to each of these two questions by different approaches or schools of thought in evolutionary biology.

For many evolutionary biologists, especially those close to the core of neo Darwinism, this lack of understanding about development has not always been perceived as a limit on progress in understanding morphological evolution (Haldane 1932, Mayr 1982). For some authors (Haldane 1932) this does not derive from lack of understanding development. Instead, natural selection is seen as the main or unique force determining how the phenotype changes (Fisher 1930,

Evolving Pathways: Key Themes in Evolutionary Developmental Biology, ed. Alessandro Minelli and Giuseppe Fusco. Published by Cambridge University Press.

Charlesworth *et al.* 1982). The question of which phenotypic variation arises and what its role is in determining how phenotypic distributions change over time is either not addressed or assumed to be unanswerable. Which variation is produced is essentially seen, according to the neo-Darwinian view, as unpredictable, other than that it should be abundant, small (gradual) and, in principle, possible in any trait.

These views about the role of development and variation in evolutionary theory have found many detractors during the past century, especially since the 1980s (Alberch 1980, Goodwin 1994, Newman and Müller 2000). This research claims that development has an important effect in evolution because it produces variation that differs from the variation assumed by the neo-Darwinian view. These studies introduce concepts such as developmental constraint, developmental bias, evolutionary novelty and robustness (Müller, Chapter 1 of this book) to describe aspects of evolution that are not conveniently approached from the Modern Synthesis. Often these approaches are not perceived as an integrated or consensus alternative to the Modern Synthesis nor as being in total agreement with it. Most of this research, which will be referred to here as the constraint school, involves conceptual (Alberch 1980, Horder 1989, Goodwin 1994, Fusco 2001, Arthur 2004) and experimental embryology studies (Richardson and Chipman 2003, Newman and Müller 2005) considering how development can affect morphological evolution, but not (with some exceptions: Newman and Müller 2000) how developmental mechanisms themselves evolve.

The advances in developmental genetics towards the end of the twentieth century have been perceived (Gilbert and Sarkar 2000) as a unique opportunity to integrate development into evolutionary theory. Many of these studies are concentrated on the search for the ultimate molecular differences underlying morphological differences between species. This does not directly explain which morphological variation is possible by genetic variation in development (question 1). Most of the research in developmental genetics is experimentally driven. The extraction of general theoretical insights to explain which morphological variation development can produce is just at its beginning, and it is still not clear how these insights will modify evolutionary theory or the relationship between the developmental constraint school and the Modern Synthesis.

This chapter critically reviews some advantages and disadvantages of these three schools of thought, neo-Darwinism, the developmental constraints school and developmental genetics, in understanding the way in which morphology changes in evolution (as described). It will be discussed how some assumptions of neo-Darwinism contrast with what is currently known about morphological variation and its developmental bases, how the constraint school and developmental genetics share or require some neo-Darwinian assumptions, and how research agendas in developmental genetics can be slightly rephrased so that they can be more easily used in the other two schools and, in general, for the study of how development affects evolution. A new set of concepts and studies that try to avoid these disadvantages will be explained. These will be used to argue that some aspects of the evolution of morphology and development are predictable if some information is available about development and selective pressures in previous generations. These new concepts allow us to approach questions that are not within reach for these three schools. Inferences will be made about the relative involvement of different types of developmental mechanisms in the evolution of morphology under different selective pressures. In general the aim is to explain how the study of pattern formation can help in understanding which morphological variation is possible in development (question 1) and, at the same time, how development itself evolves.

THE NEO-DARWINIAN APPROACH TO MORPHOLOGICAL EVOLUTION

Morphological variation is produced by the process of development. In order for natural selection to be the main force in determining how morphology changes it is required that (1) any small morphological variant is possible and (2) that genetic and morphological variation are linked by a simple relationship. If the first assumption does not hold true, then the direction of morphological change is also determined by which morphological variants are possible and/or more frequent. If the relationship between genotype and phenotype is not simple, then evolving from one morphology to another one that is more adaptive may not always be possible. This is either because morphologies produced by similar genotypes might be very different and then have very different fitness values, or because similar morphologies are not easily accessible through changes in development. Consequently, there is not always a path of permanently increasing fitness leading from one morphology to another.

Studies in population and quantitative genetics do not claim that any variation is possible. Instead, it is claimed that there is abundant additive genetic variation, that variation is of small magnitude and that it can be accumulated to eventually produce any trait value.

These conditions are necessary but not sufficient for selection to be the only force determining the direction of evolution. For that, it is also required that any combination of trait values is possible. In other words, if the morphology of an individual can be represented as a point in a hyperspace of trait values, any point should be reachable by accumulation of small mutational steps. This implies, since traits are arbitrary, that in each generation, variation is possible in any direction (any small change in trait value combinations). Defining what is a small or gradual change is relative and there has been lengthy discussion about it (Mayr 1982, Gould and Eldredge 1977). The question of the nature of morphological variation (question 1) has been directly addressed much less frequently (Alberch 1980, Goodwin 1994, Newman and Müller 2000, Salazar-Ciudad 2006b). These assumptions may be useful when studying simple univariate traits and, in fact, many artificial selection studies show that there is abundant, small, cumulative, heritable variation for these traits (Weber 1992, Broni-kowski *et al.* 2004). However, when morphology is described by several measurements, the variation observed is neither isotropic nor possible for all trait combinations. For some trait value combinations, there is no variation (the traits are highly correlated) while for others, there is variation.

There is no evidence that the relationship between phenotype and genotype is simple. In fact, one of the few general perceptions acquired from the study of development, and already claimed before the advent of modern developmental biology, is that this relationship is anything but simple (Horder 1993). Most genes affect several traits in a way that depends on the environment and/or the rest of the genotype. Some small genetic changes can give rise to dramatic phenotypic effects (Kangas *et al.* 2004) whereas sometimes relatively large genetic changes (such as gene deletions) can produce no obvious phenotypic changes in laboratory conditions (Wilkins 2001). Moreover, the same new pheno-type can be produced by different genetic changes (Alberch 1980, Horder 1989). As such, even when the question is only about the kinetics of replacement between morphological variants, developmental dynamics cannot be ignored. In other words, the adaptive landscape where each genotype has a fitness according to the morphology it produces is often complex because the mapping between genotype and morphology is often complex.

Studies of multivariate quantitative genetics do not rely on these assumptions about morphological variation. The relationship between genotype and phenotype is not assumed, and neither are the magnitude and nature of variation; instead, they are estimated statistically as the matrix of additive genetic covariance between traits. Again, however, the question is mainly about the kinetics of replacement between variants. Which variation appears is not explained but merely described. This approach cannot explain anything about the evolution of development or about how, and how often, G-matrices evolve. In addition, there are some methodological assumptions that are worth considering. Morphology is often measured on the basis of a set of traits that are (arguably) homologous between the individuals compared. With this approach, morphological variation changing the features that are used to define the traits themselves is not contemplated. The same situation applies to morphological changes that have no obvious homology with previous morphologies. This problem is relevant because this kind of variation is likely to be classified as novelty (Müller and Wagner 1991), and novelty is the kind of phenomenon that has been claimed to be difficult to explain from the neo-Darwinian approach. The mathematical apparatus of quantitative genetics requires that each trait value is determined by the sum of many genes or loci of small and invariable phenotypic effect. In other words, the effect of a gene does not depend on the other genes in the genotype. Quantitative genetic models with epistasis also exist but also assume phenotypic values determined by the summation of genetic effects. In addition, since all measurements are based on variances, these methods work better when phenotypic distributions are normal and in fact, the summing up of many genes of small effect warrants normal distributions. However, these assumptions contrast with what is known about development. The coordination of the interactions of gene products during development is hardly describable as the addition of the effects of genes that are independent of the genetic background. On the contrary, the role of a gene, and which traits it affects, normally depends on the rest of the genes in a genome. In other words, epistasis is more often the rule than the exception. As a result, G-matrix estimations are often variable with the genetic background (Carlborg and Haley 2004). This problem is likely to be more important for complex morphologies (Salazar-Ciudad 2006a). In fact, in those few cases in which genetic and cellular interactions have been mathematically modelled to reproduce accurately, through development, complex adult morphologies (Salazar-Ciudad and Jernvall 2002, Harris *et al.* 2005, Newman and Müller 2005) the genetic architecture of morphological variation has turned out to be totally different from the one assumed by quantitative genetic inferences (phenotypic traits are not determined by summation of fixed genetic effects).

THE CONSTRAINT SCHOOL

Many studies in the early 1980s and afterwards (Alberch 1980, Goodwin 1994, Newman and Müller 2000) have argued that, even if natural selection cannot always act on how the phenotypes are produced (Mayr 1982), understanding developmental dynamics is important because it determines which kind of morphological variation can appear from genetic variation (question 1). This is normally stated as development constraining (Maynard Smith *et al.* 1985) or biasing morphological evolution. These concepts are devised to contrast with the previous view in which any small morphological variation is possible. However, all morphological variation is produced by development and because of development. This means that even the neo-Darwinian assumptions on variation logically require development to work in a specific way (as described in the previous section and in Salazar-Ciudad 2006a). However, the question is not whether development constrains evolution (or whether there are developmental constraints) because development is always there. In fact, contrary to what has been proposed, developmental constraints cannot be tested (Beldade *et al.* 2002; see Salazar-Ciudad 2006a for a discussion). The question is, instead, how different kinds of development affect evolution. This question should be addressed by looking at which variation each kind of development can produce (and not which imaginable variation they cannot produce): these are called here the variational properties of a type of development (Salazar-Ciudad *et al.* 2003).

Another central concept in the criticisms to the neo-Darwinian approach to morphological evolution is novelty. Many definitions consider novelty (Müller and Wagner 1991) as phenotypic change that does not fit into the scheme of small quantitative changes in existing traits (for example, the appearance of new traits). It is often proposed that novelty occurs rarely, while more gradual quantitative changes occur often. Novelty research often points out that the qualitative aspects of morphological variation may be related to developmental dynamics. However, since neo-Darwinism does not consider development, nor any intrinsic patterns of morphological variation appearing from it, the distinction between rare qualitative changes and common quantitative changes may not be the best way to describe morphological variation. If what is known about the morphological variation possible by development is taken into account (Salazar-Ciudad 2006b), then there are several kinds of possible morphologies, each of them with its specific multivariate pattern of variation. In a sense, the concept of novelty is useful only as far as the gradual totipotent view of variation is accepted.

THE DEVELOPMENTAL GENETICS SCHOOL

The understanding of the genetic bases of development has increased dramatically during the past 30 years. Interest in development had not always come, as in the case of the constraint school, from the perception that there is a conflict between our current understanding of morphological variation and development and the conceptual framework of the Modern Synthesis. The aim of the developmental genetics school of evo-devo is often perceived as the identification of the genetic bases underlying morphological differences in evolution (Wilkins 2001). Here I argue that most studies in evo-devo are not designed to address question 1.

Many researchers in developmental genetics have become interested in evolutionary issues after the discovery that many important genes in development are conserved across animals. Often, the same genes, gene interactions and similar patterns of gene expression are involved in the development of similar morphological structures in different species. Many research papers in evo-devo consist of standard research in developmental genetics with some comparative discussions at the end. By comparing developmental genetics between different species, essential data about variation in development are acquired. These studies identify differences in the patterns of expression of homologous genes (Patel 1994, Abzhanov and Kaufman 1999), differences in the target genes of homologous transcriptional or signalling pathways (Akam 1998), or even differences in the *cis*-regulatory elements of homologous genes (Carroll 2000), that correlate with morphological differences between species. These studies show correlations between genetic differences and morphological differences between species but do not show which were the original genetic changes that gave rise to these differences. This is not even the case when experimental mutations in one species produce morphologies reminiscent of the other species (Solé *et al.* 1999, Nijhout 2002). Even between closely related species the genetic structure of development may have changed since the time the morphological differences under study appeared. Then the genetic differences currently found may differ from the ones originally involved in the formation of a morphological structure. It has even been suggested that in many cases (Salazar-Ciudad *et al.* 2001a,b, Newman and Müller 2000) the developmental mechanisms underlying a specific morphology can change during evolution.

Although identifying these genetic differences may be a first necessary step, it does not in itself explain which patterns of morphological variation are produced by development or how development evolves. A gene difference, a difference of gene expression or a difference in a target gene does not explain a morphological difference unless the network of genetic, cellular and epigenetic interactions in which it is acting is causally understood. In other words, single genes or even small groups of gene interactions do not constitute mechanisms to explain morphological variation.

In fact, the problem of morphological variation can be seen, by using adequate definitions, to be the same as the problem of pattern formation: that is, how relatively simple spatial distributions of cell types and extracellular components (hereafter called a pattern or previous pattern) are transformed, during development time, into different (normally more complex) spatial distributions of cell types and extracellular components (hereafter called a final pattern). From these definitions development can be seen as a sequence of these transformations between patterns; morphology and its variation as these patterns and their variation. Causal explanations of development require a description of (1) the network of gene interactions involved in a specific pattern transformation, (2) how these affect basic cell behaviours (e.g. proliferation, apoptosis, adhesion, signal secretion and reception, etc.), (3) how these networks and cell behaviours interact with the epigenetic information existing in the previous pattern and in other parts of the embryo (i.e. the spatial distributions of extracellular signals and components, and the spatial distribution and mechanical properties of cells) and (4) how the patterns produced will change when (1), (2) or (3) change. These networks of interactions are herein called developmental mechanisms (Salazar-Ciudad *et al.* 2003) while the pattern variations they can produce are called the variational properties of a developmental mechanism. Without an understanding of how pattern formation takes place in a morphological structure, the identification of a genetic difference underlying a morphological difference between species cannot be meaningfully interpreted. However, hypotheses about pattern formation that are based on partially known gene networks and previous patterns can be used to give tentative explanations about the morphological differences between species. These hypotheses can be refuted or refined by comparing their predictions with morphological variation between mutants and between species. This approach, here called 'variational', has been used in some previous studies (Goodwin 1994, Holloway and Harrison 1999, Salazar-Ciudad and Jernvall 2002, Harris *et al.* 2005, Jaeger *et al.* 2004, Newman and Müller 2005) and has the advantage of explaining which morphological variation is possible by development, rather than only which are the genetic changes underlying particular morphological variants.

Even if pattern formation is one of the more apparent and important phenomena in development, explicit hypotheses about the mechanisms for specific pattern transformations are rare. Even more rarely is it demonstrated how proposed hypotheses can explain the production of a pattern and its variation from previous patterns and networks of genetic and cellular interactions. For complex pattern transformations, verbal arguments are unlikely to be able to describe precisely how the patterns will change according to a given hypothesis when mutations or previous pattern changes occur. In fact, the most explicitly formulated hypotheses that exist include mathematical models that give precise morphological quantitative descriptions about the patterns produced (Holloway and Harrison 1999, Salazar-Ciudad and Jernvall 2002, Harris *et al.* 2005, Newman and Müller 2005). Research with explicit pattern formation hypotheses based on networks of gene, cell and epigenetic context interactions are relatively rare. Most of the research is more focused on identifying genes, and gene or tissue interactions (and their timing), that are required for the development of some morphological structure. Even when the focus is on pattern formation, it is rarely stated from which previous pattern a pattern appears. Moreover, the reasons for the formation of a pattern are sometimes indicated as being due to the existence of a previous pattern. Although most pattern transformations occur from spatially heterogeneous patterns, the question of pattern formation cannot be passed back to previous patterns in some kind of preformationist chain (Horder 1993). In each developmental stage some spatial information (in the form of spatial distributions of cells) is produced from the previous one. How that happens is the question of pattern formation. Inevitably the explanation of the transformation between a previous pattern and a later pattern requires a precise description of both patterns. Accurate descriptions of both final morphology and intermediate embryonic morphologies are, however, relatively rare (see for exceptions Jernvall *et al.* 2000, Streicher and Müller 2001, Kuszak *et al.* 2004). The same occurs for patterns of gene expression, even though it is well known that they change in complex ways in space and time. Somehow, in the same way that some researchers (Dobzhansky 1937) tried to reduce evolution to changes in gene frequencies, some developmental biologists see development as reducible to gene interactions, or even *cis*-regulatory regions (Carroll 2000), both at the causal level and at the level of explanation.

THE EVOLUTION OF DEVELOPMENT

For many developmental biologists there may not be any general principles for the evolution of development (Akam 1998). The claimed opportunistic nature of evolutionary change (Sander and Schmidt-Ott 2004) may be one of the reasons for that. Although for short timescales some genetic changes may seem more likely than others, evolution will, over time, use whatever genetic variation is at hand to produce adaptive variation. As a consequence, previous history about genetic changes may easily fade away over time, precluding predictions about development evolution. Moreover, it has been suggested that, since selection cannot see how the phenotype is produced, mutations that do not alter how development works but do increase its genetic complexity may accumulate over time. This would result in the logic of development becoming more and more baroque over time (True and Haag 2001, Salazar-Ciudad 2006a). Here it will be argued, however, that precisely because of the opportunistic nature of evolution, general predictions about the evolution of development and the effect of development in evolution are possible. These predictions do not involve the exact molecular nature of changes in development but rather involve more general aspects about the logic and topology of genetic networks and how they interact within the developing epigenetic context of the embryo or intermediate phenotype.

In the same way that development can be described as a sequence of pattern transformations, it can also be described as a sequence of action of developmental mechanisms (each mechanism responsible for a specific pattern transformation). Then the evolution of development can be described as changes in the developmental mechanisms used in a lineage over time. These changes can occur because new developmental mechanisms are recruited or because existing developmental mechanisms are replaced by other ones. In this chapter two developmental mechanisms are considered to be different if their gene network topologies are different or if they affect different cell behaviours. This definition is merely a choice of convenience that reflects the fact that gene network topology is among the best-known aspects of developmental mechanisms. Moreover, topological changes have been suggested to have, on average, more dramatic effects on variational properties than other kinds of change (Salazar-Ciudad *et al.* 2000). From this perspective, the morphological outcome of development can change for two reasons: (1) because mutations (or environment) change which subset of the variational properties of some developmental mechanisms are produced, or

(2) because mutations (or environment) change some developmental mechanisms (meaning topology and then, probably, variational properties).

The chances of a de novo formation of a developmental mechanism depend on its genetic structure. The more genes and gene interactions a developmental mechanism involves, or in other words the more base changes in more genes are required, the more unlikely a mechanism will appear through random mutation. The same applies for the number of genetic changes required to evolve one developmental mechanism from another. The likelihood of an existing developmental mechanism being recruited in a new place and time in the developing embryo depends, instead, on the proportion of genes in the network that can lead to the activation of the whole mechanism.

Developmental mechanisms are not selected by themselves nor by their variational properties but by the specific morphology they produce in specific individuals (which is only a part of the variational properties). If the genetic and variational properties of developmental mechanisms are known, then the evolution of development can be understood by how likely it is that different developmental mechanisms appear or are recruited by mutation, and how likely it is that the morphologies they produce are adaptive in specific environments. Only a limited number of developmental mechanisms are known to that extent. However, it has been suggested on theoretical grounds that there is a limited number of types of gene network topologies that can produce pattern transformations (Salazar-Ciudad *et al.* 2000, 2003). Pattern formation capacity and opportunism (in the form of likelihood of appearance by mutation) confine the spectrum of developmental mechanisms that can possibly be involved in different situations in development and evolution. This variational approach can be useful because, at least, all developmental mechanisms can be classified into exclusive types that share some aspects of their genetic and variational properties. Predictions about the relative involvement of each type at different stages of development and under different selective pressures are then possible. The next sections offer examples of how developmental mechanisms can be classified and how this approach can give interesting evolutionary insights. This 'variational' approach can be used on different classifications (for example Salazar-Ciudad *et al.* 2001a) and is not dependent on the specific one presented in the next section.

MORPHODYNAMIC MECHANISMS VERSUS POSITIONAL
INFORMATION

One widespread view about how pattern formation takes place is positional information (Wolpert 1989): cells in a territory acquire information (positional information) about their fate (or differentiation state) according to the specific concentration of a diffusible signal they receive from a source at the border of a territory. In that sense two diffusible signals (morphogens) can establish a two-dimensional coordinate system according to which cells determine their positional values. These morphogens determine the genetic program that a cell undertakes. From this view the exact spatial distribution of the morphogens is not important (provided that each cell or group of cells receives different concentrations). What is important is how this positional information is interpreted inside the cells. Originally it was even proposed that the differences in morphology among species would simply arise from different interpretation of a universal coordinate system (this idea has since been discarded). This perspective could explain how the fate of cells in a territory is determined, but not how the form of the territory can change. Morphogenetic movements are proposed to occur later as part of the interpretation of positional information (Wolpert 1989). How positional information is interpreted has not been described by these authors. Then the whole concept simply states that cells make developmental decisions according to signal concentration differences. The rest of pattern formation is relegated to the as yet unexplained interpretation of these differences. In practice, most developmental biologists use positional information just to state that cells in different places have different fates because they receive different signals or combinations of signals.

Nowadays it is well known that cells often communicate while they are engaged in morphogenetic movements. In addition, communication between cells is often reciprocal. Cells respond to received signals by sending other signals, expressing signal receptors, changing their adhesive properties (and then moving or changing shape), proliferating, dying, secreting extracellular matrix, etc. This results in a constant complex dynamic change in the position of cells and patterns of gene expression. This gives limited applicability to the positional information metaphor, even for the systems where it was first proposed: see the chick limb (Hinchliffe and Horder 1993) and *Drosophila* segmentation (Jaeger *et al.* 2004). In spite of that, this metaphor can be used as an extreme from which to exemplify how the variational approach can be used to compare developmental mechanisms. Precise signal concentration sensing can be done, as in early dorso-ventral patterning in *Drosophila* (Markstein *et al.* 2004), by having enhancers of different affinity (or different number of enhancers) for the same transcriptional factor in the promoter regions of different genes (if there is a molecular pathway transducing signal concentration into active transcriptional factor concentration). At least one distinct enhancer is required for any different fate choice based on morphogen concentration. By having two non-spatially overlapping sources of different morphogens, any distribution of cell types can be produced in a given two-dimensional spatial distribution of cells (three sources of different morphogens are required in three dimensions). This is because in this condition every cell receives a unique combination of concentrations of each morphogen (Figure 2.1) and then by appropriate combinations of enhancers of differential affinity any pattern can be produced. The number of different enhancers required increases by at least one with the number of cells having a state other than the default (not to be confused with the number of different cell states; Figure 2.1). Thus, a large genetic complexity is required to produce complex or large patterns. Parts in a pattern can change independently by genetic mutation only if they do not share an enhancer. In general, genetic variation is likely to produce relatively gradual pattern changes. In other words, the morphological variation produced and its relationship with genetic variation is more consistent with the neo-Darwinian view.

This mechanism does not make use of the form of the spatial distribution of signals (it is enough that each cell receives a different concentration). In contrast, there are other developmental mechanism that can use this spatial information opportunistically to produce relatively complex patterns without requiring many mutational changes and complex genetic properties. These involve simple gene networks where cells directly respond to signals by sending other signals (Figure 2.2). In that way, additional and more complex spatial distributions of signals are produced by combining the spatial distributions of several signals from sources in the previous pattern. If in addition cells are changing their location while signalling (in what has been called morphodynamic mechanisms) additional spatial distributions of signals or cells can be produced. This is because signals can be sent and received from groups of cells that have the forms possible by morphogenetic movements (rods, invaginations, condensation nodes, etc.; Newman and Müller 2000, Salazar-Ciudad 2006b). Essentially a larger

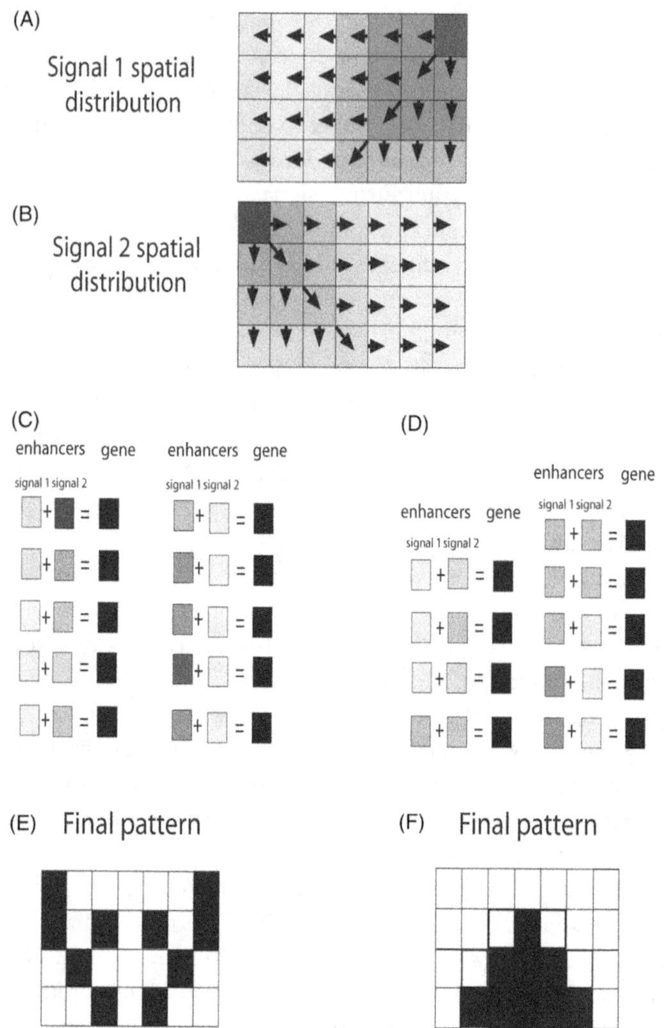

Figure 2.1 The diagram exemplifies the morphodynamic mechanism described in the text. In A and B the same field of cells is represented with the grey tones representing the concentrations of two signals. Signal 1 in A and signal 2 in B. Each matrix cell represents a cell. The spatial distribution of the signals is idealised (it should be more curved). In C and D a schema shows how different combinations of each signal concentration cause a cell to differentiate to black type. The interpretation in C gives rise to the pattern in E and the interpretation in D to the pattern in F.

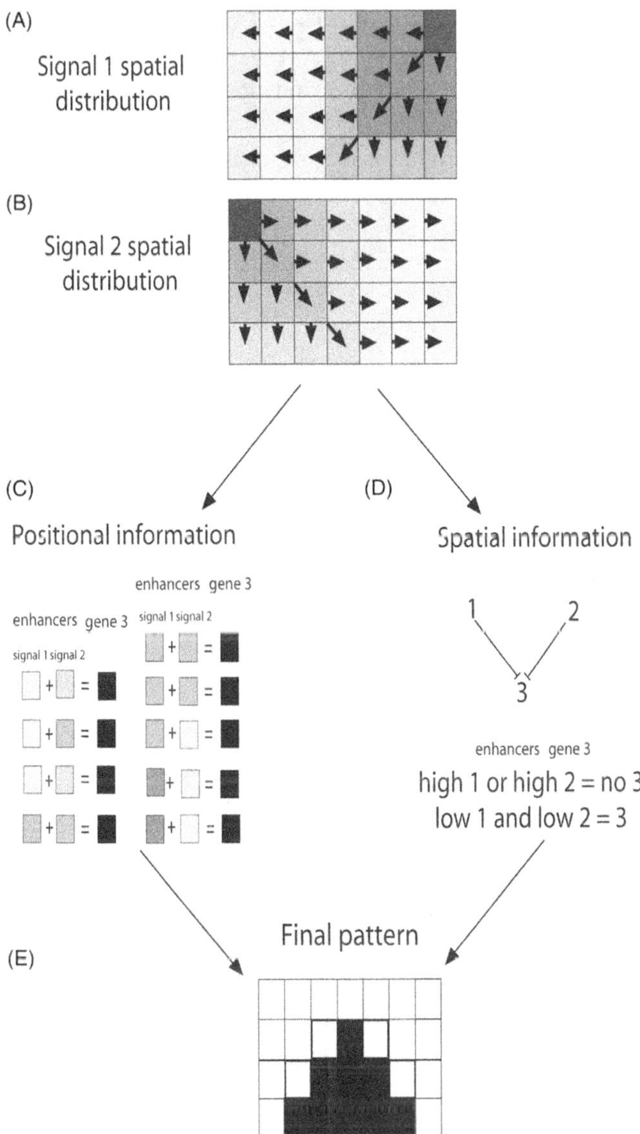

Figure 2.2 A and B as in Figure 2.1, and C as D in Figure 2.1. The interpretation in C of the signals in A and B gives rise to the pattern in E. In D it is shown how the same final pattern (in E) can be produced by gene 3 (expressed in black cells in E) being repressed where signal 1 or 2 has high concentration and active anywhere else.

spectrum of forms of groups of cells and signals can be combined to produce a pattern. This does not allow the production of more patterns but the production of more complex and more diverse patterns from the same amount of genetic variation (Salazar-Ciudad and Jernvall 2004, Salazar-Ciudad 2006b). What is important is not only signal interpretation but also the spatial distributions of signals and the collective dynamic behaviour of groups of cells.

This positional information mechanism is an example of hierarchic (Salazar-Ciudad *et al.* 2000) morphostatic mechanism (in which morphogenetic movements happen after and because of signalling; Salazar-Ciudad *et al.* 2003). Simulations of both kinds of mechanisms have shown that morphodynamic mechanisms do indeed produce patterns that are more complex, more distinct and related in more complex ways to genetic variation (Salazar-Ciudad and Jernvall 2004).

This has led to suggestions that morphodynamic mechanisms are more often involved in the formation of a pattern the first time it appears in evolution (Salazar-Ciudad and Jernvall 2004). This is because many more patterns easily appear by random mutation in morphodynamic mechanisms (while both mechanisms are equally likely to appear). Thus, evolution may often proceed opportunistically by first recruiting morphodynamic mechanisms. Over time, some of these mechanisms may be replaced by morphostatic mechanisms for the production of the same patterns because they allow more gradual variation and a simpler relationship between phenotype and genotype. This allows, depending on selective pressures, a faster and more efficient adaptation. Note that this is not opposed to the abovementioned baroque trends in the evolution of development because a morphostatic mechanism that can replace a morphodynamic mechanism necessarily involves many more genes and gene interactions. If new patterns are more likely to be added in later development, then morphodynamic mechanisms should be more frequent in later development (Salazar-Ciudad and Jernvall 2004).

CONCLUDING REMARKS

The previous examples show how evolutionary considerations (in the form of expectations about the likelihood of origin by genetic variation) can produce expectations about which competing hypothesis for developmental mechanisms is more likely for specific pattern transformations (for example the one in Figure 2.2). In the same way it has been explained how consideration of the genetic and variational properties of developmental mechanisms allows predictions about the evolution of morphology and development in different situations and stages. These inferences are modest in scope because relatively little is yet known about developmental mechanisms. This is likely to change soon. In general, this chapter advocates for the unique role of pattern formation and related concepts as a bridge to relate what is known in the different schools and in evolution and development in general. In that sense the concept of variational properties offers a definition of how hypotheses about how development functions should be tested and, at the same time, a basic conceptualisation of how different developmental mechanisms variously affect morphological evolution. The tandem genetic/variational properties, on the other hand, are used to infer ways in which development can evolve. These concepts help in explaining morphological change not only by selective arguments but also on the basis of which morphological variation is more likely to appear. Overall, these concepts aim to help give more explicit theoretical grounds for the gradual switch in evolutionary theory from selective explanations to explanations based on the interplay between selective forces and developmental capacities.

ACKNOWLEDGEMENTS

I thank Jukka Jernvall, Alistair Evans, Kathy Kavanagh and an anonymous referee. This work was supported by the Academy of Finland.

REFERENCES

Abzhanov, A. & Kaufman, T. C. 1999. Homeotic genes and the arthropod head: expression patterns of the *labial, proboscipedia,* and *Deformed* genes in crustaceans and insects. *Proceedings of the National Academy of Sciences of the USA* **96**, 10224–10229.

Akam, M. 1998. Hox genes: from master genes to micromanagers. *Current Biology* **24**, 676–678.

Alberch, P. 1980. Ontogenesis and morphological diversification. *American Zoologist* **20**, 653–667.

Arthur, W. 2004. The effect of development on the direction of evolution: toward a twenty-first century consensus. *Evolution & Development* **6**, 282–288.

Beldade, P., Koops, K. & Brakefield, P. M. 2002. Developmental constraints versus flexibility in morphological evolution. *Nature* **416**, 844–847.

Bronikowski, A. M., Rhodes, J. S., Garland, T. *et al.* 2004. The evolution of gene expression in mouse hippocampus in response to selective breeding for increased locomotor activity. *Evolution* **58**, 2079–2086.

Carlborg, S. & Haley, C. S. 2004. Epistasis: too often neglected in complex trait studies? *Nature Reviews Genetics* **5**, 618–625.

Carroll, S. B. 2000. Endless forms: the evolution of gene regulation and morphological diversity. *Cell* **101**, 577–580.

Charlesworth, B., Lande, R. & Slatkin, M. 1982. A neo-Darwinian commentary on macroevolution. *Evolution* **36**, 474–498.

Dobzhansky, T. 1937. *Genetics and the Origin of Species*. New York: Columbia University Press.

Fisher, R. A. 1930. *The Genetical Theory of Natural Selection*. Oxford: Clarendon Press.

Fusco, G. 2001. How many processes are responsible for phenotypic evolution? *Evolution & Development* **3**, 279–286.

Gilbert, S. F. & Sarkar, S. 2000. Embracing complexity: organicism for the 21st century. *Developmental Dynamics* **219**, 1–9.

Goodwin, B. 1994. *How the Leopard Changed its Spots, the Evolution of Complexity*. New York: Charles Scribner's Sons.

Gould, S. J. & Eldredge, N. 1977. Punctuated equilibria: the tempo and mode of evolution reconsidered. *Paleobiology* **3**, 115–151.

Haldane, J. B. S. 1932. *The Causes of Evolution*. London: Longmans. Reprinted by Princeton University Press (1990).

Harris, M. P., Williamson, S., Fallon, J. F., Meinhardt, H. & Prum, R. O. 2005. Molecular evidence for an activator-inhibitor mechanism in development of embryonic feather branching. *Proceedings of the National Academy of Sciences of the USA* **16**, 11734–11739.

Hinchliffe, J. R. & Horder, T. J. 1993. Testing the theoretical models for limb patterning. In H. G. Othmer, P. K. Maini & J. D. Murray (eds.) *Experimental and Theoretical Advances in Biological Pattern Formation*. New York: Plenum Press, pp. 105–119.

Holloway, D. M. & Harrison, L. G. 1999. Algal morphogenesis: modelling interspecific variation in *Micrasterias* with reaction–diffusion patterned catalysis of cell surface growth. *Philosophical Transcriptions of the Royal Society of London, Series B* **1382**, 417–433.

Horder, T. J. 1989. Syllabus for an embryological synthesis. In D. B. Wake & G. Roth (eds.) *Complex Organismal Functions*. New York: John Wiley, pp. 315–348.

Horder, T. J. 1993. The chicken and the egg. In H. G. Othmer, P. K. Maini & J. D. Murray (eds.) *Experimental and Theoretical Advances in Biological Pattern Formation*. New York: Plenum Press, pp. 121–148.

Jaeger, J., Blagov, M., Kosman, D. *et al.* 2004. Dynamical analysis of regulatory interactions in the gap gene system of *Drosophila melanogaster*. *Genetics* **167**, 1721–1737.

Jernvall, J., Keranen, S. V. E. & Thesleff, I. 2000. Evolutionary modification of development in mammalian teeth: quantifying gene expression patterns and topography. *Proceedings of the National Academy of Sciences of the USA* **97**, 14444–14448.

Kangas, A., Evans, A. R., Thesleff, I. & Jernvall, J. 2004. Non-independence of mammalian dental characters. *Nature* **432**, 211–214.

Kuszak, J. R., Zoltoski, R. K. & Tiedemann, C. E. 2004. Development of lens sutures. *International Journal of Developmental Biology* **48**, 889–902.

Markstein, M., Zinzen, R., Markstein, P. *et al.* 2004. A regulatory code for neurogenic gene expression in the *Drosophila* embryo. *Development* **131**, 2387–2394.

Maynard Smith, J., Burian, R., Kauffman, S. A. *et al.* 1985. Developmental constraints and evolution. *Quarterly Review of Biology* **60**, 265–287.

Mayr, E. 1982. *The Growth of Biological Thought*. Cambridge, MA: Harvard University Press.

Müller, G. B. & Wagner, G. P. 1991. Novelty in evolution: restructuring the concept. *Annual Review of Ecology and Systematics* **22**, 229–256.

Newman, S. A. & Müller, G. B. 2000. Epigenetic mechanisms of character origination. *Journal of Experimental Zoology* **288**, 304–317.

Newman, S. A. & Müller, G. B. 2005. Origination and innovation in the vertebrate limb skeleton: an epigenetic perspective. *Journal of Experimental Zoology B (Molecular and Developmental Evolution)* **304**, 593–609.

Nijhout, H. F. 2002. The nature of robustness in development. *BioEssays* **24**, 553–563.

Patel, N. H. 1994. The evolution of arthropod segmentation: insights from comparisons of gene expression patterns. *Development* **94** (supplement), 201–207.

Richardson, M. K. & Chipman, A. D. 2003. Developmental constraints in a comparative framework: a test case using variations in phalanx number during amniote evolution. *Journal of Experimental Zoology B (Molecular and Developmental Evolution)* **296**, 8–22.

Salazar-Ciudad, I. 2006a. Developmental constraints vs. variational properties: how pattern formation can help to understand evolution and development. *Journal of Experimental Zoology B (Molecular and Developmental Evolution)* **306**, 107–125.

Salazar-Ciudad, I. 2006b. On the origins of morphological disparity and its diverse developmental bases. *BioEssays* **28**, 1112–1122.

Salazar-Ciudad, I., Garcia-Fernàndez, J. & Solé, R. V. 2000. Gene networks capable of pattern formation: from induction to reaction–diffusion. *Journal of Theoretical Biology* **205**, 587–603.

Salazar-Ciudad, I. & Jernvall, J. 2002. A gene network model accounting for development and evolution of mammalian teeth. *Proceedings of the National Academy of Sciences of the USA* **99**, 8116–8120.

Salazar-Ciudad, I. & Jernvall, J. 2004. How different types of pattern formation mechanisms affect the evolution of form and development. *Evolution & Development* **1**, 6–16.

Salazar-Ciudad, I. & Jernvall, J. 2005. Graduality and innovation in the evolution of complex phenotypes: insights from development. *Journal of Experimental Zoology B (Molecular and Developmental Evolution)* **304**, 619–631.

Salazar-Ciudad, I., Jernvall, J. & Newman, S. A. 2003. Mechanisms of pattern formation in development and evolution. *Development* **130**, 2027–2037.

Salazar-Ciudad, I., Newman, S. A. & Solé, R. V. 2001a. Phenotypic and dynamical transitions in model genetic networks. I. Emergence of patterns and genotype–phenotype relationships. *Evolution & Development* **3**, 84–94.

Salazar-Ciudad, I., Solé, R. V. & Newman, S. A. 2001b. Phenotypic and dynamical transitions in model genetic networks. II. Application to the evolution of segmentation mechanisms. *Evolution & Development* **3**, 95–103.

Sander, K. & Schmidt-Ott, U. 2004. Evo-devo aspects of classical and molecular data in a historical perspective. *Journal of Experimental Zoology B (Molecular and Developmental Evolution)* **15**, 69–91.

Solé, R. V., Salazar-Ciudad, I. & Garcia-Fernàndez, J. 1999. Landscapes, gene networks and pattern formation: on the Cambrian explosion. *Advances in Complex Systems* **2**, 313–337.

Streicher, J. & Müller, G. B. 2001. 3D modelling of gene expression patterns. *Trends in Biotechnology* **19**, 145–148.

True, J. R. & Haag, E. S. 2001. Developmental system drift and flexibility in evolutionary trajectories. *Evolution & Development* **3**, 109–119.

Weber, K. E. 1992. How small are the smallest selectable domains of form? *Genetics* **130**, 345–353.

Wilkins, A. S. 2001. *The Evolution of Developmental Pathways.* Sunderland, MA: Sinauer Associates

Wolpert, L. 1989. Positional information revisited. *Development* **107** (supplement), 3–12.

3

Conflicting hypotheses on the nature of mega-evolution

WALLACE ARTHUR

Here is a question of the utmost importance for our understanding of what has been called the 'big picture' of evolution (Simpson 1944, 1953): are the divergences that lead ultimately to high-level sister groups, such as those that would typically be labelled as orders, classes and phyla, qualitatively or quantitatively different from those that lead to low-level sister groups, such as races, species and genera? In other words, is mega-evolution more than just accumulated micro/macro-evolution, or alternatively is evolution effectively 'scale-independent' (Leroi 2000)?

This question can be approached in three ways. We can choose to compare the *magnitude* of changes involved in high- and low-level divergences, the *type* of changes, or the *timing* (in development) of changes. Here, I argue that previous work on the first of these has been unproductive and has generated more heat than light; but that the second and third offer better prospects for shedding light on this important issue. However, in an unusual strategy, I also play devil's advocate with my own argument at the end of the chapter. This helps to take us in an interesting, final (for now) direction.

Because the designation of high-level sister groups as, for example, orders or classes, is a subjective rather than an objective process, I will, wherever possible, use specific examples rather than general levels of taxon.

MAGNITUDES OF CHANGE

Throughout the history of evolutionary biology, there have been assertions that large-scale evolution involves individually bigger changes in

Evolving Pathways: Key Themes in Evolutionary Developmental Biology, ed. Alessandro Minelli and Giuseppe Fusco. Published by Cambridge University Press.

the developmental trajectory, and hence in the adult phenotype, than does its small-scale, micro-evolutionary, equivalent. Such a view has been labelled macromutational or saltational; it had several well-known proponents (Bateson 1894, De Vries 1910, Goldschmidt 1940). Although macromutational theories of evolution have been rejected by the majority of biologists (most prominently, perhaps, by Dawkins 1986), they still occasionally surface (Whiting and Wheeler 1994) and are worthy of brief consideration.

At the phenotypic level, most change that occurs in the divergence of intraspecific variants, and of congeneric species, is continuous rather than discrete. A classic example is the divergence of beak size and shape in Darwin's finches, which is now beginning to be understood at the molecular level (Abzhanov et al. 2006). However, when higher-level divergences are considered, these often seem to be explicable only by invoking qualitative changes instead of, or as well as, quantitative ones. For example, in the Diptera there is only one pair of wings, in contrast to the four-wing condition from which this clade emerged. This famously led Goldschmidt to propose that the divergence of taxa at this level involved homeotic mutations (Goldschmidt 1952).

One of the reasons that Goldschmidt's ideas were not accepted is that homeotic mutations usually confer a massive loss in fitness; another is that the reduction of a pair of wings to halteres (Diptera) or their specialisation for some other purpose (e.g. the evolution of the elytra from the forewing of coleopteran ancestors) can be achieved by a series of small changes. In other words, the invoking of evolution by homeotic mutation in these cases would appear to be both problematic and unnecessary. However, it is possible that the argument of major fitness decrease being *necessarily* associated with homeotic mutation is context-dependent. For example, the discovery of an apparently healthy adult homeotic centipede in a natural population (Kettle *et al.* 1999, 2000) contrasts with the *Drosophila* situation. Also, a role for homeotic mutations in the evolution of angiosperm flowers now seems possible (De Craene 2003).

Although segment identity (the realm of homeotic mutations) can change gradually, segment number, as a meristic character, cannot. In this realm, we can only have one integer value or another. So perhaps there is a stronger case to be made here for a macromutational theory of the divergence of orders and classes, especially within the Arthropoda? Actually this is not so. There are indeed some high-level arthropod sister-groups characterised by a segment number difference between them but a lack of (or negligible) segment number variation within either of them, especially in the Crustacea (Schram 1986). However, there are also well-known cases of variation in segment number within arthropod species, such as the geophilomorph centipedes (Minelli and Bortoletto 1988, Arthur and Kettle 2001). There is no reason why normally invariant groups should not exhibit transient periods of intraspecific variability, perhaps environmentally induced but partially heritable, which provide the source of a later high-level difference in segment number.

Although segmentation has provided fertile ground for controversies about possible saltational evolution, we should not forget that it is itself an evolutionarily derived condition – regardless of whether it has been derived on one, two or three occasions (Davis and Patel 1999). A feature of broader phylogenetic scope than segmentation is symmetry/asymmetry. This again provides an opportunity for saltational theories of mega-evolution, perhaps most importantly in relation to the reversal of dorso-ventral orientation that appears to have accompanied the divergence of protostomes and deuterostomes (Geoffroy Saint-Hilaire 1822, Holley *et al.* 1995).

As with segment number, however, there are examples of intraspecific variation in the direction of asymmetries, not just rare cases such as the medical condition known as *situs inversus* in humans, but also polymorphisms in natural populations, such as for dextrality/sinistrality in the gastropod *Partula suturalis* (Murray and Clarke 1966).

The most reasonable conclusion at present with respect to the magnitudes of changes at the phenotypic level that contribute to evolutionary divergence is as follows. All levels of divergence typically involve accumulation of minor continuous variants; and all levels of divergence involve the occasional incorporation of changes that are 'large' and discrete. Some of these latter changes are based on a single big-effect mutation – as in chirality, but perhaps with some complications (Gould *et al.* 1985). Others probably have a more complex genetic basis. In any event, consideration of the *magnitude* of effect of changes in development and in the adult phenotype provides us with no clear evidence for a qualitative difference between mega-evolution and its smaller-scale (micro/macro) equivalent.

TYPES OF CHANGE

In the long term, this may be the best approach to the issue of scale-dependence versus scale-independence in evolution, but we still lack adequate information to come up with a rational and complete classification of types.

There is a connection here with the concept of an evolutionary novelty – an exciting and important concept but one that has been, and continues to be, hard to define. Different authors have adopted very different approaches (Mayr 1963, Müller and Wagner 1991). Given the problems with defining novelties, it is probably best to proceed by considering examples.

Most high-level clades are characterised by possession of a novelty – so novelties can be considered as a subset of the more general category of synapomorphies. Examples include chelonians (shells), mammals (hair), centipedes (forcipules) and dipterans (halteres). At the highest taxonomic levels, we can perhaps equate novelties with body plans (vertebrate endoskeleton, arthropod exoskeleton, and so on); while at middle levels (as in Diptera), the novelties appear to be less deep-rooted in the body architecture, although whether this subjective notion can be quantified is another matter. In fact, 'novelties' can be thought of not as a discrete category but rather as a hierarchy that ranges from the very conspicuous (e.g. the origin of skeletons) to the much less conspicuous (e.g. the redeployment of certain bones of the skeleton, as in the case of the origin of mammalian ear ossicles), thus intergrading with more minor evolutionary changes that should probably not be described as novelties at all (e.g. the segment number increase that characterises centipedes of the order Geophilomorpha when compared with their sister-order, the Scolopendromorpha).

Given this intergradation, there would seem to be no strong argument arising from consideration of novelties that mega-evolution is fundamentally different from its lower-level counterparts. It is true, perhaps, that *some* novelties would appear to involve a very specific and rare sort of variant – e.g. the chelonian shell with its fused ribs and internal shoulder girdle (Gilbert *et al.* 2001) – but this is hardly the basis for a *general theory* of the origin of novelties.

TIMING OF CHANGE

This may, at least for the moment, represent the best opportunity for making headway with the issue of the relationship between mega-evolution on the one hand and micro/macro-evolution on the other.

The key question now becomes: are those divergences that ultimately lead to the origin of higher taxa characterised by a different distribution of changes, in *developmental* time, from those which lead only to the origin of congeneric species? More specifically, are 'big' divergences characterised by *earlier* (on average) ontogenetic changes than 'small' ones? This is one of those so-called deceptively simple questions – i.e. it is not simple at all. I will approach it from a historical angle.

The obvious starting point is von Baer (1828) and his idea of divergence in embryonic trajectories. The most famous of the many comparisons he made are those involving different vertebrate classes and orders. For example, the ontogenies of birds and mammals start off being rather similar, but end up being very different (Figure 3.1A). Although von Baer did not interpret such patterns in evolutionary terms, Haeckel (1866) subsequently did, and he combined von Baerian divergence with a more enlightened version of recapitulation than that of the earlier 'nature-philosophers,' as can be seen in the following quote from the English translation of his *Anthropogenie* (Haeckel 1896):

> examination of the human embryo in the third or fourth week of its evolution shows . . . that it exactly corresponds to the undeveloped embryo-form presented by the Ape, the Dog, the Rabbit and other Mammals, at the same stage of their Ontogeny.

It has subsequently been emphasised that the very earliest developmental stages do not fit into this apparently neat picture. For example, the ontogenies of birds and mammals have very different starting points before converging to their point of maximum similarity (the pharyngula stage), beyond which von Baerian divergence does indeed occur. Thus we have come to recognise the 'phylotypic stage', and this applies not just to vertebrates but to other groups as well. In fact, the phrase was first used to describe the germ-band stage in insect development (Sander 1983).

The recognition of the phylotypic stage led to the egg-timer or hour-glass model of comparative embryology (Duboule 1994); see Figure 3.1B. And a further refinement of this was to acknowledge that, when enough species were compared, the idea of a phylotypic 'stage' was too neat, and a better model was based on the idea of an extended phylotypic *period* (Richardson 1995). In any event, it must be stressed that the 'egg-timer' of comparative embryology is a very asymmetric one: the point of constriction is much closer to the start of ontogeny, the fertilised egg, than it is to the end-point, the adult. Of course, I use 'start' and 'end' here in a pragmatic way; it is not my intention to take an overly 'adultocentric' (Minelli 2003) view of development.

Another complication to our overall picture is the existence, indeed the prevalence, of complex life-cycles, such as those of insects, amphibians, parasitic platyhelminthes and many marine invertebrates with assorted types of larvae, including trochophores and pluteuses.

(A)

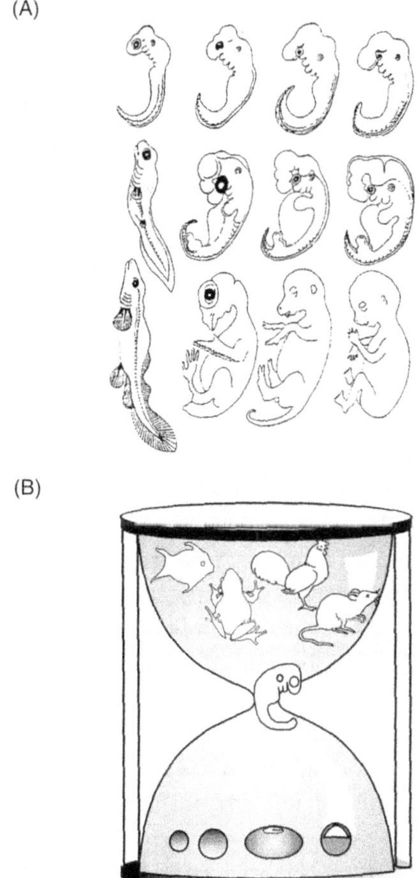

(B)

Figure 3.1 A, von Baerian divergence, as illustrated by the early similarity of four vertebrate embryos giving way to later differences. B, The egg-timer or hour-glass model of comparative embryology, showing convergence from different starting points to a point of maximum similarity prior to von Baerian divergence from that point onwards. Note that the time axis is distorted. The point of maximum similarity, or phylotypic stage, occurs very soon after the start of development. Reproduced, with permission, from two sources (Raff and Kaufman 1983, Richardson *et al.* 1997).

These complex life cycles clearly cannot be described in the same terms as the simple ones of direct developers such as birds and mammals. However, they do not require a fundamentally different model than von Baerian divergence, or its more sophisticated (and more accurate) egg-timer equivalent. Rather, they require that such a model be applied separately to each life stage rather than to the whole of ontogeny throughout the life-cycle. Because the questions at issue here are difficult enough without having to think simultaneously about different life stages, in what follows I will base the discussion largely on direct developers; the argument can be extended later to deal with indirect developers.

There is something that so far has remained implicit; now is the time to make it explicit. The comparisons of two or more ontogenies that we have been making have been of phylogenetically distant animals: that is, those belonging to different classes or orders (Figure 3.1A). What picture of similarity and difference through developmental time would we see if, instead, we made a comparison between the ontogenies of a 'typical' pair of congeners, in as much as there is such a thing?

We are hindered here because of a shortage of comparative developmental studies at this level, particularly those that include an embryonic component rather than merely comparing post-embryonic growth patterns (Andersson 1990). So it may be useful to begin with a 'thought experiment', as follows. Suppose that a research project is undertaken over the next few months, specifically to make a series of comparisons of the ontogeny of pairs of direct-developing congeners, sampling from a range of different phyla. For example, in the molluscs, we might compare the ontogenies of the land-snails *Cepaea hortensis* and *Cepaea nemoralis* that are well known from an ecological genetics perspective (Jones *et al.* 1977) but not from a developmental perspective. And we would make similar inter-congener comparisons within other phyla. What overall pattern would we see if many such studies were undertaken?

My suspicion is that, after the appropriate phylotypic stage (where there is one) we would see von Baerian divergence, with the point at which the two ontogenies begin to diverge being much *later* than in the case of comparisons between species from different orders or classes.

Suppose that this result – congeners diverging later than more distantly related animals (say from different classes) – is a general one. This leads us to an interesting argument about the possibility (or otherwise) of extrapolating from micro/macro-evolution of development to its mega-evolutionary counterpart. The argument goes like this. When we think only of *magnitudes* of morphological change in evolution, the accumulation of many small differences (as typically found between congeners) over long periods of time, will, with an inescapable logic, produce the much larger differences that we observe between more phylogenetically distant forms. This view of evolution has been disseminated to a wide audience (Dawkins 1986) and is generally accepted. However, when we think in explicitly developmental terms, and especially in terms of the timing of changes in ontogeny, the equivalent argument does not work: *lots of late changes in development do not add up to an early one.*

At this juncture, it is worth stressing the contrast between absolute and statistical differences, for example in the 'evolvability' (Kirschner and Gerhart 1998) of different developmental stages, because this point has been neglected in some previous discussions.For example, consider the following (Scholtz 2005):

> The comparative view of developmental processes is that of a series of potentiallly independent patterns (stages) which can be altered individually ... The recognition of this potential freedom of developmental stages results in a refutation of a special importance of any stage.

In my view, Scholtz is stating the obvious when he says that the various stages of development can *all* be altered in the course of evolution. But the relative probabilities of altering different stages are not the same. The fact that different phylotypic stages, such as vertebrate pharyngula and insect germband, are manifestly different from each other attests to the former; while the fact that these stages are recognizable *within* higher taxa attests to the latter.

These considerations lead us in the following direction: both early (e.g. phylotypic) and late (e.g. allometric) developmental stages can change in evolution, but the latter change more often, and so are typically found in all comparisons of taxa at whatever level; while the former are rarer and so are *not often* found in comparisons involving congeneric or confamilial species. Although this argument might be a circular one in the case of congeneric species defined as such solely on developmental grounds, it would not be so if the species concerned were defined by other means, such as DNA sequence data.

Note, however, that 'not often' and 'never' are different. Just as we saw earlier that *big-effect* changes sometimes occur between congeners, *early* changes sometimes do likewise. Indeed, sometimes these are the same changes looked at in a different way. Not only do chirality reversals in gastropods invert the symmetry of the whole animal, but they do so by modifying, through the maternal-effect genes involved, the very earliest stage of development, namely cleavage (Verdonk and van den Biggelaar 1983). Also, evolutionary switches between the planktotrophic and lecithotrophic larvae of congeneric sea urchin species have major morphological effects (loss of feeding arms) and involve *early* developmental stages (Wray and Raff 1989).

DISCUSSION

Consideration of the timing of changes in development rather than their magnitudes of effect on the adult phenotype is helpful in one way but not in another. It is helpful because there are no temporal equivalents of the philosophically loaded terms micro- and macromutation, and no history of ill-tempered debate equivalent to that between Goldschmidt and the architects of the 'Modern Synthesis.' However, it is unhelpful because although it suggests that mega-evolution is unique in its temporal distribution of changes through developmental time, it is not conclusive in this respect.

Let us approach this problem in two ways: first, by playing devil's advocate with the 'lots of late changes do not add up to an early one' argument; and second, by re-examining the concept of an evolutionary novelty.

The devil's advocate argument can be quite a simple one, if the right context is chosen, as follows. Consider two extant high-level sister groups of direct developers, for example two mammalian orders, chosen so that their phylotypic stages are very similar. Ignore the pre-phylotypic phase of their development. With regard to the post-phylotypic phase, ask the following question. Is there any time-point, t_x, between the phylotypic stage, t_p, and the completion of development, t_c, beyond which no inter-congener embryonic divergences (within either order) go back, while the inter-order divergence itself does? If there *is* such a time-point, then accumulated 'ordinary' speciations cannot account for the origin of the two orders concerned, because the argument 'lots of late changes do not add up to an early one' applies. On the other hand, if there is *not* such a time-point, scale-independence is possible. In most cases, we lack enough relevant comparative developmental data to know which of these alternative possibilities is true.

Another problem of an argument based solely on timing is that, while mega-evolutionary divergences usually, perhaps always, include early changes in the developmental pathways in one or both of the diverging lineages, earliness is not in itself a sufficient descriptor of what is going on. A novelty characterising a major clade often involves a change in development that is in some sense 'special' as well as 'early'. Returning to the example of the chelonian shell: getting the pectoral girdle to develop inside the ribcage rather than outside, as in the (unknown) chelonian ancestor and in all other tetrapod groups, would appear to involve a quite distinct form of developmental reprogramming (Arthur 2000) from the quantitative changes in the size, shape and position of this girdle within either chelonians or typical tetrapods.

This is just one specific example of the general difference between qualitative and quantitative changes. Another angle on this difference was taken by D'Arcy Thompson in relation to his method of geometric transformations (Thompson 1917). This quantitative method could adequately (and beautifully) describe the evolutionary changes that produced the different members of a particular group. But it could not be extended beyond a certain level of divergence. D'Arcy Thompson himself pointed this out, and argued 'that discontinuous variations are a natural thing, that "mutations" or sudden changes, greater or less – are bound to have taken place, and new "types" to have arisen now and then.'

So, for now we must be satisfied with the following tentative conclusion about conflicting hypotheses on the nature of the small minority of divergences that lead ultimately to the production of high-level sister-clades characterised by different novelties – i.e. to mega-evolutionary change. There are not just two such hypotheses but three. The first, that mega-evolutionary changes are something quite apart from 'routine' ones, in the sense of non-overlapping sets, can in my view be rejected on the basis of the available evidence. The second, that mega-evolutionary divergences are statistically different from their lower-level counterparts, cannot. The third, that all levels of evolution are the same in both absolute and statistical senses, also cannot be rejected. Therefore, the question of whether evolution is a scale-dependent process, or a scale-independent one, remains open.

ACKNOWLEDGEMENTS

I would like to thank Alessandro Minelli and Giuseppe Fusco for the invitation to speak at the Venice meeting, and to contribute to this book deriving from it; also Chris Klingenberg for his very helpful comments on an earlier version of the manuscript.

REFERENCES

Abzhanov, A., Kuo, W. P., Hartmann, C., Grant, B. R., Grant, P. R. & Tabin, C. 2006. The calmodulin pathway and evolution of elongated beak morphology in Darwin's finches. *Nature* **442**, 563–567.

Andersson, G. 1990. About the duration of the different stadia in the post-embryonic development of some Lithobiomorph species. In A. Minelli (ed.) *Proceedings of the 7th International Congress of Myriapodology*. Leiden: Brill, pp. 323–335.

Arthur, W. 2000. The concept of developmental reprogramming and the quest for an inclusive theory of evolutionary mechanisms. *Evolution & Development* **2**, 49–57.

Arthur, W. & Kettle, C. 2001. Geographic patterning of variation in segment number in geophilomorph centipedes: clines and speciation. *Evolution & Development* **3**, 34–40.

Bateson, W. 1894. *Materials for the Study of Variation, Treated with Especial Regard to Discontinuity in the Origin of Species*. London: Macmillan.

Davis, G. K. & Patel, N. H. 1999. The origin and evolution of segmentation. *Trends in Genetics* **15**, M68–M72.

Dawkins, R. 1986. *The Blind Watchmaker*. London: Longman.

De Craene, L. P. R. 2003. The evolutionary significance of homeosis in flowers: a morphological perspective. *International Journal of Plant Science* **164**, S225–S235.

De Vries, H. 1910. *The Mutation Theory: Experiments and Observations on the Origin of Species in the Vegetable Kingdom*. London: Kegan Paul, Trench, Trübner & Co.

Duboule, D. 1994. Temporal collinearity and the phylotypic progression: a basis for the stability of a vertebrate Bauplan and the evolution of morphologies through heterochrony. *Development 1994 supplement*, 135–142.

Geoffroy Saint-Hilaire, E. 1822. Considérations générales sur la vertèbre. *Mémoires du Muséum d'Historie Naturelle* **9**, 89–119.

Gilbert, S. F., Loredo, G. A., Brukman, A. & Burke, A. C. 2001. Morphogenesis of the turtle shell: the development of a novel structure in tetrapod evolution. *Evolution & Development* **3**, 47–58.

Goldschmidt, R. 1940. *The Material Basis of Evolution*. New Haven: Yale University Press.

Goldschmidt, R. 1952. Homeotic mutants and evolution. *Acta Biotheoretica* **10**, 87–104.

Gould, S. J., Young, N. D. & Kasson, B. 1985. The consequences of being different: sinistral coiling in *Cerion*. *Evolution* **39**, 1364–1379.

Haeckel, E. 1866. *Generelle Morphologie der Organismen*. Berlin: Georg Reimer.

Haeckel, E. 1896. *The Evolution of Man: A Popular Exposition of the Principal Points of Human Ontogeny and Phylogeny*. New York: Appleton.

Holley, S. A., Jackson, P. D., Sasai, Y. *et al.* 1995. A conserved system for dorsal-ventral patterning in insects and vertebrates involving *sog* and *chordin*. *Nature* **376**, 249–253.

Jones, J. S., Leith, B. H. & Rawlings, P. 1977. Polymorphism in *Cepaea*: a problem with too many solutions? *Annual Review of Ecology and Systematics* **8**, 109–143.

Kettle, C., Arthur, W., Jowett, T. & Minelli, A. 1999. Homeotic transformation in a centipede. *Trends in Genetics* **15**, 393.

Kettle, C., Arthur, W., Jowett, T. & Minelli, A. 2000. A homeotically-transformed specimen of *Strigamia maritima* (Chilopoda, Geophilomorpha), and its morphological, developmental and evolutionary implications. In J. Wytwer & S. Golovatch (eds.) *Progress in Studies on Myriapoda and Onychophora*. *Fragmenta Faunistica* **43** (supplement), 105–112. Warsaw.

Kirschner, M. & Gerhart, J. 1998. Evolvability. *Proceedings of the National Academy of Sciences of the USA* **95**, 8420–8427.

Leroi, A. M. 2000. The scale independence of evolution. *Evolution & Development* **2**, 67–77.

Mayr, E. 1963. *Animal Species and Evolution*. Cambridge, MA: Harvard University Press.

Minelli, A. 2003. *The Development of Animal Form: Ontogeny, Morphology and Evolution*. Cambridge: Cambridge University Press.

Minelli, A. & Bortoletto, S. 1988. Myriapod metamerism and arthropod segmentation. *Biological Journal of the Linnean Society* **33**, 323–343.

Müller, G. B. & Wagner, G. P. 1991. Novelty in evolution: restructuring the concept. *Annual Review in Ecology and Systematics* **22**, 229–256.

Murray, J. & Clarke, B. 1966. The inheritance of polymorphic shell characters in *Partula* (Gastropoda). *Genetics* **54**, 1261–1277.

Raff, R. A. & Kaufman, T. C. 1983. *Embryos, Genes and Evolution: The Developmental Genetic Basis of Evolutionary Change*. New York: Macmillan.

Richardson, M. K. 1995. Heterochrony and the phylotypic period. *Developmental Biology* **172**, 412–421.

Richardson, M. K., Hanken, J., Gooneratne, M. L. *et al.* 1997. There is no highly conserved embryonic stage in the vertebrates: implications for current theories of evolution and development. *Anatomy & Embryology* **196**, 91–106.

Sander, K. 1983. The evolution of patterning mechanisms: gleaning from insect embryogenesis and spermatogenesis. In B. C. Goodwin, H. Holder & C. C. Wyllie (eds.) *Development and Evolution*. Cambridge: Cambridge University Press, pp. 137–159.

Scholtz, G. 2005. Homology and ontogeny: pattern and process in comparative developmental biology. *Theory in Biosciences* **124**, 121–143.

Schram, F. R. 1986. *Crustacea*. New York: Oxford University Press.

Simpson, G. G. 1944. *Tempo and Mode in Evolution*. New York: Columbia University Press.

Simpson, G. G. 1953. *The Major Features of Evolution*. New York: Columbia University Press.

Thompson, D'A. W. 1917. *On Growth and Form*. Cambridge: Cambridge University Press.

Verdonk, N. H. & van den Biggelaar, J. A. M. 1983. Early development and the formation of the germ layers. In N. H. Verdonk, J. A. M. van den Biggelaar & A. S. Tompa (eds.) *The Mollusca*, Vol. 3, *Development*. New York: Academic Press, 91–122.

von Baer, K. E. 1828. *Über Entwicklungsgeschichte der Tiere: Beobachtung und Reflexion*. Königsberg: Bornträger.

Whiting, M. F. & Wheeler, W. C. 1994. Insect homeotic transformation. *Nature* **368**, 696.

Wray, G. A. & Raff, R. A. 1989. Evolutionary modification of cell lineage in the direct-developing sea urchin *Heliocidaris erythrogramma*. *Developmental Biology* **132**, 458–470.

4

Prospects of evo-devo for linking pattern and process in the evolution of morphospace

PAUL M. BRAKEFIELD

Evolutionary developmental biology or evo-devo will make crucial contributions over the coming years to exploring the occupancy of morphospace. Why do species show the patterns of diversity and disparity they do, and to what extent do such patterns reflect the ways in which phenotypic variation is generated as well as the processes of natural selection? Evo-devo in appropriate study systems is providing the means to explore fully how phenotypic variation is generated by the processes intrinsic to individuals, especially those of development. Substantial progress will be made in understanding the occupancy of morphospace if the results of this type of evo-devo can then be combined with analyses of how this same variation is influenced by natural selection and other extrinsic processes to result in the patterns of evolution. Use can also be made here of recent advances in developing appropriate null models for testing adaptive versus neutral or random-walk explanations of evolution (Pie and Weitz 2005).

In combination, this type of broad evo-devo and evolutionary biology can begin to unravel how evolvability, the capacity to evolve at the genetical and developmental levels, contributes to shaping the evolution of patterns of diversity and disparity in phenotypic space (Brakefield 2006). We will then be able to examine the extent to which such phenomena as genetic channelling, developmental bias and developmental drive are reflected in patterns of evolution, whether involving change or stasis (e.g. Maynard Smith *et al.* 1985, Schluter 1996, Wagner and Altenberg 1996, Arthur 2001, Blows and Hoffman 2005).

Evolving Pathways: Key Themes in Evolutionary Developmental Biology, ed. Alessandro Minelli and Giuseppe Fusco. Published by Cambridge University Press.
© Cambridge University Press 2008.

Developmental biology has flourished as the application of new molecular and genetical tools in several model organisms has opened up the mechanisms of development. Differences in embryonic development have been explored across such highly divergent organisms as flies, nematode worms, fish, frogs, chickens and mice. Specific genes and genetic 'toolkit' pathways central to development have been identified through the study of the consequences of major mutations obtained from mutant screens in the model organisms. It is becoming clear that much of the evolution of morphological diversity is about teaching old genes new tricks, and that this frequently does not involve changes in the encoding sequences themselves but rather evolutionary tinkering in their complex regulatory apparatus (Carroll 2005).

Development is central to evolution because the processes of development map morphologies on to genotypes, and differences in development among genotypes generate the variation in phenotype on which selection can act. Developmental biology has been slow to expand upon the work with highly divergent model organisms to examine more subtle differences in form such as are found within a single lineage and which provide the developmental basis of variation in natural populations. However, through the application of evo-devo, variation in form can now be traced on to genetic variation via developmental mechanisms and the processes of pattern formation. A full understanding of the roles of genetic variation in morphological evolution will require an effective integration of the mechanisms of development with studies of evolution. This will need to be done both for variation in form among species with known phylogenetic relationships, and for the variability within the populations of single species. The extent to which this latter level of variation within populations will inform us about the fixed differences that characterise related species is at present unclear but will only be resolved as both levels are examined within particular lineages. This will need to be achieved for many morphologies that are relevant in an ecological arena, such as interactions with potential mating partners or with predators (Brakefield *et al.* 2003, Brakefield and French 2006).

There is an undoubted power in studying natural genetic variation in developmental processes in a lineage that includes one of the model organisms of developmental biology. For example, work on butterfly wings is successfully building on developmental insights about wing development from *Drosophila* whilst at the same time taking advantage of the exciting opportunities presented by the well-known ecology in this group (Beldade and Brakefield 2002, McMillan *et al.* 2002,

Joron *et al.* 2006). Indeed, one vigorous aspect of evo-devo is that it is rapidly expanding to include new, emerging model organisms which lend themselves to studies both in the laboratory and in the wild. Studies on *Heliconius* butterflies, dung beetles, stalk-eyed flies, fruit flies, nematode worms, centipedes, and stickleback, cichlid and danio fish all spring to mind (for references see Brakefield *et al.* 2003, Brakefield and French 2006). This body of work is also ranging widely over differing morphologies from body plans and larval forms through to skeletal morphology, patterns of pigmentation and bristles, and structures such as horns, spines, segments, eyes and so on. Developmental plasticity is becoming better represented, for example in research with butterfly wing patterns, beetle horns, *Daphnia* helmets and aphid wings. Eventually, this wide net across the animal kingdom and different morphologies will capture many of the general properties of how the genetic variation that underlies developmental change influences patterns in the evolution of animal forms. Such generalisations will become easier as the comparative approaches to variation in form and function for specific morphologies become more integrated with the detailed studies of variation in genetical and developmental mechanisms in the laboratory. One case study that illustrates an integrative approach to linking the evolution of developmental mechanisms with understanding the role of development in evolution is work on wing eyespots and other traits in *Bicyclus* butterflies.

EYESPOTS MATTER

Butterfly wings are decorated by mosaics made up of scale cells in patterns resembling the overlapping tiles of a roof. An eyespot is one example of a wing pattern element, and is made up of concentric rings of modified epithelial cells. The scale cells in the eyespot rings synthesise different colour pigments in the course of their development. Each eyespot in *Bicyclus* butterflies usually has a central white pupil surrounded by a middle black ring and an outer gold ring. Eyespots in *Bicyclus* and other butterflies and moths (together, the Lepidoptera) are known to function both in interactions with predators (Stevens 2005, Brakefield and Frankino 2007) and during courtship and mate choice (Breuker and Brakefield 2002, Robertson and Monteiro 2005). Thus both natural selection and sexual selection are relevant to understanding functional differences in eyespot patterns among species.

Field experiments with *Bicyclus* butterflies in Malawi have taken advantage of the phenomenon of seasonal polyphenism to show that eyespot patterns can strongly influence adult survival (Brakefield and Frankino 2007). Species of *Bicyclus* (Nymphalidae: Satyrinae) in regions of Africa with wet–dry seasonal cycles have evolved developmental plasticity. They exhibit seasonal forms differing dramatically in colour pattern on their ventral wing surfaces, especially with respect to the expression of the marginal eyespots (Figure 4.1). The ventral wings are exposed to predators when the butterflies are at rest on the forest floor or feeding on fallen fruit with their wings closed above the body. In each polyphenic species, both of *Bicyclus* and many other satyrine genera, the dry-season form (DSF) is a more or less uniformly brown insect whereas butterflies of the wet-season form (WSF) have a series of marginal eyespots and a pale medial band across both wings (Brakefield and Larsen 1984, Windig *et al.* 1994). The essential idea about the adaptive significance of the developmental plasticity is that natural selection in the dry season favours a comparatively inactive or quiescent life style in which butterflies spend most of their time at rest on brown leaf litter. They rely on background matching and crypsis or camouflage (i.e. looking like dead leaves) to survive the long dry season before they can reproduce at the onset of rains, laying eggs on newly green and growing grasses. In contrast, butterflies of the wet-season form are active, reproduce quickly, and rely on marginal eyespots as active deflection devices against vertebrate predators (Brakefield and Larsen 1984, Lyytinen *et al.* 2004, Stevens 2005); if an attack is aimed at a 'target' eyespot, a butterfly can escape albeit with the loss of some wing tissue grabbed by the predator.

Cohort analyses using mark–release–recapture experiments were performed with *Bicyclus* butterflies in Malawi to test this hypothesis (Brakefield and Frankino 2007). Here, the results are summarised for the dry season when a colour pattern made more conspicuous by markings like eyespots is predicted to be disadvantageous. Initial experiments used releases of reared butterflies with phenotypes ranging from extreme DSF with no ventral eyespots through to WSF with very large eyespots (Figure 4.2A). Butterflies were released in a forest environment and recaptured using about 40 fruit-baited traps. Patterns of movement over the traps were similar for the different phenotypes. In the dry season, butterflies of the WSF had a much lower probability of recapture than those of the DSF. However, experiments using releases of bred individuals do not demonstrate that it is the ventral eyespots per se that account for higher mortality of the WSF in the dry season since we know that the seasonal forms differ for a suite of traits including physiology and life history.

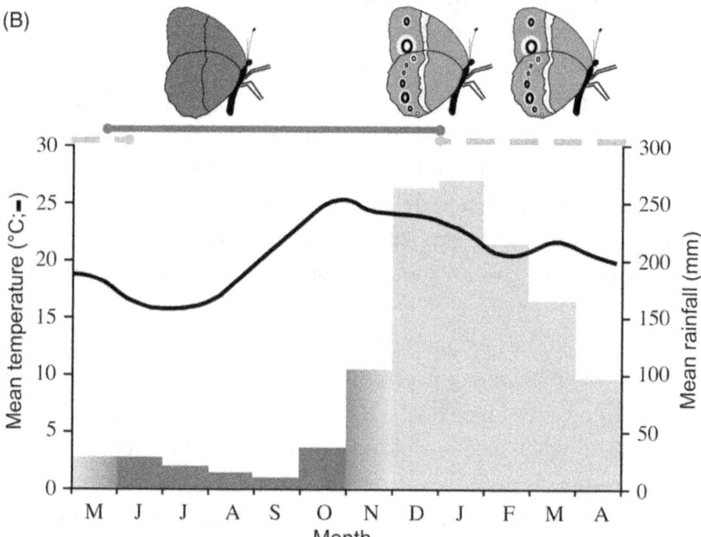

Figure 4.1 Developmental plasticity in *Bicyclus* butterflies. A, The seasonal forms illustrated by two sisters of *B. safitza*. The specimen to the left is of the wet-season form (WSF) and was reared at 27 °C, whereas the one to the right, which is of the dry-season form (DSF), was switched to a low temperature for a few days just before pupation. B, The wet–dry seasonal cycle in Malawi. A dry season with low rainfall (histogram) is followed by a wet season. Temperatures are cooler throughout most of the dry season before increasing prior to the rains. Two generations of the WSF are produced in each rainy period followed by a single generation of the DSF that emerges in the early dry season before the larval food plants have completely dried out.

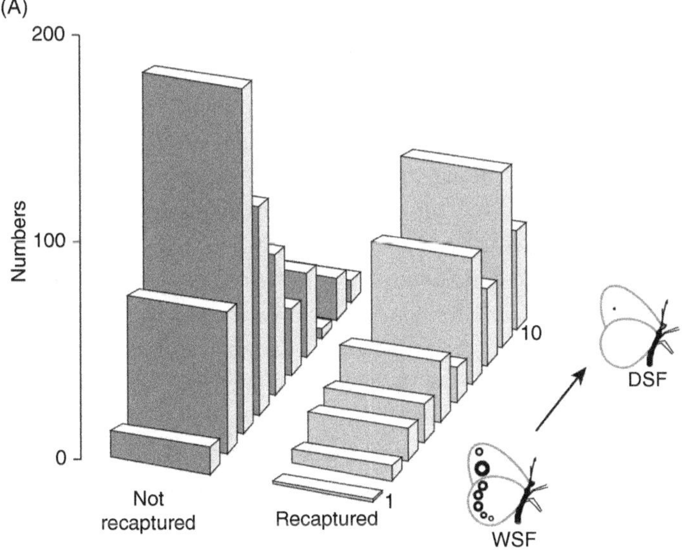

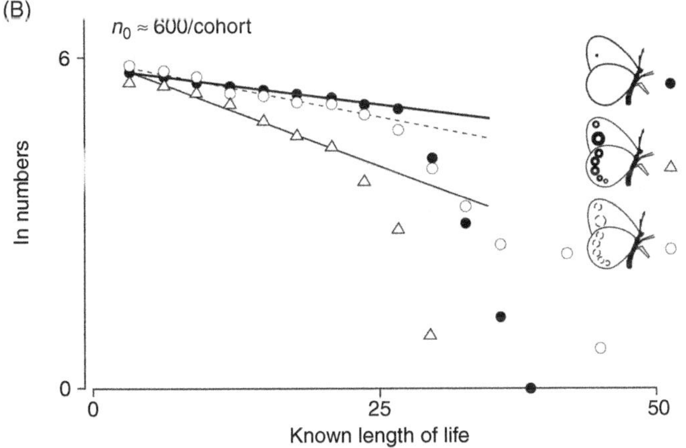

Figure 4.2 Natural selection on eyespot size in *Bicyclus safitza* butterflies demonstrated by field experiments in the dry season in Africa. A, Probability of recapture over a grid of forest traps for releases of reared butterflies of 10 phenotypic classes ranging from extreme dry-season form (DSF – no eyespots) to extreme wet-season form (WSF – very large eyespots). The WSF shows much higher mortality. B, Survivorship curves for releases of about 1800 DSF butterflies collected in another forest and divided among three treatments: unpainted controls; painted with conspicuous eyespots; and painted with inconspicuous eyespots. Butterflies painted with conspicuous eyespots show a dramatically higher mortality consistent with the eyespots making them easier to find by predators in the dry season. Lines show periods of age-independent survival for each cohort. From Brakefield and Frankino (2007).

Additional field experiments using butterflies with manipulated wing patterns tested directly the hypothesis that in the dry season the eyespots result in higher mortality by increasing butterfly conspicuousness (Figure 4.2B). DSF butterflies collected in the wild were marked individually on their dorsal wings that are hidden when at rest, and then randomly divided among three cohorts: (1) no further treatment (control); (2) painted using marker pens with full series of black–gold eyespots on each ventral wing surface (treatment with conspicuous eyespots); and (3) painted in the same way as treatment butterflies but using marker pens of a similar colour to the brown background of the wings (sham control with inconspicuous eyespots). All butterflies were released and recaptured in the same trapping grid as in the previous experiment. Figure 4.2B shows that the butterflies with inconspicuous eyespots (to our eyes) and the unpainted controls had closely similar survivorship curves. In sharp contrast, treated butterflies with conspicuous eyespots showed a much higher mortality. These results demonstrate that the eyespots themselves contribute to the lower relative fitness of the WSF phenotype in the dry season. Whatever the details of exactly how natural selection works through the seasonal cycles in the field (see also Lyytinen *et al.* 2004), such experiments show that eyespot patterns can matter.

Serial repeats and artificial selection experiments on eyespots

The wing patterns of butterflies in the family Nymphalidae are made up of combinations of different pattern elements including colour bands, stripes, and marginal ocelli or eyespots (Nijhout 1991). A reconstruction of a nymphalid 'groundplan' shows repeated series of the different types of element arranged in anterior–posterior columns on each surface of the fore- and hindwings. Each repeated series of a particular element can be considered as a module. A wing surface is subdivided by wing veins into a series of wing cells, each with its own combination of the different pattern elements. The development of one set of such elements, the marginal eyespots, is becoming understood both in terms of cell–cell signalling mechanisms and candidate genetic pathways (Beldade and Brakefield 2002). Essentially, understanding development involves discovering how different populations of epithelial cells in the wings – the scale cells to be – gain information during wing growth in the late larval and early pupal stages and, thus, become fated to lay down different colour pigments just before adult eclosion. Our understanding of morphogenesis and pattern formation can now be used alongside artificial selection experiments targeted on the pattern of eyespots to explore how the mechanisms involved in generating phenotypic variation may influence evolutionary trajectories.

The eyespots of our laboratory model species, *B. anynana*, are all formed in the late larval stage and the early pupa by the same developmental pathway (Brunetti *et al.* 2001, Beldade and Brakefield 2002, Reed and Serfas 2004, Beldade *et al.* 2005). Transplantation experiments performed in early pupae show that each eyespot is formed around a group of organising cells known as an eyespot focus; transplanting a focus to a novel site in the pupal wing yields an ectopic eyespot around the grafted tissue in the adult wing (French and Brakefield 1995). Establishment of the foci occurs in late larvae, and then in early pupae each focus sets up a gradient in the surrounding epithelial cells, presumably via one or more diffusible morphogens. These cells then respond to the concentration gradient of the signal, and become fated to synthesise a particular colour pigment just before emergence of the adult.

All the eyespots of *B. anynana* express the same developmental genes, and at comparable stages in eyespot formation (Brunetti *et al.* 2001, Beldade *et al.* 2005, Reed and Serfas 2004). These genes include *Distal-less, hedgehog, engrailed, spalt* and *Notch*. Typically, mutant alleles established in laboratory stocks also affect all eyespots. Moreover, artificial selection experiments targeted on the size or colour of a single eyespot yield highly correlated responses for the target trait in other eyespots, especially on the same wing surface (Monteiro *et al.* 1994, 1997). The shared morphogenesis in terms of genetic variation and developmental mechanisms led us to design experiments with *B. anynana* to examine the potential developmental flexibility of the repeated eyespot elements to evolve in different directions in trait space or developmental morphospace (Brakefield 1998).

Beldade *et al.* (2002) explored whether a phenotype in which one eyespot was smaller and the other larger could be produced as readily by artificial selection as one in which both eyespots were either larger or smaller than in the wild type. We had predicted that morphological change would, in some sense, be more limited when eyespots were selected in opposite directions (Brakefield 1998). We targeted the pattern on the dorsal forewing with a smaller anterior eyespot and a larger posterior eyespot (see Figure 4.3); these eyespots show no

(A)

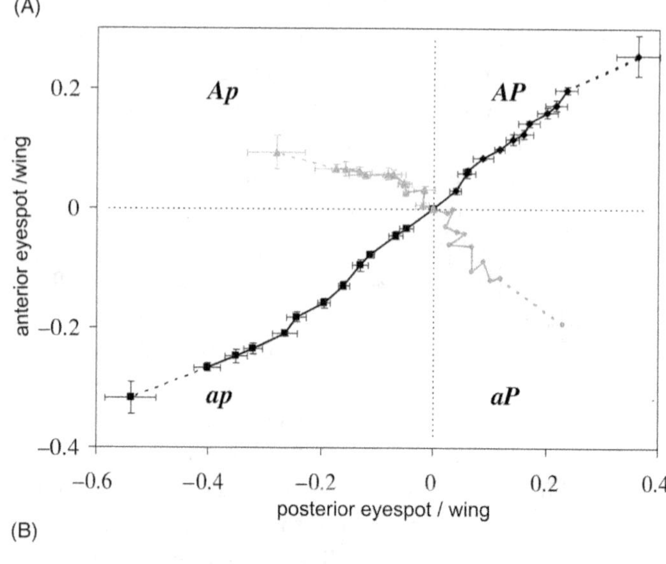

(B)

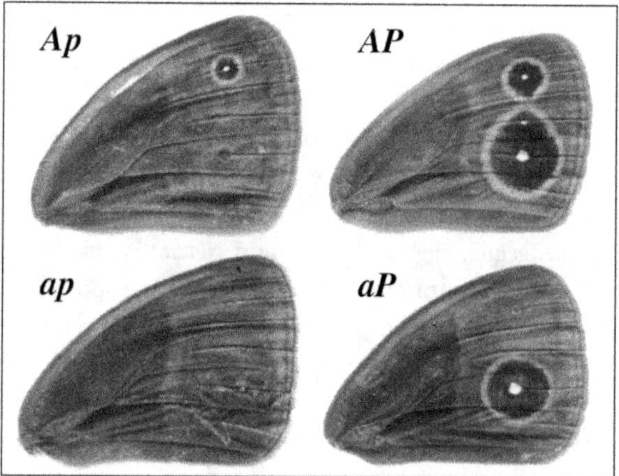

Figure 4.3 Responses drawn in morphospace for artificial selection experiments on the size of the anterior (large, A, or small, a) and posterior (large, P, or small, p) forewing eyespots on the dorsal wings of *B. anynana*. A, Changes from generation to generation in mean eyespot diameter relative to wing size along the four different directions of selection starting from the wild type in the centre of morphospace. Broken lines join the mean values for generation 11 and 25 phenotypes and all values are given relative to unselected control values. B, Representative wings of females for generation 25 phenotypes. Note that most ap females have no eyespots whereas many AP specimens have extra, satellite eyespots. From Beldade *et al.* 2002.

developmental plasticity. Replicated lines were established from the same base population and selected towards each corner of trait space; that is, along the 'coupled' axis towards two small or two large eyespots as is consistent with the shared genetics and development, and along the 'uncoupled' axis where the two eyespots are selected in opposite directions. This latter axis, therefore, represents one orthogonal to the proposed line of genetic channelling and the plane of developmental bias.

Artificial selection occurred over 25 generations (Figure 4.3). As expected, selection either 'up' or 'down' the 'coupled' axis of shared genetics and development produced rapid responses, with butterflies eventually either having no eyespots (ap) or two very large ones (AP). These morphologies differ widely from any produced in the base population, and are, therefore, highly novel. However, populations along the other 'uncoupled' axis, orthogonal to the axis of shared development, also responded well to selection, eventually producing phenotypes in which one eyespot was very large and the other absent or very small (Ap or aP). Again, these represent highly novel phenotypes. We concluded that this pattern in relative size of the two eyespots behaved in a developmentally flexible manner with a high evolvability in all directions through morphospace. There is no reason to imagine that other directions of change, such as ones in which one eyespot changes and the other does not, would be any more resistant to change under the influence of equally intense and targeted selection. We suggested that this capacity for independent evolution was the product of a long legacy of natural selection and evolutionary tinkering leading to morphological diversity among species and to corresponding evolvabilities for eyespot size at different sites in the wings (Beldade *et al.* 2003). We have now begun to examine whether the properties of the responses to artificial selection for different eyespot traits in our laboratory model species are reflected in the patterns of morphological disparity across the 80 or so species of the *Bicyclus* lineage (Beldade *et al.* 2002, Brakefield and Roskam 2006).

Current work is focused on a different eyespot trait, namely colour composition. There is some variability in our laboratory stock with respect to whether particular eyespots have a comparatively narrow outer gold ring or a broader one relative to the size of the inner black disc. Work on the colour of the posterior forewing eyespot had previously demonstrated a comparably high heritability and correlations among eyespots as for eyespot size, but a low genetic correlation between the two traits suggesting different genetic bases (Monteiro *et al.* 1994, 1997). Again, artificial selection along the 'coupled' axis for a specific pair of eyespots rapidly yields novel morphologies in which both targeted eyespots (and others) have narrower or broader gold rings. However in contrast to eyespot size, morphological change is much more strongly limited along the 'uncoupling' axis in which eyespots are selected in opposite directions in morphospace (C. E. Allen *et al.*, in preparation).

These experiments explore the potential roles of flexibility in genetic variation and developmental mechanisms in shaping evolution, but they do not examine directly evolvability in the context of the performance of the same morphologies in an ecological arena. Although, as outlined above, other work shows that eyespot size in *Bicyclus* can be a target for both natural and sexual selection (Breuker and Brakefield 2002, Robertson and Monteiro 2005, Brakefield and Frankino 2007), measurements of fitness remain to be made on the different eyespot phenotypes yielded by artificial selection in our experiments. A new study on the evolution of patterns of allometry involving the wings of *Bicyclus* butterflies has, however, attempted to combine artificial selection experiments with examinations of the consequences of changes in morphology for fitness.

The evolution of allometry involving wings

Diversification in the patterns of relative growth of different appendages in an organism must also involve the uncoupling of traits that originally shared all of their genetic pathways and developmental mechanisms. Species of butterfly can differ greatly in the size of the forewings relative to the hindwings, or in the size of both pairs of wings relative to the body (or 'wing-loading'). Again, artificial selection in *B. anynana* is exploring the potential flexibility in short-term responses for these scaling relationships.

Artificial selection has resulted in divergence in the scaling relationship for wing size relative to body size, and for forewing to hindwing size (Frankino *et al.* 2005, and in press), in each case producing novel morphologies relative to those in the base population. Following selection, the populations with divergent scaling relationships were each crossed to produce single populations with wide phenotypic variance for each scaling relationship. These latter populations were then used to compare the mating success for males showing changed allometry with males of wild-type allometry. This was done by performing competition experiments in a spacious tropical greenhouse, and by using the transfer of fluorescent dusts of different colours during copulation to trace male mating success (Joron and Brakefield 2003). For each scaling relationship, the wild-type males had an approximately three times higher mating success than either of the divergent phenotypes (Frankino *et al.* 2005, and in press). These studies detected strong stabilising selection in specific environments, but also indicate the necessary evolvability to account for the evolution of diversity in scaling relationships in new environments. Patterns of diversity and disparity with respect to these types of traits in butterfly morphospace may, therefore, be shaped mainly by natural selection acting differentially among environments and lifestyles.

Interactions between morphologies and life histories: the roles of hormones

The developmental plasticity and seasonal polyphenism in *Bicyclus* butterflies has provided an opportunity to examine the consequences of interactions between morphologies and life histories mediated by a common hormonal mechanism. The size difference in the marginal eyespots between the seasonal forms of *B. anynana* is mediated by circulating ecdysteroid hormones in early pupae that regulate the expression of the developmental genes specifying eyespot formation (Brakefield *et al.* 1996). Micro-injections or diffusion of ecdysone into young pupae destined to be adults of the dry-season form results in development of larger marginal eyespots and a broader medial band as characterise the wet-season form (Koch *et al.* 1996, Brakefield *et al.* 1998, Zijlstra *et al.* 2004). Thus, the seasonal polyphenism is regulated by the dynamics of ecdysteroid hormones immediately following the larval–pupal moult with a later build-up of 20-hydroxyecdysone in pupae of the dry-season form.

The seasonal forms (Figure 4.1A) are induced by environmental temperature in the larval stage especially in the period closest to pupation (but before the release of ecdysone hormone and subsequent wing pattern determination). In the field, temperature is high throughout the early and middle parts of the wet season when the two generations of the wet-season form of *B. anynana* develop but then declines as larvae that will produce the generation of the dry-season form grow at the transition from the warm wet season to the cool dry season (Figure 4.1B). Similarly in the laboratory, the alternative wing pattern phenotypes are obtained using split families reared either at high or low temperatures, respectively (Kooi and Brakefield 1999). However, our overall data suggest that developmental time rather than temperature per se is the key parameter such that any factor slowing pre-adult development leads to butterflies with a more exaggerated dry-season form phenotype. The same ecdysteroid hormone that mediates eyespot plasticity also regulates growth and metamorphosis suggesting that important interactions could exist among the wing pattern and life history traits.

In summary, the wing pattern with small eyespots in the cool dry season goes hand in hand with a long development time and delayed metamorphosis, whereas the reverse pattern is observed at higher temperatures. We examined the consequences of potential interactions by applying two-trait artificial selection. This included selection along the 'uncoupling' axis, that is for smaller eyespots and a shorter development time or the reverse combination, as well as for both traits along the same axis as found for the seasonal forms (Zijlstra *et al.* 2004). In a similar manner to the artificial selection experiments on the size of two eyespots, selection produced responses along both of these two axes.

Ecdysteroid titers and sensitivity to ecdysone injections were assayed for pupae from these two-trait selected lines (Zijlstra *et al.* 2004). Although the selected lines had diverged more for eyespot size than for developmental time, the widest differences in timing of ecdysteroid titers were observed between the development time selection regimes. Thus, fast selected lines, whether with small or large eyespots, had an earlier hormonal increase after pupation than either type of slow selected lines. Furthermore, sensitivity to ecdysone injection, as measured by a subsequent decrease in pupal time, was significantly lower for slow selected lines than for fast or unselected lines. We concluded that the observed response in eyespot size to artificial selection must have been achieved via alteration of mechanisms other than the circulating hormones since the dynamics of the hormone were apparently strongly dictated by the selection on developmental time. These alternative mechanisms probably involved changes directly in the expression of the developmental genes of eyespot morphogenesis. The overall developmental and physiological system is thus flexible enough to allow evolution in directions opposing the correlation between wing pattern and developmental time, and responses to selection are not constrained or strongly biased by a shared hormonal system.

DEVELOPMENT AND THE FUNCTIONAL PHENOTYPE

The latter example of work in *Bicyclus* butterflies indicates the potential for a broad evo-devo and evolutionary genetics to address not only morphological evolution but other traits including life histories.

Developmental biology is focused on morphology. Whilst exciting strides have been made in understanding the making of animal forms, if evolutionary biology is to benefit in a wide context such work will need to be extended to whole, functional phenotypes rather than form alone. Natural selection screens variation in the reproductive performance of individuals whose phenotypes are made up of complex suites of morphological, physiological, metabolic and behavioural traits. As in evo-devo, areas of research, including functional genomics, gene mapping, comparative physiology, epigenetics and behavioural biology, are making headway in opening up different aspects of the genotype to functional phenotype map. Eventually, the melding of information sourced from all these areas will reveal how the different types of traits interact with one another and the environment during development to make and maintain functional phenotypes, as well as to generate phenotypic variation.

Progress will be made increasingly in understanding the generation of subtle differences in phenotype. The challenge will then be more in the direction of producing a comparably sophisticated understanding of how natural selection screens such subtle phenotypic variants. This can again be illustrated by reference to the butterfly eyespot system. Artificial selection in *B. anynana* has yielded phenotypes where the relative size of two particular eyespots is changed (Figure 4.3). We will eventually be able to map and identify the genes or quantitative trait loci responsible for the response to selection, as well as demonstrate how the underlying developmental mechanisms have been modified to produce the changes in eyespot pattern. We may also be able to examine whether other species with patterns similar to those produced by the artificial selection in *B. anynana* (Brakefield and Roskam 2006) share some degree of genetical and developmental basis, or whether other options for morphological change were involved in their evolution. At the same time we have shown how capture–recapture experiments in the field can detect overall differences in survival in a dry-season environment between large cohorts of butterflies with conspicuous eyespots and those with no eyespots (Figure 4.2). Whilst this is undoubtedly important, it will be much more challenging to design comparable experiments to detect the effects in a natural selection arena of the phenotypes produced by the artificial selection because any differences in fitness, especially for the dorsal eyespots, are likely to be (much) smaller. When the complexities of spatial and temporal variation in the natural environment are also considered, the asymmetry of the challenges with regard on the one hand to the intrinsic processes of making the phenotype, and on the other to the extrinsic processes of natural selection becomes even clearer. Even so, to explore successfully why species occupy phenotypic space in the way they do will require such a two-pronged approach.

PERSPECTIVES AND FUTURE CHALLENGES

Evo-devo is clearly extending its scope beyond the traditional model organisms of developmental biology. It is gaining momentum in the exploration of the roles of genetic variation segregating within and among populations in the changes in development that underlie the evolution of morphologies in nature. Future progress will surely reveal how genetic change in the processes of development has contributed to the patterns of diversity in form from the differences across phyla and the origins of key innovations in body plans through to subtle variation among individuals within populations. In turn, this will bring us closer to understanding more about how developmental processes contribute to the patterns of occupancy of morphospace.

A successful fusion of work on different types of traits and the use of varied approaches from genomics to gene mapping will undoubtedly teach us exactly how very subtle changes in phenotypes can be made. Such phenotypes will extend from morphologies to metabolic traits, behaviours and life histories, and their interactions. Both epigenetic phenomena and developmental plasticity will become clear in mechanistic terms. Looking beyond such success in understanding the evolvability of complex traits suggests that the challenges for the future will be more in terms of measuring the effects of subtle changes in phenotype in the natural selection arena. This will be required to reach any firm conclusions about how both the processes of generating variation in the phenotype and those of natural selection and functional performance contribute together to the evolution of the occupancy of trait space. However, comparative methodologies using information about phylogenetic relationships and patterns of diversity among populations or taxa will provide increasingly sophisticated tools and data sets at the indirect level. It is the direct measurements of fitness curves that will remain highly challenging; small differences in fitness, including those arising via variation in the environment and in genotype × environment interactions, will always be difficult to tie down. The move towards integrating evo-devo and ecology will become increasingly dominated by the challenges of taking laboratory-based studies of how variation in the phenotype is made into the more functional domain of how such variants perform in natural environments.

ACKNOWLEDGEMENTS

I thank the organisers of the meeting in Venice for providing me with the opportunity to attend and to prepare this manuscript. The ideas presented here are the outcome of a program of work supported by all the past and present members of the *Bicyclus* laboratory in Leiden. I am extremely grateful to them all.

REFERENCES

Arthur, W. 2001. Developmental drive: an important determinant of the direction of phenotypic evolution. *Evolution & Development* **3**, 271–278.

Beldade, P. & Brakefield, P. M. 2002. The genetics and evo-devo of butterfly wing patterns. *Nature Reviews Genetics* **3**, 442–452.

Beldade, P., Brakefield, P. M. & Long, A. D. 2005. Generating phenotypic variation: prospects from "evo-devo" research on *Bicyclus anynana* wing patterns. *Evolution & Development* **7**, 101–107.

Beldade, P., Koops, K. & Brakefield, P. M. 2002. Developmental constraints versus flexibility in morphological evolution. *Nature* **416**, 844–847.

Beldade, P., Koops, K. & Brakefield, P. M. 2003. Modularity, individuality, and evo-devo in butterfly wings. *Proceedings of the National Academy of Sciences of the USA* **99**, 14262–14267.

Blows, M. W. & Hoffmann, A. A. 2005. A reassessment of genetic limits to evolutionary change. *Ecology* **86**, 1371–1384.

Brakefield, P. M. 1998. The evolution-development interface and advances with the eyespot patterns of *Bicyclus* butterflies. *Heredity* **80**, 265–272.

Brakefield, P. M. 2006. Evo-devo and constraints on selection. *Trends in Ecology and Evolution* **21**, 362–368.

Brakefield, P. M. & Frankino, W. A. 2007. Polyphenisms in Lepidoptera: multidisciplinary approaches to studies of evolution. In D. W. Whitman & T. N. Ananthakrishnan (eds.) *Phenotypic Plasticity in Insects. Mechanisms and Consequences.* Plymouth, UK: Science Publishers Inc., pp. 121–151.

Brakefield, P. M. & French, V. 2006. Evo-devo focus issue. *Heredity* **97**, 137–138.

Brakefield, P. M., French, V. & Zwaan, B. J. 2003. Development and the genetics of evolutionary change within species. *Annual Review in Ecology, Evolution, and Systematics* **34**, 633–660.

Brakefield, P. M., Gates, J., Keys, D. *et al.* 1996. Development, plasticity and evolution of butterfly eyespot patterns. *Nature* **384**, 236–242.

Brakefield, P. M., Kesbeke, F. & Koch, P. B. 1998. The regulation of phenotypic plasticity of eyespots in the butterfly *Bicyclus anynana*. *American Naturalist* **152**, 853–860.

Brakefield, P. M. & Larsen, T. B. 1984. The evolutionary significance of dry and wet season forms in some tropical butterflies. *Biological Journal of the Linnean Society* **22**, 1–12.

Brakefield, P. M. & Roskam, J. C. 2006. Exploring evolutionary constraints is a task for an integrative evolutionary biology. *American Naturalist* **168**, S4–S18.

Breuker, C. J. & Brakefield, P. M. 2002. Female choice depends on size but not symmetry of dorsal eyespots in the butterfly *Bicyclus anynana*. *Proceedings of the Royal Society of London B* **269**, 1233–1239.

Brunetti, C. R., Selegue, J. E., Monteiro, A. S. B. *et al.* 2001. The generation and diversification of butterfly eyespot color patterns. *Current Biology* **11**, 1578–1585.

Carroll, S. B. 2005. Evolution at two levels: on genes and form. *PloS Biology* **3**, 1159–1166.

Frankino, W. A., Zwaan, B. J., Stern, D. L. & Brakefield, P. M. 2005. Natural selection and developmental constraints in the evolution of allometries. *Science* **307**, 718–720.

Frankino, W. A., Zwaan, B. J., Stern, D. L. & Brakefield, P. M. in press. Internal and external constraints in the evolution of allometries among morphological traits in a butterfly. *Evolution*.

French, V. & Brakefield, P. M. 1995. Eyespot development on butterfly wings: the focal signal. *Developmental Biology* **168**, 112–123.

Joron, M. & Brakefield, P. M. 2003. Captivity masks inbreeding effects on male mating success in butterflies. *Nature* **424**, 191–194.

Joron, M., Jiggins, C. D., Papanicolaou, A. & McMillan, W. O. 2006. *Heliconius* wing patterns: an evo-devo model for understanding phenotypic diversity. *Heredity* **97**, 157–167.

Koch, P. B., Brakefield, P. M. & Kesbeke, F. 1996. Ecdysteroids control eyespot size and wing color pattern in the polyphenic butterfly *Bicyclus anynana* (Lepidoptera: Satyridae). *Journal of Insect Physiology* **42**, 223–230.

Kooi, R. E. & Brakefield, P. M. 1999. The critical period for wing pattern induction in the polyphenic tropical butterfly *Bicyclus anynana* (Satyrinae). *Journal of Insect Physiology* **45**, 201–212.

Lyytinen, A., Brakefield, P. M., Lindström, L. & Mappes, J. 2004. Does predation maintain eyespot plasticity in *Bicyclus anynana*? *Proceedings of the Royal Society of London B* **271**, 279–283.

Maynard Smith, J., Burian, R., Kaufman, S. *et al.* 1985. Developmental constraints and evolution. *Quarterly Review of Biology* **60**, 265–287.

McMillan, W. O., Monteiro, A. & Kapan, D. D. 2002. Development and evolution on the wing. *Trends in Ecology and Evolution* **17**, 125–133.

Monteiro, A., Brakefield, P. M. & French, V. 1994. The evolutionary genetics and developmental basis of wing pattern variation in the butterfly *Bicyclus anynana*. *Evolution* **48**, 1147–1157.

Monteiro, A., Brakefield, P. M. & French, V. 1997. Butterfly eyespots: the genetics and development of the color rings. *Evolution* **51**, 1207–1216.

Nijhout, H. F. 1991. *The Development and Evolution of Butterfly Wing Patterns*. Washington: Smithsonian Institution Press.

Pie, M. R. & Weitz, J. S. 2005. A null model of morphospace occupation. *American Naturalist* **166**, E1–E13.

Reed, R. D. & Serfas, M. S. 2004. Butterfly wing pattern evolution is associated with changes in a Notch/Distal-less temporal pattern formation process. *Current Biology* **14**, 1159–1166.

Robertson, K. A. & Monteiro, A. 2005. Female *Bicyclus anynana* butterflies choose males on the basis of their dorsal UV-reflective eyespot pupils. *Proceedings of the Royal Society of London B* **272**, 1541–1546.

Schluter, D. 1996. Adaptive radiation along genetic lines of least resistance. *Evolution* **50**, 1766–1774.

Stevens, M. 2005. The role of eyespots as anti-predator mechanisms, principally demonstrated in the Lepidoptera. *Biological Review* **80**, 573–588.

Wagner, G. P. & Altenberg, L. 1996. Complex adaptations and the evolution of evolvability. *Evolution* **50**, 967–976.

Windig, J. J., Brakefield, P. M., Reitsma, N. & Wilson, J. G. M. 1994. Seasonal polyphenism in the wild: survey of wing patterns in five species of *Bicyclus* butterflies in Malawi. *Ecological Entomology* **19**, 285–298.

Zijlstra, W. G., Steigenga, M. J., Koch, P. B., Zwaan, B. J. & Brakefield, P. M. 2004. Butterfly selected lines explore the hormonal basis of interactions between life histories and morphology. *American Naturalist* **163**, E76–E87.

5

The molecular biology underlying developmental evolution

CLAUDIO R. ALONSO

Stephen Jay Gould opens the Prospectus of his influential *Ontogeny and Phylogeny* (Gould 1977) with the following quotation from Van Valen (1973): 'A plausible argument could be made that evolution is the control of development by ecology. Oddly, neither area has figured importantly in evolutionary theory since Darwin, who contributed much to each. This is being slowly repaired for ecology . . . but development is still neglected.'

As accurate as these comments may have been in 1977, today, 30 years later, they no longer hold true: two new fields centred on the study of organismal development have now emerged in modern biology. One of them, which has successfully married the traditional fields of embryology and genetics, is the field of *developmental genetics*. The other one is known as *developmental evolution, evolutionary developmental biology* or simply *evo-devo*, and is the primary subject of this book and this chapter.

The evo-devo field has set as its ultimate goal to provide a mechanistic explanation of how developmental mechanisms changed during evolution, and how these alterations are causally linked to modifications in morphological patterns (Holland 1999). These questions are most relevant, as, so far, the formal structure of the evolutionary theory has been based upon the dynamics of alleles, individuals and populations under selective pressures and genetic drift 'assuming' the prior existence of these entities (Fontana and Buss 1993). The problems related to this assumption were already recognised more than a century ago by Hugo DeVries, who closes his famous book *Species and Varieties: Their Origin by Mutation* (DeVries 1904) with the comments of Arthur Harris on the

Evolving Pathways: Key Themes in Evolutionary Developmental Biology, ed. Alessandro Minelli and Giuseppe Fusco. Published by Cambridge University Press.
© Cambridge University Press 2008.

limits of natural selection: 'natural selection may explain the survival of the fittest, but it cannot explain the arrival of the fittest.' We could probably say that the central business of the evo-devo field is to determine the mechanisms that lead to *the arrival of the fittest.*

As it is now evident that the answer to the mechanistic question on *the arrival of the fittest* involves changes in the function of genes controlling developmental programs, it is opportune and important to reflect on the nature of the elements and systems controlling developmental gene function using an updated molecular background. I dedicate this chapter to precisely this task.

MOLECULAR BIOLOGY AND EVOLUTIONARY DEVELOPMENTAL
BIOLOGY: HISTORICAL LINKS

It would be hard to argue that should you need to select the two pieces of work that started off the field of molecular biology, one of them should not be the publication in 1953 of the three-dimensional model for the structure of DNA, based on the work of Rosalind Franklin at King's College London, and James Watson and Francis Crick at the University of Cambridge (Watson and Crick 1953a,b). The other contribution should indisputably be the description in 1961 of the operon model by François Jacob, Jacques Monod and André Lwoff (Jacob and Monod 1961). This second piece of work not only provided the first mechanistic model for gene regulation involved in a complex physiological process (Figure 5.1), but also explained the nature of the mechanisms for information transfer during the formation of proteins.

It was in this bubbling atmosphere of exciting discoveries about transcriptional regulation that Roy Britten and Eric Davidson proposed a first explicit theoretical model for gene regulation in 'higher' (eukaryotic) cells (Britten and Davidson 1969). The essence of the Britten–Davidson model can be summarised as a theory of how eukaryotic cells may achieve multiple changes in gene activity from relatively simple signals, on the assumption that a given state of differentiation depends on the coordinated activity of a number of biochemical systems. Britten and Davidson (1969) proposed that the transcription of batteries of 'producer' genes (i.e. protein-coding genes) is regulated by 'integrator' genes (i.e. genes encoding transcription factors). The main effect of the integrator genes is therefore to induce the transcription of many producer genes in response to a signal, which will be sensed via the 'sensor' DNA elements controlling the transcription of the integrators (Figure 5.2A). The concerted activation of one or more gene batteries

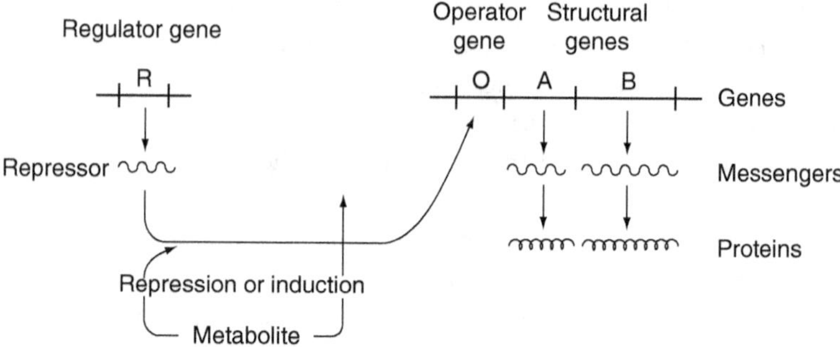

Figure 5.1 The operon model as originally published in 1961 (Jacob and Monod 1961). The simple and powerful logic of this bacterial gene regulatory circuit suggested that gene activity might be primarily regulated at the transcriptional level in most organisms. However, the discovery of several molecular post-transcriptional regulatory systems in the late 1970s invited a revision of this view.

would underlie the existence of diverse states of differentiation. Not least, the article also suggests the idea that in what the authors call 'higher grades of organization' or 'higher organisms,' evolution might be considered principally the result of changes in the regulatory systems encoded in the genome(Britten and Davidson 1969).

Leaving aside insignificant details, the similarities between the concepts presented in the article by Britten and Davidson, and those now used to describe the function of gene regulatory networks controlling transcriptional patterns during development reflect the enormous impact of the ideas of Britten and Davidson on mainstream views in the evo-devo field (Wray 2003, Levine and Davidson 2005). To illustrate this remarkable level of similarity, Figure 5.2 displays one of the diagrams in the original paper by Britten and Davidson (Figure 5.2A) together with a modern description of the gene regulatory interconnections controlling aspects of the development of the sea urchin (Figure 5.2B) (Davidson *et al.* 2003).

There are two possible interpretations for such parallels between the views of Britten and Davidson and the prevalent views of gene regulation within the evo-devo field at present. One of them is that the authors were well ahead of their time in setting the stage and the logic of what was to emerge in the field of developmental evolution in the following four decades. The other is that the emerging field of developmental evolution, after absorbing this influential package of ideas, has never again taken the time to explore the problem of gene regulation

(A)

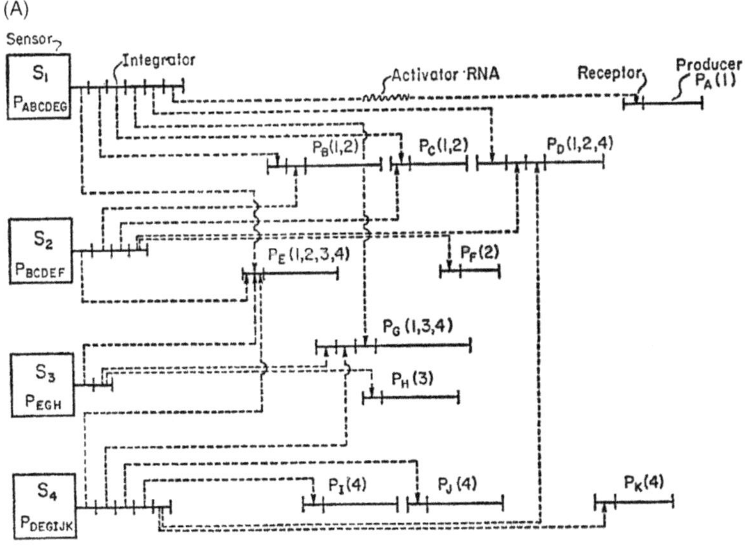

(B)

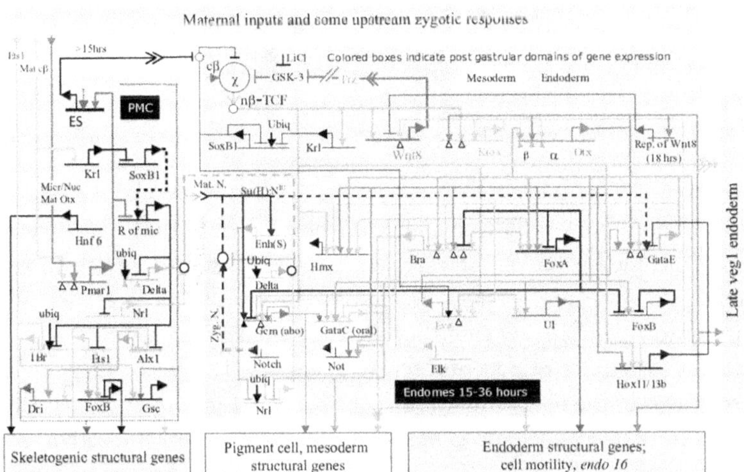

Figure 5.2 A, Original diagrams describing a theory for gene regulation in 'higher cells' (Britten and Davidson 1969), the diagram indicates the transcriptional circuits that coordinate the activity of gene batteries in cells. B, A modern representation of the gene regulatory networks controlling different developmental programs in the sea urchin (Davidson *et al.* 2003). A great degree of similarity exists between the two panels; for possible interpretations of these parallels see main text.

from a broader or updated molecular context. I see no reason why these two interpretations should be mutually exclusive.

A central message of this chapter is that the ideas of Britten and Davidson (1969) were directly absorbed by the evo-devo field at the time of its birth, and that they have never been revised in the light of the conceptual changes that took place in the field of molecular biology since the late 1960s. This accidental ostracism kept the evo-devo field from looking anywhere other than within the province of transcriptional regulation at the time of looking for the mechanisms affecting developmental evolution. Unsurprisingly, the field developed rather extreme positions on the mechanistic basis for developmental change, such as the general notion that the evolution of development largely reflects the evolution of transcriptional regulation (Arnone and Davidson 1997, Davidson 2001, Wray 2003, Wray *et al.* 2003).

Before we leave this section, it is perhaps opportune to bring in a key quotation from Britten and Davidson's (1969) article: 'Undoubtedly important regulatory processes occur at all levels of biological organiz-ation. We emphasize that this theory is restricted to processes of cell regulation at the level of genomic transcription.'

We are now ready to look at the importance of regulatory pro-cesses at non-transcriptional levels.

THE RISE OF POST-TRANSCRIPTIONAL GENE REGULATION

The pioneer work by Britten and Davidson in the late 1960s could never have included elements of what we now call *post-transcriptional* regu-lation. In those years, these levels of gene regulation were not well understood, and in any case they were not considered significant even within the circles of pure molecular biologists. This was partly due to the power of the operon model and the beautiful mechanics of its oper-ation to ensure gene- and context-specific patterns of gene activity via transcriptional control.

With the physical structure of the gene clearly established from the work in bacteria and bacteriophages, the appreciation that the sequences of the gene (DNA), RNA and protein were organised in a colli-near manner, and evidence emerging from the field of genetics suggesting that eukaryotic genes would by and large work similarly to those in prokaryotic organisms, it soon was assumed that most aspects of gene organisation in bacterial systems were likely to be universal.

With common gene structures, most regulatory mechanisms were bound to be similar, and therefore what was true for *Escherichia coli*, the λ

phage and the likes, was equally true for an elephant or a fly (Sharp 1993). As the operon model was regulated at the transcriptional level, in those years (and those years only!) the field of molecular biology considered gene regulation as a synonym of transcriptional regulation.

This reality began to change by the mid 1970s with a better appreciation of the functional constraints imposed on gene regulation by the existence of a nuclear structure in eukaryotes and the discovery of new ways of modulation affecting eukaryotic genes. For instance, mRNAs transcribed from nuclear loci are physically distant from the protein translational machinery, which is mostly located in the cytoplasm; nuclear compartmentalisation could then offer a site for specific mRNA processing and transport. Also, it was noted that the DNA content of eukaryotic germ cells varied dramatically across organisms without an apparent variation in the total number of genes. Certain organisms appeared to have 10 times as much DNA as was required to encode all their proteins; what was then the function of those large (and energeti cally expensive) tracts of DNA? Perhaps these extra DNA sequences were there to guide gene activity in an unknown way. Third, the phenomenon of heterogeneous nuclear RNA (hnRNA) suggested that long RNAs were transcribed from diverse nuclear sequences (Sharp 1993). These hnRNAs had very brief half-lives relative to the stable cytoplasmic mRNAs suggesting that they could potentially be precursors to mRNAs. Furthermore, both the long and unstable hnRNAs and the shorter and better-known mRNAs appeared to have similar chemical modifications in their 5′ and 3′ termini (i.e. what are now termed the CAP structure and poly-adenylation tails, respectively).

Although the important issue of how comparable were the structures of genes in prokaryotic and eukaryotic cells was not really explored or questioned at that time, the field headed towards the determination of the exact biochemical pathway linking a gene in the nucleus, its mRNA in the cytoplasm and its functional protein in the cell.

The regulation of gene expression, setting the logical circuits of all interesting biology, including cancer, infection and, not least, development, started to be viewed as the salient result from changes in the rates or efficiencies *of every one of the various steps* in the information pathway running from the gene to the protein. The understanding of the functions of the whole pathway of eukaryotic gene expression (related and unrelated to transcriptional modulation) began to be seen as the route to understanding important biological and biomedical phenomena (Sharp 1993).

These general considerations about the importance of gene regulation at different hierarchical levels were suddenly supported by a whole range of unexpected findings in post-transcriptional regulation made in the late 1970s. Because of space limitations, I will focus on just one of such findings: the discovery of alternative splicing.

The discovery of 'split' genes in adenovirus by Phillip Sharp and his team at the Massachusetts Institute of Technology in 1977 (Berget *et al.* 1977) swiftly led to the realisation that cellular genes suffered splicing reactions to remove introns or intervening sequences from their precursor RNAs (Breathnach *et al.* 1977, Tilghman *et al.* 1978). Protein coding sequences could thus arise from the joining of segments of mRNAs (exons) derived from DNA loci sitting far apart in the genome. In turn, the identification of conserved sequences at intron boundaries (Breathnach and Chambon 1981) and the observation that these sequences were common to vertebrate, plant and yeast cells (Padgett *et al.* 1986) suggested that the splicing process was evolutionarily conserved (Sharp 1993).

A year after the discovery of splicing, the team of Phillip Sharp described the first cases of alternative splicing in adenovirus and SV40 (Berk and Sharp 1978a,b). Shortly afterwards, several cellular mRNAs were shown to undergo alternative splicing, proving that post-transcriptional regulation was not a curiosity, but a powerful molecular strategy to change the function of gene products of a given gene (Kornblihtt *et al.* 1984, Smith *et al.* 1989).

Therefore, since the late 1970s the field of molecular biology has seen gene regulation as a complex matter that involves transcriptional as well as post-transcriptional regulatory processes. This realisation, however, has never broken through the integument of transcriptional domination that characterises the evo-devo field.

On this setting, I shall review what is now known about the processes of gene regulation, emphasising the regulatory potential for developmental evolution that resides in the many molecular elements controlling gene activity in eukaryotic organisms.

THE MOLECULAR BIOLOGY OF GENE REGULATION:
A POSTGENOMIC VIEW

As mentioned earlier in this chapter, prevalent ideas in the evo-devo field sustain the view that transcriptional regulation is the most important level of regulation underlying the evolution of developmental gene expression.

Now, most attention has actually been focused on a particular type of transcriptional regulatory element, the *enhancers*, rather than on transcription factor genes, promoter core elements or other transcriptionally relevant elements. Enhancers can be broadly defined as *cis*-acting DNA sequences containing a series of binding sites for transcription factors, which increase the net rate of transcriptional activity of genes located upstream or downstream of them.

Mutations affecting enhancers have been treated as the principal component of all evolutionarily relevant mutations (Stern 2000), and indeed, over time, evidence has accumulated in line with this view (Ludwig *et al.* 1998, Rockman and Wray 2002, Wray *et al.* 2003), bringing enhancers to the focal point of the mechanistic analysis of developmental evolution (Carroll 1995, Akam 1998, Stern 2000, Davidson 2001).

However, enhancers are not the only molecular sites able to modulate gene activity. In fact, a rather wide spectrum of molecular elements regulates the function of eukaryotic genes. To appreciate the richness provided by such *alternative regulatory levels* (ARLs) to enhancer regulation (Alonso and Wilkins 2005), let us begin by looking at how the expression of genetic information is initiated.

The transcription of all protein-coding genes in the cell requires a complex biochemical machine known as the basal transcriptional apparatus. This machine consists of the enzyme RNA polymerase II (RNApolII) and many other factors collectively known as general transcription factors. These factors mediate the physical interactions of the RNApolII with core promoter elements (e.g. TATA box, initiator, etc.) and regulatory transcription factors operating enhancers and silencers. The recruitment of the basal transcriptional apparatus to specific DNA sites is determined through physical and biochemical interactions with chromatin structures (Szutorisz *et al.* 2005), which rely on the state of chemical modification of chromatin proteins, as well as on interactions with factors bound to the gene's enhancers and silencers. The identity of core promoter elements and tissue-specific auxiliary factors also contributes to define the kinetics of transcription initiation (Smale and Kadonaga 2003). Other transcriptional *cis*-regulatory elements include insulators, which prevent enhancers and silencers in one gene from regulating a neighbouring gene, and the recently discovered global control regions, which are able to regulate gene transcription over large chromosomal domains (hundreds of kilobases). Therefore, when considering transcriptional regulation, enhancer elements represent only one of the many elements that influence this process.

In any case, transcription constitutes just the beginning of a long series of highly regulated biochemical events: once made, RNA transcripts suffer a wide range of alterations (Figure 5.3). Notably, the nature of such modifications or 'processing' events (e.g. splicing,

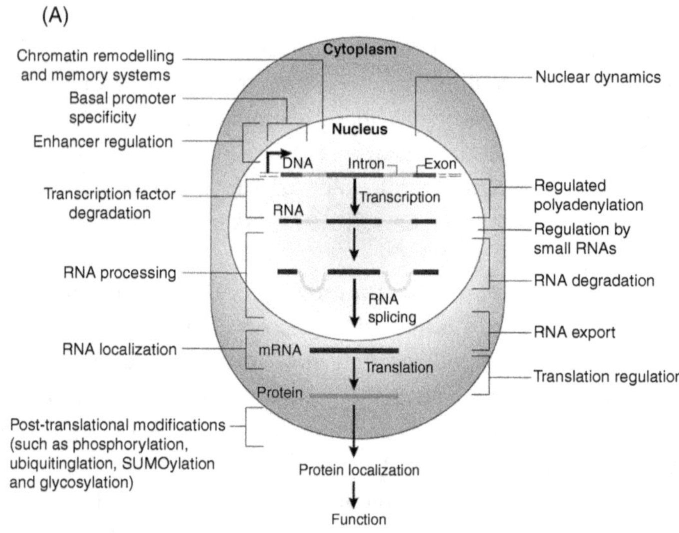

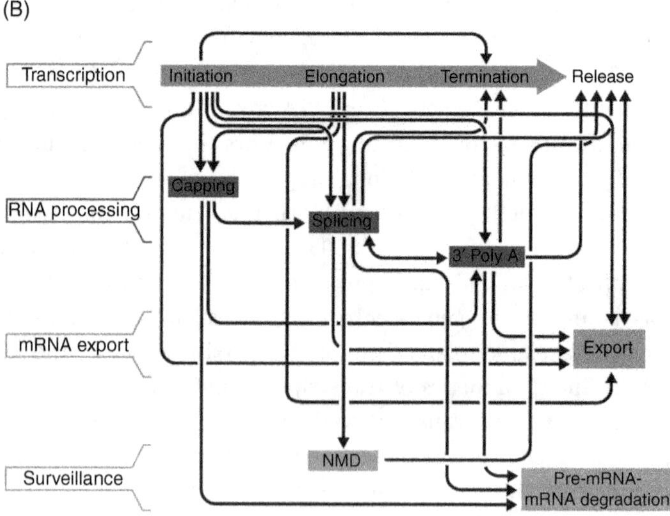

Figure 5.3 A, Multiple regulatory mechanisms regulate the final output of a given gene: enhancer-mediated gene regulation is only one of these mechanisms (Alonso and Wilkins 2005). B, The biochemical machines controlling gene expression interact with one another, creating a network topology that may offer yet another niche for genetic variation affecting contact surfaces between the machines (Maniatis and Reed 2002).

capping, poly-adenylation, editing, trans-splicing) affects the quality and quantity of the resulting mature RNA message. Processed messages are subjected to yet another regulatory layer by the quality control/degradation systems, which may label the message for chemical destruction or allow its export from the nucleus; the latter process (RNA export) could itself be subject to regulation (Darzacq *et al.* 2005).

Many messages contain sequence 'signatures' that convert them into targets of small regulatory RNA molecules, known as micro-RNAs (miRNA). These miRNAs are 22-nucleotide single-stranded non-coding RNAs that bind to target mRNAs and silence/reduce their expression (Plasterk 2006) via effects on protein synthesis (Carthew 2006) and/or effects on target RNA stability (Jing *et al.* 2005, Wu *et al.* 2006). Other sequence elements in messenger RNAs determine their localisation in particular sub-cellular foci, where they will await further signals for translational release. Translation and post-translational modifications including phosphorylation, glycosylation, ubiquitinylation and SUMOylation offer further regulatory opportunities to adjust protein quality, quantity and localisation (Alonso and Wilkins 2005).

The rate, rhythm, gene- and environment-specificity of all these processes are determined via sequence and structural labels present in particular mRNAs and proteins. Thus, it is the physical presence of these molecular tags, and the processes triggered by them, that will determine the qualitative and quantitative output of a given gene in the cell at a particular moment.

The brief consideration of the many biochemical steps and gene-specific tags involved in the information transfer from DNA to function puts enhancer sequences into a true minority within the molecular codes that specify controlled gene activity.

However, the functional properties of all gene-specific tags need not be identical. Enhancers do possess a series of attributes that make them particularly suitable to act as control nodes for developmental evolution. For example, because of the modular structure and combinatorial function of enhancers, mutational changes in their sequences may result in highly specific changes of gene expression causing developmental alterations that escape lethality. Thus, enhancer structure and *modus operandi* make enhancers especially apt for the generation of functional diversity. Nonetheless, a crucial point in our discussion is the realisation that these attributes are not exclusive to enhancers: the inspection of ARL properties reveals that they can affect gene function in a gene-specific, context-specific manner comparable to that of enhancers. To illustrate this, I will return to our previous case of alternative splicing with examples of its biological functions during the development of the fruit fly *Drosophila melanogaster.*

The *Drosophila Dscam* gene illustrates the modular nature and extreme structural diversity that is achievable by alternative splicing. The gene encodes a cell surface protein, member of the immunoglobulin (Ig) superfamily, with several protein domains including ten Ig and six fibronectin domains (Schmucker *et al.* 2000, Bharadwaj and Kolodkin 2006). The precise architecture of the Dscam protein is determined through the regulation of alternative splicing: the *Dscam* gene includes 95 variable exons that have the potential to produce 38 016 distinct alternatively spliced isoforms (Schmucker and Flanagan 2004). Phenotypic studies in mutant flies indicate that Dscam is involved in the regulation of axonal and dendritic branching, and axonal targeting and fasciculation (the bundling of axons into tracts) (Schmucker *et al.* 2000, Hummel *et al.* 2003, Zhu *et al.* 2006). Notably, a recent study (Chen *et al.* 2006) shows that the isoform diversity of the Dscam protein in *Drosophila* is required to establish stereotypical axonal branching patterns, demonstrating that the selective expression of *Dscam* alternative splice variants in particular cells determines neural connectivity.

At the other end of the complexity spectrum is the alternative splicing of the *Drosophila Sex-lethal* (*Sxl*) gene, which is a key regulatory gene in sexual differentiation (Bell *et al.* 1988, Smith *et al.* 1989). Here, male- and female-specific transcripts differ by the inclusion in males of a specific exon introducing a premature termination codon. Thus, *Sxl* sex-specific splicing patterns give rise to a functional protein product in females, while in males no functional protein is produced (reviewed in Lopez 1998), illustrating the utility of alternative splicing as a developmental control mechanism.

Another example of the developmental importance of alternative splicing is given by one of the members of the family of *Hox* genes: the *Drosophila Hox* gene *Ultrabithorax (Ubx)*. *Ubx* is expressed in the posterior thorax and anterior abdomen regions, where it determines segment-specific characteristics of many different cell lineages, including epidermis, central and peripheral nervous system, and mesodermal tissues (Bender *et al.* 1983, White and Wilcox 1984). Notably, the *Ubx* gene produces a family of six protein isoforms through alternative splicing. Isoforms differ from one another by the presence of optional microexons, which alter the distance between the homeodomain (the DNA-binding unit) and a cofactor-interaction module termed the hexapeptide. *Ubx* isoforms display different expression patterns during embryonic development: isoforms containing both microexons account for most of the *Ubx* expressed in epidermis, mesoderm, and peripheral nervous system (Lopez and Hogness 1991). In contrast, isoforms lacking the microexons are expressed exclusively in the central nervous system (Lopez and Hogness 1991, Bomze and Lopez 1994). The complex and quantitative nature of this regulation is unlike that of any other well-studied model systems in *Drosophila*; in addition, *Ubx* splicing patterns are conserved in drosophilids with 60 million years of independent evolution (Bomze and Lopez 1994). My laboratory has recently explored the biological relevance of *Ubx* alternative splicing, analysing the in vivo effects of the *Ubx* isoforms on the activation of a natural *Ubx* molecular target: the regulatory region of the gene *decapentaplegic* (*dpp*). These experiments showed that when *Ubx* isoforms are ectopically expressed in embryos they differ in their abilities to activate *dpp* in different tissues. In vitro studies, also in my laboratory, show that *Ubx* isoforms vary in their ability to bind target DNA elements in the presence of cofactors (H. C. Reed, M. Akam and C. R. Alonso, in preparation).

Thus, alternative splicing patterns dictate the functional specificity of several developmental genes including the *Hox* genes, one of the best-studied gene families in the evo-devo field. At this point it seems safe to conclude that gene regulation at ARLs such as alternative splicing is of clear relevance during development. However, the biological significance of this type of regulation is further highlighted by its prevalence: more than 15% of all *Drosophila* genes suffer alternative splicing. In addition, analyses of human expressed sequence tags (EST) and cDNA datasets have conservatively estimated that about 40–60% of human genes are alternatively spliced (reviewed in Modrek and Lee 2002); notably, this number increased to 73% when alternative splicing microarray data were combined with ESTs (Johnson *et al.* 2003). This has been independently corroborated by 'genome tiling' microarrays across human chromosomes 21 and 22, which indicated that alternative splicing occurs in >80% of genes (Kampa *et al.* 2004). In other words, alternative splicing is much more the rule than the exception.

If ARLs can affect development, can we think of ways in which these alternative levels of regulation may have changed during evolution? For this the first requisite is the accumulation of genetic variation affecting ARLs function.

In our familiar example of alternative splicing, this is illustrated by the fact that at least 15%, and perhaps as many as 50%, of human genetic diseases arise from mutations either in consensus splice site sequences, or in the more variable elements known as exon and intron splicing enhancers (ESEs and ISEs) and silencers (ESSs and ISSs) (Caceres and Kornblihtt 2002, Faustino and Cooper 2003, Pagani and

Baralle 2004), which are involved in defining both constitutive and alternative exons.

Now, let us return to our general discussion on the importance of enhancers in the evo-devo field. A possible reason that enhancer-dependent developmental change could predominate over that mediated by ARLs is that sites for the former might be far more abundant than for the latter, at the level of individual genes. This is essentially an argument about relative mutational 'target sizes'. Unfortunately, the information to resolve this matter, even for one selected organism, is not yet available in full. However, to gain an approximation for the dimensions of both enhancer-associated and ARL-associated DNA target sizes, a recent study has compiled information from different genomic databases and established a preliminary estimate; this shows that the mutational target sites offered by ARLs are actually comparable to those offered by enhancers, if not larger than the latter (Alonso and Wilkins 2005), implying that ARLs are likely to be hit by mutations more frequently than enhancers.

Finally, the importance of the multilayered regulatory systems controlling developmental gene activity as an evolutionary substrate is further stressed by recent studies proposing the view that, in contrast to a simple linear assembly line, a complex and extensively coupled network has evolved to coordinate the activities of the individual gene expression machines controlling the pathway from DNA to protein to function (Maniatis and Reed 2002). Such extensive coupling can be accommodated in a model in which the machines are tethered to each other via contact surfaces so that they form what one may call 'gene expression factories' which maximise the efficiency and specificity of each step in gene expression (Figure 5.3). With the landscape of gene regulation emerging as multidimensional and highly interconnected, the importance of ARLs involving discrete contact surfaces and signature sequences in mRNAs and proteins becomes more apparent.

In sum, we may need to look at gene expression in a rather different way than we are used to. For this, the conventional linear 'assembly line' representation seems no longer adequate. A more appropriate description of gene expression might instead involve a multidimensional space, composed of a series of planes, each one of them representing the set of functional interactions taking place between an element originated from a given gene sequence (e.g. pre-mRNA, mRNA, protein, modified protein) and relevant elements within that gene regulatory level (e.g. transcription, RNA-processing, RNA export, RNA degradation). I see no perfect way of illustrating this; nonetheless, an attempt is shown in Figure 5.4. There, positive and negative inter-actions between elements are represented through conventional acti-vation/repression symbols; the diagram also incorporates quantitative aspects of these interactions which are commonly left aside in most evo-devo studies. These quantitative aspects are depicted by arrows with different thicknesses and lengths to accommodate strengths and duration of the interactions between the different elements (Figure 5.4A). This representation of interactions can be applied to the expression of genetic information from DNA to RNA to protein, as stated in the central dogma of molecular biology (Figure 5.4B). However, future developmental gene function analysis may need to understand the biological significance of the particular gene expression path followed by a gene element as development proceeds. The diagram in Figure 5.4C illustrates an example of a 'gene expression path' followed by a given gene and its products through the different gene regulatory planes.

According to this view, the functional output of a given gene in a particular cellular and developmental context will be determined by the efficiency with which each one of the gene's products (e.g. primary mRNA, spliced mRNA, exported mRNA, etc.) moves through the series of gene regulatory planes that compose the multidimensional space of gene regulation. It thus follows that, as important as transcriptional net-works might be, they represent one level and one level only within the multilayered architecture of gene regulation.

EPILOGUE: THE PHILOSOPHY OF GENE REGULATION AND THE
FUTURE OF THE EVO-DEVO FIELD

Six years after Britten and Davidson's (1969) paper on the theory of oper-ation of transcriptional regulatory circuits in 'higher' cells, Mary-Claire King and Allan C. Wilson published a seminal article in which they reported their findings on the genetic differences between humans and chimpanzees (King and Wilson 1975). In brief, their study applied a combination of amino acid sequencing and immunological and electro-phoretic techniques to compare a large set of proteins from humans and chimpanzees. All three approaches yielded consistent results indicating that an average human polypeptide is more than 99% similar to its chimpanzee counterpart. Based on their data, King and Wilson suggested that the evolutionary changes in anatomy and 'way of life' (i.e. physiology and behaviour; Carroll 2005) are more likely to be the result of changes in the mechanisms controlling the expression of

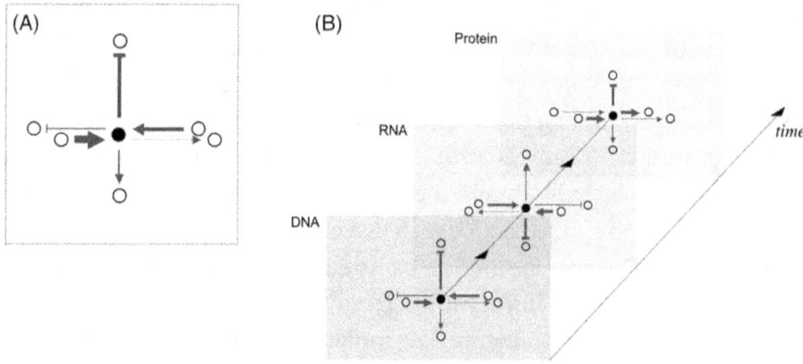

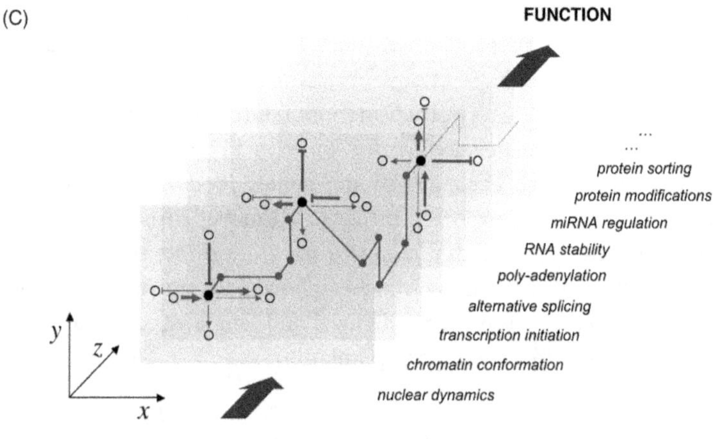

Figure 5.4 Possible representations of the multidimensional space controlling gene regulation in eukaryotic cells. A, Individual genetic elements (black circles) interact with many other elements through positive and negative interactions (activation and repression arrows, respectively); quantitative aspects of these interactions are represented by variable thicknesses (strength of the interaction) and lengths (duration of the interaction) of the interaction arrows. B, As a concept diagram, interaction network motifs as in A have been plotted on successive planes representing the dogma of molecular biology. C, A holistic view of gene expression in eukaryotic cells. The diagram illustrates a specific 'gene expression circuit' followed by a given gene and its products through the different regulatory planes. The functional output of a given gene will be determined by the efficiency with which each one of the gene's products (e.g. primary mRNA, spliced mRNA, exported mRNA, etc.) moves through the series of gene regulatory planes that compose the multidimensional space of the gene expression path.

genes, rather than the product of sequence changes affecting protein composition. This beautiful and powerful study grasped the attention of all of those interested in the genetic underpinnings of developmental evolution, taking them to the dilemma between structural (protein-coding) versus regulatory mutations; King and Wilson's study left, however, little margin for doubts about the power of regulatory changes.

Now, returning to our previous discussions on the prevalent molecular thinking in the mid 1970s that gene regulation was almost exclusively controlled at the level of transcription, and regardless of King and Wilson's (1975) careful writing, their article ended up being largely interpreted as supporting the view that transcriptional regulation was the key molecular level of regulation for organismal evolution.

Not surprisingly, the field searched, and searched again, and after three decades of intense work managed to find a handful of clean examples in which enhancer variation is indeed involved in morphological differences between species (Belting *et al.* 1998, Wang and Chamberlin 2004, Wang *et al.* 2004, Gompel *et al.* 2005, Jeong *et al.* 2006, Prud'homme *et al.* 2006). Do these examples prove that enhancers are the principal nodes for change in developmental evolution? As we have seen above, given that the complexity of gene regulation exceeds transcriptional control, the answer is, probably, no: to distinguish the relative contribution of genetic variation at one regulatory level or other, we must look equally thoroughly at all regulatory levels, and this has not been done until now.

Yet another relevant piece of work needs to be mentioned to close this discussion: one that was to come two years after King and Wilson's article was published. This is François Jacob's 'Evolution and Tinkering' paper (Jacob 1977). Here, Jacob put forward his views about how evolution proceeds, contrasting the process of evolution with the jobs of an engineer and that of a tinkerer. In his own words:

> Natural selection has no analogy with any aspect of human behaviour. However, if one wanted to play with a comparison, one would have to say that natural selection does not work like an engineer. It works like a tinkerer who does not know exactly what he is going to produce but uses whatever he finds around him whether it be pieces of string, fragments of wood, or old cardboards; in short it works as a tinkerer who uses everything at his disposal to produce some kind of workable object.

The experiments of King and Wilson, together with the ideas of Jacob, nicely complement each other telling us that the business of the evolutionary tinkerer is likely to be a regulatory one, and thus, chances are that he is going to use everything at hand, be it alternative splicing, RNA localisation, protein degradation or enhancer modules in order to manipulate developmental programs over time. The fact that we have found a few cases of variation at enhancer modules must not be confused with evidence in support of the idea that enhancers are the only elements involved in developmental evolution.

Although the history of ideas can itself be fascinating, the prime goal of this text is scientific utility. The exploration of the molecular elements used for the generation of developmental diversity during evolution is a venture of paramount importance in modern biology; to tackle this problem, the comparative study of variation at *all* levels of gene regulation emerges as the only unbiased strategy to establish the principal avenues of molecular change used during developmental evolution. This is, in my view, the only impartial path to follow for the true understanding of the *arrival of the fittest*.

ACKNOWLEDGEMENTS

I thank Adam Wilkins for key discussions on some of the ideas expressed here. I also wish to thank Michael Akam for his support and valuable feedback. Many thanks also to my wife, Juliane Mossinger, and baby daughter Sofia Alonso-Mossinger for their love and support during the writing of this chapter.

REFERENCES

Akam, M. 1998. Hox genes, homeosis and the evolution of segment identity: no need for hopeless monsters. *International Journal of Developmental Biology* **42**, 445–451.

Alonso, C. R. & Wilkins, A. S. 2005. The molecular elements that underlie developmental evolution. *Nature Reviews Genetics* **6**, 709–715.

Arnone, M. I. & Davidson, E. H. 1997. The hardwiring of development: organization and function of genomic regulatory systems. *Development* **124**, 1851–1864.

Bell, L. R., Maine, E. M., Schedl, P. & Cline, T. W. 1988. *Sex-lethal*, a *Drosophila* sex determination switch gene, exhibits sex-specific RNA splicing and sequence similarity to RNA binding proteins. *Cell* **55**, 1037–1046.

Belting, H. G., Shashikant, C. S. & Ruddle, F. H. 1998. Modification of expression and cis-regulation of *Hoxc8* in the evolution of diverged axial morphology. *Proceedings of the National Academy of Sciences of the USA* **95**, 2355–2360.

Bender, W., Akam, M., Karch, F. *et al.* 1983. Molecular genetics of the bithorax complex in *Drosophila melanogaster*. *Science* **221**, 23–29.

Berget, S. M., Moore, C. & Sharp, P. A. 1977. Spliced segments at the 5′ terminus of adenovirus 2 late mRNA. *Proceedings of the National Academy of Sciences of the USA* **74**, 3171–3175.

Berk, A. J. & Sharp, P. A. 1978a. Spliced early mRNAs of simian virus 40. *Proceedings of the National Academy of Sciences of the USA* **75**, 1274–1278.

Berk, A. J. & Sharp, P. A. 1978b. Structure of the adenovirus 2 early mRNAs. *Cell* **14**, 695–711.

Bharadwaj, R. & Kolodkin, A. L. 2006. Descrambling *Dscam* diversity. *Cell* **125**, 421–424.

Bomze, H. M. & Lopez, A. J. 1994. Evolutionary conservation of the structure and expression of alternatively spliced *Ultrabithorax* isoforms from *Drosophila*. *Genetics* **136**, 965–977.

Breathnach, R. & Chambon, P. 1981. Organization and expression of eucaryotic split genes coding for proteins. *Annual Review of Biochemistry* **50**, 349–383.

Breathnach, R., Mandel, J. L. & Chambon, P. 1977. Ovalbumin gene is split in chicken DNA. *Nature* **270**, 314–319.

Britten, R. J. & Davidson, E. H. 1969. Gene regulation for higher cells: a theory. *Science* **165**, 349–357.

Caceres, J. F. & Kornblihtt, A. R. 2002. Alternative splicing: multiple control mechanisms and involvement in human disease. *Trends in Genetics* **18**, 186–193.

Carroll, S. B. 1995. Homeotic genes and the evolution of arthropods and chordates. *Nature* **376**, 479–485.

Carroll, S. B. 2005. Evolution at two levels: on genes and form. *PLoS Biology* **3**, e245.

Carthew, R. W. 2006. Gene regulation by microRNAs. *Current Opinion in Genetics and Development* **16**, 203–208.

Chen, B. E., Kondo, M., Garnier, A. *et al.* 2006. The molecular diversity of *Dscam* is functionally required for neuronal wiring specificity in *Drosophila*. *Cell* **125**, 607–620.

Darzacq, X., Singer, R. H. & Shav-Tal, Y. 2005. Dynamics of transcription and mRNA export. *Current Opinion in Cell Biology* **17**, 332–339.

Davidson, E. H. 2001. *Genomic Regulatory Systems*. San Diego: Academic Press.

Davidson, E. H., McClay, D. R. & Hood, L. 2003. Regulatory gene networks and the properties of the developmental process. *Proceedings of the National Academy of Sciences of the USA* **100**, 1475–1480.

DeVries, H. 1904. *Species and Varieties: Their Origin by Mutation*. Chicago: The Open Court Publishing Company.

Faustino, N. A. & Cooper, T. A. 2003. Pre-mRNA splicing and human disease. *Genes & Development* **17**, 419–437.

Fontana, W. & Buss, L. W. 1993. "The arrival of the fittest": towards a theory of biological organization. *Bulletin of Mathematical Biology* **56**, 1–64.

Gompel, N., Prud'homme, B., Wittkopp, P. J., Kassner, V. A. & Carroll, S. B. 2005. Chance caught on the wing: *cis*-regulatory evolution and the origin of pigment patterns in *Drosophila*. *Nature* **433**, 481–487.

Gould, S. J. 1977. *Ontogeny and Phylogeny*. Cambridge, MA: Harvard University Press.

Holland, P. W. 1999. The future of evolutionary developmental biology. *Nature* **402**, C41–C44.

Hummel, T., Vasconcelos, M. L., Clemens, J. C. *et al.* 2003. Axonal targeting of olfactory receptor neurons in *Drosophila* is controlled by *Dscam*. *Neuron* **37**, 221–231.

Jacob, F. 1977. Evolution and tinkering. *Science* **196**, 1161–1166.

Jacob, F. & Monod, J. 1961. Genetic regulatory mechanisms in the synthesis of proteins. *Journal of Molecular Biology* **3**, 318–356.

Jeong, S., Rokas, A. & Carroll, S. B. 2006. Regulation of body pigmentation by the Abdominal-B Hox protein and its gain and loss in *Drosophila* evolution. *Cell* **125**, 1387–1399.

Jing, Q., Huang, S., Guth, S. *et al.* 2005. Involvement of microRNA in AU-rich element-mediated mRNA instability. *Cell* **120**, 623–634.

Johnson, J. M., Castle, J., Garrett-Engele, P. *et al.* 2003. Genome-wide survey of human alternative pre-mRNA splicing with exon junction microarrays. *Science* **302**, 2141–2144.

Kampa, D., Cheng, J., Kapranov, P. *et al.* 2004. Novel RNAs identified from an in-depth analysis of the transcriptome of human chromosomes 21 and 22. *Genome Research* **14**, 331–342.

King, M. C. & Wilson, A. C. 1975. Evolution at two levels in humans and chimpanzees. *Science* **188**, 107–116.

Kornblihtt, A. R., Vibe-Pedersen, K. & Baralle, F. E. 1984. Human fibronectin: molecular cloning evidence for two mRNA species differing by an internal segment coding for a structural domain. *EMBO Journal* **3**, 221–226.

Levine, M. & Davidson, E. H. 2005. Gene regulatory networks for development. *Proceedings of the National Academy of Sciences of the USA* **102**, 4936–4942.

Lopez, A. J. 1998. Alternative splicing of pre-mRNA: developmental consequences and mechanisms of regulation. *Annual Review of Genetics* **32**, 279–305.

Lopez, A. J. & Hogness, D. S. 1991. Immunochemical dissection of the Ultrabithorax homeoprotein family in *Drosophila melanogaster*. *Proceedings of the National Academy of Sciences of the USA* **88**, 9924–9928.

Ludwig, M. Z., Patel, N. H. & Kreitman, M. 1998. Functional analysis of eve stripe 2 enhancer evolution in *Drosophila*: rules governing conservation and change. *Development* **125**, 949–958.

Maniatis, T. & Reed, R. 2002. An extensive network of coupling among gene expression machines. *Nature* **416**, 499–506.

Modrek, B. & Lee, C. 2002. A genomic view of alternative splicing. *Nature Genetics* **30**, 13–19.

Padgett, R. A., Grabowski, P. J., Konarska, M. M., Seiler, S. & Sharp, P. A. 1986. Splicing of messenger RNA precursors. *Annual Review of Biochemistry* **55**, 1119–1150.

Pagani, F. & Baralle, F. E. 2004. Genomic variants in exons and introns: identifying the splicing spoilers. *Nature Reviews Genetics* **5**, 389–396.

Plasterk, R. H. 2006. Micro RNAs in animal development. *Cell* **124**, 877–881.

Prud'homme, B., Gompel, N., Rokas, A. *et al.* 2006. Repeated morphological evolution through *cis*-regulatory changes in a pleiotropic gene. *Nature* **440**, 1050–1053.

Rockman, M. V. & Wray, G. A. 2002. Abundant raw material for cis-regulatory evolution in humans. *Molecular Biology and Evolution* **19**, 1991–2004.

Schmucker, D., Clemens, J. C., Shu, H. *et al.* 2000. *Drosophila Dscam* is an axon guidance receptor exhibiting extraordinary molecular diversity. *Cell* **101**, 671–684.

Schmucker, D. & Flanagan, J. G. 2004. Generation of recognition diversity in the nervous system. *Neuron* **44**, 219–222.

Sharp, P. A. 1993. Split genes and RNA splicing. In N. Ringertz (ed.) *Nobel Lectures, Physiology or Medicine 1991–1995*. Singapore: World Scientific Publishing Co., pp. 145–174.

Smale, S. T. & Kadonaga, J. T. 2003. The RNA polymerase II core promoter. *Annual Review of Biochemistry* **72**, 449–479.

Smith, C. W., Patton, J. G. & Nadal-Ginard, B. 1989. Alternative splicing in the control of gene expression. *Annual Review of Genetics* **23**, 527–577.

Stern, D. L. 2000. Evolutionary developmental biology and the problem of variation. *Evolution* **54**, 1079–1091.

Szutorisz, H., Dillon, N. & Tora, L. 2005. The role of enhancers as centres for general transcription factor recruitment. *Trends in Biochemical Sciences* **30**, 593–599.

Tilghman, S. M., Tiemeier, D. C., Seidman, J. G. *et al.* 1978. Intervening sequence of DNA identified in the structural portion of a mouse beta-globin gene. *Proceedings of the National Academy of Sciences of the USA* **75**, 725–729.

Van Valen, L. 1973. Festschrift. *Science* **180**, 488.

Wang, X. & Chamberlin, H. M. 2004. Evolutionary innovation of the excretory system in *Caenorhabditis elegans*. *Nature Genetics* **36**, 231–232.

Wang, X., Greenberg, J. F. & Chamberlin, H. M. 2004. Evolution of regulatory elements producing a conserved gene expression pattern in *Caenorhabditis*. *Evolution & Development* **6**, 237–245.

Watson, J. D. & Crick, F. H. 1953a. Genetical implications of the structure of deoxyribonucleic acid. *Nature* **171**, 964–967.

Watson, J. D. & Crick, F. H. 1953b. Molecular structure of nucleic acids; a structure for deoxyribose nucleic acid. *Nature* **171**, 737–738.

White, R. A. & Wilcox, M. 1984. Protein products of the bithorax complex in *Drosophila*. *Cell* **39**, 163–171.

Wray, G. A. 2003. Transcriptional regulation and the evolution of development. *International Journal of Developmental Biology* **47**, 675–684.

Wray, G. A., Hahn, M. W., Abouheif, E. *et al.* 2003. The evolution of transcriptional regulation in eukaryotes. *Molecular Biology and Evolution* **20**, 1377–1419.

Wu, L., Fan, J. & Belasco, J. G. 2006. MicroRNAs direct rapid deadenylation of mRNA. *Proceedings of the National Academy of Sciences of the USA* **103**, 4034 4039.

Zhu, H., Hummel, T., Clemens, J. C. *et al.* 2006. Dendritic patterning by *Dscam* and synaptic partner matching in the Drosophila antennal lobe. *Nature Neuroscience* **9**, 349–355.

6

Evo-devo's identity: from model organisms to developmental types

RONALD A. JENNER

EVO-DEVO'S IDENTITY

Evo-devo studies the evolution of development, and how changes in development influence phenotypic evolutionary change. The evolution of novelties and body plans are considered as the most distinctive research areas of evo-devo (Wagner 2000, 2001, Wagner *et al.* 2000, Müller and Newman 2005). Nevertheless, there seems to be little consensus about evo-devo's disciplinary identity. It has been regarded as a branch of developmental biology, part of evolutionary biology, a revision of evolutionary theory or an independent new synthetic discipline (Gilbert *et al.* 1996, Arthur 2000, 2002, 2004a,b, Hall 2000, Raff 2000, Wagner 2000, Robert *et al.* 2001, Gould 2002, Wilkins 2002, Baguñá and Garcia-Fernàndez 2003, Gilbert 2003, Kutschera and Niklas 2004, Amundson 2005, Müller and Newman 2005). Similarly, there has been skepticism about evo-devo's promise in both the literature (Wagner 2000, 2001, Richardson 2003, Wagner and Larsson 2003, Coyne 2005) and at meetings such as the one in 2006 in Venice, at which the present book was conceived.

Although various factors are at play, I think that current skepticism partly results from a failure to articulate evo-devo's conceptual foundation properly. This issue comes into focus when it is observed that the papers outlining evo-devo's research agenda almost exclusively link the promise of evo-devo to discovering general concepts and rules.

Arthur (2002: 757), for example, expresses concern when he writes that we are currently in a situation 'where it almost seems that anything goes, that is, any developmental gene, its expression pattern and the resultant ontogenetic trajectory can evolve in any way. If this were true, no generalisations would be possible, let alone universally applicable laws'. A senior developmental biologist in my institution expresses it thus: 'I am left thinking that there are no rules, hence nothing for evo-devo to discover'.

Here I discuss the identity of evo-devo from the perspective of an important, but neglected, epistemological dualism: idiographics vs. nomothetics. By grounding evo-devo's identity in this framework I show that the above pessimism is misguided, and bring into focus a bias in what evo-devo is generally expected to contribute to biology. This perspective is also vital for understanding the role of model organisms in evo-devo (Jenner 2006a), and for salvaging the status of developmental types, which have increasingly fallen into disrepute in the recent literature.

GENERALITY AND UNIQUENESS IN SCIENCE: NOMOTHETICS AND IDIOGRAPHICS

An epistemological distinction can be made between idiographic and nomothetic aspects of science. The description of unique and historically contingent particulars is the domain of idiographics, while the discovery of law-like regularities or generalities falls under the rubric of nomothetics. This epistemological dualism is predicated on the individuality thesis, which distinguishes between *individuals* and *classes* (Ghiselin 1997). Idiographics is strictly concerned with the description of unique, concrete individuals, while nomothetics is concerned with formulating generalisations for abstract classes of which individuals may be members. Such generalisations can be formulated with respect to traits shared between the members of a class. Evo-devo embodies both principles, and like other historical sciences such as anthropology, paleobiology and evolutionary biology, evo-devo strives to relate the detailed description of particulars to law-like regularities (Gould 1980, Ghiselin 1997, 2005, Lyman and O'Brien 2004). Specifically, evo-devo aims to understand how the unique evolutionary histories of particular body plans, or origins of novelties, relate to the involvement of different classes of evolutionary developmental mechanisms that are embodied in the nomothetic conceptual categories at the core of evo-devo's research agenda (Table 6.1).

Table 6.1 *A sample of evolutionary developmental mechanisms and the evo-devo concepts based on them, as well as developmental types with examples of model organisms.*

Although listed separately in this table, developmental types can be formulated for any class of organisms sharing particular developmental traits, including specific phenotypes, or exemplifying particular evolutionary developmental mechanisms.

Evo-devo concept	Example	Model system	References
Gene level evolutionary developmental mechanisms	Gene regulatory networks (GRN)	Sea urchin endomesoderm development	Davidson and Erwin (2006)
	Gene/genome duplication and divergence	*Drosophila* engrailed family genes	Peel *et al.* (2006)
Epigenetic evolutionary developmental mechanisms	Gene regulation by transposable elements	Mouse coat colour, teleost *Fundulus heteroclitus* differential gene expression	Richards (2006), Biémont and Vieira (2006)
Cell level evolutionary developmental mechanisms	Competence	Nematode vulva development	Rudel and Sommer (2003), Hong and Sommer (2006)
	Cell condensations	*Drosophila* (imaginal discs)	Hall (2003b)
Tissue/organ level evolutionary developmental mechanisms	Induction	Cavefish *Astyanax mexicanus* eye loss	Jeffery *et al.* (2003)
	Segmentation	Arthropods, onychophorans, annelids, chordates	Minelli and Fusco (2004)
Organism level evolutionary developmental mechanisms	Developmental bias	*Caenorhabditis elegans* body size variation	Arthur (2004b)
General evolutionary developmental mechanisms	Redundancy	*Drosophila* antero-posterior axis development	Rudel and Sommer (2003)

General themes addressed by study of evolutionary developmental mechanisms	Modularity	Cavefish *Astycnax mexicanus* sense organs	Franz-Odendaal and Hall (2006)
	Co-option	Head and thoracic horns in *Onthophagus* beetles	Moczek (2006)
	Evolvability	In silico cell lineage evolution	Azevedo *et al.* (2005)
	Developmental constraint	Gastropod *Cerion*	Gould (2002)
	Evolutionary novelties	Sea anemone *Nematostella vectensis* (bilateral symmetry, triploblasty)	Darling *et al.* (2005)
	Relationship between micro and macroevolution	Three-spined stickleback skeletal evolution, *Heliconius* butterfly wing patterns	Rudel and Sommer (2003), Joron *et al.* (2006)
	Genetic assimilation	*Bicyclus* butterfly seasonal wing pattern differences	Pigliucci (2005)
	Canalisation and cryptic genetic variation	*Drosophila* phenotypic variation increase during Hsp90 impairment	Flatt (2005)

Continued

Table 6.1 (cont.)

Evo-devo concept	Example	Model system	References
	Developmental/phenotypic plasticity, polyphenism	Ant caste polyphenism, caste determination by primordial germ cells in parasitic wasp *Copidosoma floridanum*	Abouheif and Wray (2002), Extavour (2004)
Developmental types	Organisms with reduced characters	Wing loss in sterile ant castes, pelvic reduction in fish, eye loss in cavefish	Abouheif and Wray (2002), Jeffery *et al.* (2003), Tanaka *et al.* (2005), Shapiro *et al.* (2006)
	Animals with set-aside cells	Polychaetes, sea urchins	Peterson *et al.* (1997), Ransick *et al.* (1996), Blackstone and Ellison (2000)
	Animals with type I development	Nematodes, tunicates	Davidson (1991)
	Animals with split or dispersed Hox clusters, and exhibiting Hox gene loss	*Drosophila*, the ascidian *Ciona*, the larvacean *Oikopleura*, *C. elegans*	Seo *et al.* (2004)

IDIOGRAPHICS AND NOMOTHETICS IN EVO-DEVO

Evo-devo's most outspoken practitioners have presented evo-devo as unabashedly nomothetic in its promise (Gilbert *et al.* 1996, Arthur 2000, 2002, 2004a,b, Hall 2000, Raff 2000, Baguñà and Garcia-Fernàndez 2003, Gilbert 2003), a view explicitly accepted by those who see evo-devo as an important contribution to, or corrective of, evolutionary theory (Gould 2002, Kutschera and Niklas 2004, Amundson 2005, Stoltzfus 2006). At the core of evolutionary biology, neo-Darwinian evolutionary theory supplies a set of nomothetic principles with respect to which evo-devo has staked out its conceptual territory. However, I think that a misleadingly biased perspective has been established in the literature by downplaying the importance of idiographics. For example, Arthur (2002: 759) labels evolutionary biology 'a conceptually driven discipline'. It is unlikely that Arthur simply means that evolutionary biology is based on hypothetico-deductive methodology characterised by the interplay of concepts and empirical evidence. Instead, it is clear that he refers specifically to nomothetics, to which evo-devo can make 'a conceptual contribution'. Yet, nomothetic insights are epistemologically accessible only through the study of idiographic details.

EVO-DEVO'S NOMOTHETIC ASPECTS

There are two issues that require detailed examination. Firstly, what is evo-devo's potential contribution to neo-Darwinian evolutionary theory? Secondly, what is the explanatory range of evo-devo's central nomothetic concepts?

Evo-devo's relation to neo-Darwinian evolutionary theory

> Developmental biology is reclaiming its appropriate place in evolutionary theory.
> Gilbert *et al.* 1996: 368

> The clamour to revise neo-Darwinism is becoming so loud that hopefully most practising evolutionary biologists will begin to pay attention.
> Pigliucci 2005: 566

> We suspect that most evo-devoists are not concerned with enhancing, completing, modifying or overturning the modern synthesis of evolution.
> Robert *et al.* 2001: 958

> Although a 'new synthesis' has been repeatedly announced in recent years, those announcements are premature.
> Wilkins 2002: 34

As the above quotes show, evo-devo's status and ambitions with respect to neo-Darwinian evolutionary theory are controversial. Which quote is most accurate? Consideration of the hierarchical organisation of biology provides an important insight.

Evo-devo's main challenge is to codify the relationship of its organism-level focus with the population-level focus of neo-Darwinian evolutionary theory (Wagner 2000, 2001, Gilbert 2003, Wagner and Larsson 2003, Amundson 2005). Evo-devo does not provide a new component to evolutionary theory, but instead draws attention to a previously neglected level. Evo-devo focuses on the origin and nature of the material substrate of evolution. Within the neo-Darwinian framework this rich topic was blackboxed under the rubric 'variation,' which was considered a mere boundary condition for the operation of the population-level processes deemed most important in determining the direction of evolutionary change, such as drift and selection. Specifically, any potential for shaping the direction of evolutionary change inherent in the nature of variation itself has been codified as 'constraint' in evolutionary theory (Maynard Smith *et al.* 1985, Arthur 2000, 2004b, Reif *et al.* 2000, Gould 2002, Stoltzfus 2006). This is the proper locale of evo-devo's empirical and theoretical contribution to evolutionary theory. As Gould (2002: 82) summarised it: 'the revolutionary empirics and conceptualisations of evo-devo [are] united by a common goal: to rebalance constraint and adaptation as causes and forces of evolution.' As far as evo-devo contributes general conceptual subthemes that can be categorised under the rubric of variation, it should be considered a genuine contribution to a previously neglected part of evolutionary theory (Arthur 2000, 2004a,b, Stoltzfus 2006).

Critically, this means that the term 'evolution' itself is understood differently by evo-devoists and neo-Darwinians. The standard neo-Darwinian understanding of evolution is a population-level process of the sorting of variation that is brought about by genetic recombination and mutation (Reif *et al.* 2000, Kutschera and Niklas 2004). Strikingly, as Stoltzfus (2006) points out, the neo-Darwinian perspective does not consider the processes of the origin of variation to be a part of evolution! As the *Encyclopedia of Evolution* states, 'Darwin's theory is peculiar in that evolution is not an extension of the mutational process' (Ridley 2002: 800). Under the evo-devo perspective of evolution, the processes generating variation are very much part of what evolution is.

The explanatory range of evo-devo's nomothetic components

The best answer to any question about evolution is the lawyer's answer to any general question about the law: 'It depends on the jurisdiction'.

Lewontin 2002: 17

The explanatory range of a concept can be defined as the class size over which it rules, i.e. the range of taxa, or events, or facts, over which generalisations can be made. For evo-devo, the relevant classes are defined with respect to the taxonomic range of organisms with particular developmental properties. Evo-devo's main criterion of theoretical importance is the extent to which evolutionary developmental mechanisms can influence the direction of phenotypic evolution. Evo-devo mechanisms are defined (Hall 2003a) as mechanisms operating during development that can be modified during evolution, thereby affecting phenotypic evolution (Table 6.1). Again, consideration of biology's hierarchical organisation is helpful.

Populations and organisms exemplify two distinct focal levels in life's constitutive hierarchy (Vrba and Gould 1986, Valentine and May 1996, Gould and Lloyd 1999, Gould 2002). By focusing on individual organisms evo-devo complements the neo-Darwinian focus on populations. A formal property of biology's hierarchy is a marked asymmetry of the interactions between levels that can be summarised as follows: the lower level proposes and the higher level disposes.

This sheds light on the relative importance of phenomena that occur on different hierarchical levels. Any phenomenon on the level of individual organisms or their parts has to be filtered through the population level if it is to have an effect on the direction of phenotypic evolutionary change. Any change on the higher level automatically has an impact on the lower levels, but the reverse is not true. Since individual organisms define the focal level of evolutionary developmental mechanisms, phenomena on this level will always be subject to the population level processes of natural and sexual selection, and drift. Natural and sexual selection are considered the most important forces governing phenotypic evolutionary change because they define the competitive economic context in which all evolution takes place (Ghiselin 1995, 1999, Vermeij 2004). Consequently, the maximum explanatory range of evo-devo mechanisms can logically never exceed the explanatory range of population-level processes. At best they can be equal partners. For this reason I disagree with authors who seem to try to overextend evo-devo's explanatory umbrella. For example, Gilbert (2003: 350): 'It may even be the case that the population genetics model turns out to be placed within a developmental framework', and Hall (2003a: 494): 'Evolutionary developmental mechanisms also include interactions between individuals of the same species, individuals of different species, and species and their biotic and/or abiotic environment'. These statements seem to imply that evo-devo's scope can encompass the traditional neo-Darwinian arena. I think this may obscure the legitimate complementary roles of organism- and population-level perspectives.

What then is the precise explanatory range of particular evo-devo concepts? According to Arthur (2004a,b) developmental bias represents evo-devo's most important challenge to a strict neo-Darwinian view of life. Developmental bias describes how the direction of evolutionary change is influenced by the non-random structure of variation. The potential explanatory range of developmental bias is enormous, because logic alone dictates the 'null model of zero bias' as 'inherently improbable anyhow' (Arthur 2004b: 284). However, the true extent to which developmental bias will direct evolutionary change is a matter of historical contingency, to be determined independently and idiographically for the evolution of each character in each population. So it is too with the explanatory range of other evo-devo mechanisms on the hierarchical level of the individual and below (Table 6.1).

Therefore the study of the theoretical importance of evo-devo concepts falls into the same category as the older study of general evolutionary trends, rules or laws of evolution, such as Cope's rule (phyletic size increase), Bergmann's rule (temperature dependence of body size), Williston's rule (reduction in number and specialisation of repeated body parts) and ecological rules regulating the evolution of r-versus K-strategies (based among others on rapid development and high fecundity versus long development and low fecundity, respectively). These rules are not universal as their realisation depends on the relative strength of different selection pressures and taxon-specific characteristics that may aid or constrain a certain outcome; and Williston's rule depends also on developmental mechanisms for the multiplication and/or specialisation of parts. For example, Kingsolver and Pfennig (2004) showed that in many populations with variable body sizes there is positive individual-level selection (both natural and sexual) for increased body size, providing a potential explanation for instances of Cope's rule. In many situations larger body size is selectively advantageous, which may lead to broad, but not universal, predictions of when Cope's rule will obtain. Similarly, with respect to evo-devo mechanisms or mechanisms of genomic change in general (Ryan 2006), generalisations may be formulated that may ascribe different probabilities to particular kinds of events, or even allow (probably much more rarely and difficult to study) predictions of what will happen if certain circumstances pertain. For example, on the principle that gene duplication can have important consequences for evolvability (Carroll 2005), assessing the relative frequencies of different fates of duplicated genes (neofunctionalisation, subfunctionalisation or loss of function by becoming pseudogenes) for different taxa may lead to general insights or broad predictions about the evolvability of different taxa.

THE ROLES OF EVO-DEVO'S IDIOGRAPHICS

Am I overly critical by claiming that the identity and promise of evo-devo has been presented in a biased way by overemphasising its nomothetic aspects? Surely, it is at least implicitly realised that any generalisations can only be built upon a rich idiographic foundation. However, as mentioned at the beginning of the chapter, some workers think that if anything goes, and if there are no general rules, then evo-devo has nothing to discover.

Extensive documentation of the unique contributions of evo-devo mechanisms (Table 6.1) to the origin of novelties and body plan evolution is a central idiographic goal of evo-devo. Yet, this goal seems almost pejoratively dismissed as 'merely filling in some missing details' (Arthur 2002: 757). Perhaps this is merely an unsurprising remnant of the pervasive tradition for the status ranking of scientific disciplines in which the arrow of arrogance unfailingly soars from the nomothetic domain to impale innocent idiographers (Jenner 2006b). Nevertheless, the documentation of evo-devo's unique phenomenology is integral to both evo-devo's idiographic and nomothetic goals. The central question then is how best to mine evo-devo's idiographics by the judicious choice of model organisms.

MODEL BIAS EQUALS MODEL STRENGTH

> . . . reasoning via model organisms, in a sense, has become the lingua franca of biologists.   Ankeny 2001: S259

Choice of evo-devo model organisms has been discussed in detail elsewhere (Jenner and Wills 2007). Here I restrict discussion to trait bias in model organisms. There are two extreme strategies for choosing new model organisms: (1) minimising character overlap between model organisms by maximising phylogenetic diversity; (2) maximising character overlap between model organisms by explicitly choosing them on the basis of shared developmental traits.

The first strategy maximises the amount of unique idiographic detail captured by models. It is generally recommended that a wider phylogenetic range of taxa, including satellite species that allow the easy transfer of experimental techniques, should be a prime guideline for choosing new models (Bolker 1995, Bolker and Raff 1997, Hughes and Kaufman 2000, Raff 2000, Simpson 2002, Wilkins 2002, Minelli 2003, Rudel and Sommer 2003, Sommer 2005). This perspective considers the trait bias (such as short generation time, rapid and stereotypical development) of established model organisms, especially those inherited from molecular developmental biology, as an important drawback because any general conclusions one might draw on the basis of these species 'are not universally true beyond our models' (Bolker and Raff 1997: 36).

Although broad phylogenetic sampling is important for assessing the extent of developmental variation, it is largely a distant idiographic goal. Given limited time and resources it is not the most efficient route to general nomothetic insights. A more pressing immediate goal is to establish the value of important evo-devo themes, such as developmental and phenotypic plasticity, canalisation, genetic assimilation and evolvability. These topics as still labelled 'controversial', or 'too esoteric for mainstream consideration' (Gibson and Dworkin 2004, Sniegowski and Murphy 2006) and allegedly supported only by 'anecdotal evidence' (Leroi 1998, Sniegowski and Murphy 2006).

For fulfilling evo-devo's ultimate idiographic goal of documenting the diversity of evo-devo mechanisms, and their unique roles in the evolution of novelties and body plans, each idiographic particular – each species – is equally valuable. In contrast, the value of organisms for empirically grounding evo-devo's nomothetic themes is for each model based on possessing particular developmental characteristics that provide independent support for a particular concept. Importantly, there is a trade-off between explanatory range and explanatory force at a given sample size. Maximising the amount of unique idiographic detail captured by new models minimises the ability to draw general conclusions from them. In contrast, by sampling taxa that share particular characters, one can maximise explanatory force (a measure of explanatory or predictive reliability), which is the basis of general insights. Thus nomothetic profits for each idiographic investment are maximised by the coordinated choice of models with the potential to shed independent light on each theme. An efficient search for general nomothetic insights on the way to fulfilling evo-devo's ultimate goals depends on a biased search for models. The general value of bias in model organisms becomes clear when they exemplify developmental types.

DEVELOPMENTAL TYPES: THE BASIS OF EVO-DEVO'S
NOMOTHETICS

Developmental types can be considered a special kind of body plan or Bauplan, differing only in the number or nature of body parts, or developmental aspects of form they refer to. Consequently, in the following discussion the terms 'body plan' and 'developmental type' can be interchanged without disturbing the logic of argument. Apart from this basic statement about the nature of developmental types, the literature is rife with confusion, and I know of no proper treatment of this important issue. Fitch and Sudhaus (2002: 243) pinpoint the problem when they note that the Bauplan suffers from 'uncertain ontology'. On the one hand, body plans are interpreted as concrete entities (Hall 1999, Amundson 2005, Rieppel 2006), while on the other hand they may be conceptualised as abstractions (Scholtz 2004, Rieppel 2006). However conceived, body plans and developmental types are widely considered detrimental to evolutionary research because of supposed typological connotations (Arthur 1997: 30, Richardson *et al.* 1999, Fitch and Sudhaus 2002: 243, Baguñá and Garcia-Fernàndez 2003: 708, Scholtz 2004: 5, Amundson 2005: 256, Hübner 2006: 379, Rieppel 2006: 531). This creates a paradoxical situation as Amundson (2005: 235) writes that 'Bauplans are taken very seriously within evo-devo'. The evolution of body plans is the overarching theme of evo-devo. It is therefore crucial to understand the nature of body plans and developmental types.

Body plans and developmental types aren't what they seem

Baupläne and developmental types both refer to phenotypic traits shared among taxa. For the sake of this discussion, they may refer to any genotypic, epigenetic or phenotypic traits, as well as the functional organisation that results from the interaction among organismal parts, on any hierarchical level, from the parts of individual organisms to monophyletic high-level taxa.

The nature of body plans can be clarified by being very clear about the fundamental ontological distinction between classes and individuals, which is the very foundation of natural science (Ghiselin 1997). Everything is either an individual or a class, and this distinction is very useful in addressing central evo-devo issues (Jenner 2006a). For the present discussion the following distinctions are important. Individuals are concrete and spatio-temporally restricted, while classes are abstract, spatio-temporally unrestricted concepts. The ontology of individuals is the part/whole relation, in which lower-level parts (cells) form a higher-level individual (multicellular animal). In contrast, the ontology of classes is the membership relation. Class membership is defined on the basis of possessing certain traits stipulated by the class's definition. In contrast, individuals do not have defining properties.

Baupläne can be conceived as both individuals and classes. The former seems widely favoured as body plans are said to evolve, and not to represent mere abstractions (Hall 1999, Amundson 2005, Rieppel 2006). For a body plan to be concrete it needs to refer to all parts that make up a higher-level whole. As soon as a body plan refers to only some abstracted parts of a whole, it is a class. This is intuitively obvious when considering the part/whole relationship of a multicellular individual. I am the sum of all my lower-level parts, so in total they make up a single concrete individual. However, if I refer to only some of my lower-level parts, for example my epithelial cells, this collection of parts together no longer constitutes a single concrete individual. Instead it specifies a class of traits of which parts of a higher-level individual are members. It represents an abstraction based on some specified characteristics.

In analogy, the Bauplan of a taxon is a concrete entity (individual) only when it refers to all parts making up the whole. The Bauplan consisting of all parts is then synonymous with the high-level whole. In contrast, when a Bauplan refers to only a selection of traits, such as major organ systems or developmental genes present in all of its organisms, or all traits that are shared by some but not all organisms of the clade, the Bauplan is an abstract class with an intensional definition stipulating the possession of certain traits. It seems that Baupläne are usually defined as classes, but construed as if they were individuals. This reification of a class as an individual is perniciously typological. The common view of body plans and developmental types strictly in terms of homologies and monophyletic taxa (Hall 1996, 1999, Arthur 1997: 29, Valentine 2004, Amundson 2005: 232, Hübner 2006: 370) seems to lend concreteness to an abstract concept, but this is unnecessarily restrictive. By properly defining a body plan as a class of characters, taxa can share a body plan even if it has independently evolved. Such a body plan can form the basis of generalisations that can go beyond particular monophyletic taxa. The recognition of Baupläne as classes is perfectly legitimate. Indeed, the recognition of classes is 'not really based on a historical analysis' (Scholtz 2004: 4); it is completely ahistorical. Importantly, exactly this ahistorical formulation of Baupläne and developmental types as classes provides the necessary basis for any nomothetic insights into evolution.

The heuristic value of developmental types

Developmental types attain importance in evolutionary research by specifying a class of organisms with shared properties of development. They may exemplify one or a combination of evo-devo mechanisms, or they may refer to the possession of a particular developmental genetic or morphological phenotype, such as animals with gene regulatory networks or larvae with set-aside cells (Table 6.1). This allows developmental types to function in generating nomothetic insights into the evolution of development. For example, animals with set-aside cells have functioned in several general evolutionary hypotheses. Set-aside cells are cells from which the adult body develops in animals with distinct larval stages (Peterson *et al.* 1997). They have been implicated in the evolution of germ-line sequestration as a mechanism to mediate conflict between cell lineages that may result in lowered organismal fitness (Blackstone and Ellison 2000, Michod and Roze 2001). Possession of set-aside cells also has general implications for the timing of germline formation (Ransick *et al.* 1996). Recently Peterson *et al.* (2005) also implicated set-aside cells as a general solution to the problem of alleviating the danger of predation in a sensitive phase of the life cycle. For these hypotheses it does not matter whether or not set-aside cells are homologous across groups.

Another developmental type comprises organisms with reduced morphologies, either uniformly across individuals of a species, or only characterising certain morphs, such as castes in social hymenopterans. It may be expected that a single evolutionary origin of organ loss may be reflected in an identical change at the level of developmental regulation, while convergent loss may be reflected in obvious differences in developmental regulation. However, the unique evolutionary loss of wings in non-reproductive castes of different ant species is reflected in a diversity of genetic regulatory changes in different species (Abouheif and Wray 2002), while in several populations of Mexican cavefish with independently degenerated eyes, the developmental mechanisms of eye reduction are surprisingly similar (Jeffery *et al.* 2003). This shows the value of studying independent instances exemplifying a common evo-devo theme.

The point here is not that new models should solely be chosen on the basis of a bias in developmental traits, because general patterns also need testing by potentially falsifying evidence. However, in cases where there is insufficient supporting evidence for the value of general concepts, it makes sense to focus first on documenting confirming cases, which is most efficiently achieved by the coordinated choice of model organisms to illuminate evo-devo's nomothetic themes.

CONCLUSION: FROM MODEL ORGANISMS TO DEVELOPMENTAL TYPES

Evo-devo is an ambitious young discipline with both idiographic and nomothetic goals. Idiographically, evo-devo aims to document the unique effects of changes in evolutionary developmental mechanisms on the origin of novelties and the evolution of body plans. Nomothetically, evo-devo attempts to establish the general effects of evolutionary developmental mechanisms on determining the overall direction of phenotypic evolution. Ultimately these aspects can be combined into an evolutionary narrative that relates the description of unique particulars to broad generalisations. On the long road towards fulfilment of evo-devo's ultimate aims, the coordinated choice of models to illuminate nomothetic evo-devo themes is a more efficient route to general insights than choosing new model organisms based solely on the criterion of maximising phylogenetic spread, which tends to maximise the amount of unique idiographic detail. Evo-devo's idiographics are most efficiently translated into nomothetic insights when model organisms are judiciously chosen to aid the discovery of developmental types, based on the models sharing certain developmental traits. The diversity of our models should therefore reflect the diversity of the general questions that interest us.

Love (2006: 95) observed that there is currently a mismatch between evo-devo's broad research agenda and 'the predominance of particular experimental tools', which are biased towards 'current consensus methodologies derived from genetic regulatory mechanisms'. This narrowing of evo-devo's research agenda is unsurprising insofar as our most important model systems were chosen to function within the genetic paradigm of developmental biology (Gilbert 2001). Consequently, it should be one of evo-devo's central goals to promote hitherto neglected research topics into fully fledged research programs by judicious choice of new model organisms. Important topics, such as the evolutionary ecology of plasticity, may be 'logistically cumbersome and tedious' (Pigliucci 2005: 485), and are therefore in urgent need of more empirical work. This necessitates the selection of appropriate model organisms, even if they are not selected to function in evo-devo's current genetic paradigm. This requires evo-devoists to communicate their needs clearly to granting agents to prevent the exclusive funneling of funds into narrow research areas. For example, the British

Biotechnology and Biological Sciences Research Council (BBSRC) currently requests evo-devo proposals thus: 'Applications are encouraged to make comparisons between the genetic basis of development in different organisms'. For the moment, to conclude that evo-devo 'hasn't quite lived up to expectations' (Richardson 2003: 351) is simply to expect too much too soon.

ACKNOWLEDGEMENTS

This chapter is dedicated to the memory of Stephen Jay Gould, whose *Ontogeny and Phylogeny* was an important factor in the conception of evo-devo 30 years ago. He thought long and hard about the relationship between evolution and development, and throughout his career he was deeply concerned with the relative roles of idiographics and nomothetics in evolutionary biology. Special thanks to Alessandro Minelli and Giuseppe Fusco for inviting this contribution and for being such understanding editors. I thank Wallace Arthur for reviewing the manuscript. The BBSRC is gratefully acknowledged for financial support.

REFERENCES

Abouheif, E. & Wray, G. A. 2002. Evolution of the gene network underlying wing polyphenism in ants. *Science* **297**, 249–252.

Amundson, R. 2005. *The Changing Role of the Embryo in Evolutionary Thought: Roots of Evo-Devo*. Cambridge: Cambridge University Press.

Ankeny, R. A. 2001. Model organisms as models: understanding the 'Lingua Franca' of the human genome project. *Philosophy of Science* **68**, S251–S261.

Arthur, W. 1997. *The Origin of Animal Body Plans. A Study in Evolutionary Developmental Biology*. Cambridge: Cambridge University Press.

Arthur, W. 2000. Intraspecific variation in developmental characters: the origin of evolutionary novelties. *American Zoologist* **40**, 811–818.

Arthur, W. 2002. The emerging conceptual framework of evolutionary developmental biology. *Nature* **415**, 757–764.

Arthur, W. 2004a. *Biased Embryos and Evolution*. Cambridge: Cambridge University Press.

Arthur, W. 2004b. The effect of development on the direction of evolution: toward a twenty-first century consensus. *Evolution & Development* **6**, 282–288.

Azevedo, R. B. R., Lohaus, R., Braun, V. et al. 2005. The simplicity of metazoan cell lineages. *Nature* **433**, 152–156.

Baguñà, J. & Garcia-Fernàndez, J. 2003. Evo-devo: the long and winding road. *International Journal of Developmental Biology* **47**, 705–713.

Biémont, C. & Vieira, C. 2006. Junk DNA as an evolutionary force. *Nature* **443**, 521–524.

Blackstone, N. W. & Ellison, A. M. 2000. Maximal indirect development, set-aside cells, and levels of selection. *Journal of Experimental Zoology B (Molecular and Developmental Evolution)* **288**, 99–104.

Bolker, J. A. 1995. Model systems in developmental biology. *BioEssays* **17**, 451–455.

Bolker, J. A. & Raff, R. A. 1997. Beyond worms, flies, and mice: it's time to widen the scope of developmental biology. *The Journal of NIH Research* **9**, 35–39.

Carroll, S. B. 2005. Evolution at two levels: on genes and form. *PLoS Biology* **3**, e245.

Coyne, J. A. 2005. Switching on evolution. How does evo-devo explain the huge diversity of life on Earth? *Nature* **435**, 1029–1030.

Darling, J. A., Reitzel, A. R., Burton, P. M. *et al.* 2005. Rising starlet: the starlet sea anemone, *Nematostella vectensis*. *BioEssays* **27**, 211–221.

Davidson, E. H. 1991. Spatial mechanisms of gene regulation in metazoan embryos. *Development* **113**, 1–26.

Davidson, E. H. & Erwin, D. H. 2006. Gene regulatory networks and the evolution of animal body plans. *Science* **311**, 796–800.

Extavour, C. 2004. Hold the germ cells, I'm on duty. *BioEssays* **26**, 1263–1267.

Fitch, D. H. A. & Sudhaus, W. 2002. One small step for worms, one giant leap for "Bauplan?" *Evolution & Development* **4**, 243–246.

Flatt, T. 2005. The evolutionary genetics of canalization. *Quarterly Review of Biology* **80**, 287–316.

Franz-Odendaal, T. A. & Hall, B. K. 2006. Modularity and sense organs in the blind cavefish, *Astyanax mexicanus*. *Evolution & Development* **8**, 94–100.

Ghiselin, M. T. 1995. Darwin, progress, and economic principles. *Evolution* **49**, 1029–1037.

Ghiselin, M. T. 1997. *Metaphysics and the Origin of Species*. Albany: State University of New York Press.

Ghiselin, M. T. 1999. Progress and the economy of nature. *Journal of Bioeconomics* **1**, 35–45.

Ghiselin, M. T. 2005. Homology as a relation of correspondence between parts of individuals. *Theory in Biosciences* **24**, 91–103.

Gibson, G. & Dworkin, I. 2004. Uncovering cryptic genetic variation. *Nature Reviews Genetics* **5**, 681–690.

Gilbert, S. F. 2001. Ecological developmental biology: developmental biology meets the real world. *Developmental Biology* **233**, 1–12.

Gilbert, S. F. 2003. Evo-Devo, Devo-Evo, and Devgen-Popgen. *Biology & Philosophy* **18**, 347–352.

Gilbert, S. F., Opitz, J. M. & Raff, R. A. 1996. Resynthesizing evolutionary and developmental biology. *Developmental Biology* **173**, 357–372.

Gould, S. J. 1980. The promise of paleobiology as a nomothetic, evolutionary discipline. *Paleobiology* **6**, 96–118.

Gould, S. J. 2002. *The Structure of Evolutionary Theory*. Cambridge Massachusetts: The Belknap Press of Harvard University Press.

Gould, S. J. & Lloyd, E. A. 1999. Individuality and adaptation across levels of selection: How shall we name and generalize the unit of Darwinism? *Proceedings of the National Academy of Sciences of the USA* **96**, 11904–11909.

Hall, B. K. 1996. Baupläne, phylotypic stages and constraint: why there are so few types of animals. *Evolutionary Biology* **29**, 215–261.

Hall, B. K. 1999. *Evolutionary Developmental Biology*. Dordrecht: Kluwer Academic Publishers.

Hall, B. K. 2000. Evo-devo or devo-evo – does it matter? *Evolution & Development* **2**, 177–178.

Hall, B. K. 2003a. Evo-devo: evolutionary developmental mechanisms. *International Journal of Developmental Biology* **47**, 491–495.

Hall, B. K. 2003b. Unlocking the black box between genotype and phenotype: cell condensations as morphogenetic (modular) units. *Biology & Philosophy* **18**, 219–247.

Hong, R. L. & Sommer, R. J. 2006. *Pristionchus pacificus*: a well-rounded nematode. *BioEssays* **28**, 651–659.

Hübner, C. 2006. *Hox* genes, homology and axis formation – the application of morphological concepts to evolutionary developmental biology. *Theory in Biosciences* **124**, 371–396.

Hughes, C. L. & Kaufman, T. C. 2000. A diverse approach to arthropod development. *Evolution & Development* **2**, 6–8.

Jeffery, W. R., Strickler, A. G. & Yamamoto, Y. 2003. To see or not to see: evolution of eye degeneration in Mexican blind cavefish. *Integrative and Comparative Biology* **43**, 531–541.

Jenner, R. A. 2006a. Unburdening evo-devo: ancestral attractions, model organisms, and basal baloney. *Development, Genes & Evolution* **216**, 385–394.

Jenner, R. A. 2006b. On cosmik debris and palaeontology. *Palaeontological Association Newsletter* **62**, 28–35.

Jenner, R. A. & Wills, M. A. 2007. The choice of model organisms in evo-devo. *Nature Reviews Genetics* **8**, 311–319.

Joron, M., Jiggins, C. D., Papanicolaou, A. & McMillan, W. O. 2006. *Heliconius* wing patterns: an evo-devo model for understanding phenotypic diversity. *Heredity* **97**, 157–167.

Kingsolver, J. G. & Pfennig, D. W. 2004. Individual-level selection as a cause of Cope's rule of phyletic size increase. *Evolution* **58**, 1608–1612.

Kutschera, U. & Niklas, K. J. 2004. The modern theory of biological evolution: an expanded synthesis. *Naturwissenschaften* **91**, 255–276.

Leroi, A. M. 1998. The burden of the bauplan. *Trends in Ecology and Evolution* **13**, 82–83.

Lewontin, R. C. 2002. Directions in evolutionary biology. *Annual Review of Genetics* **36**, 1–18.

Love, A. C. 2006. Reflections on the middle stages of evodevo's ontogeny. *Biological Theory* **1**, 94–97.

Lyman, R. L. & O'Brien, M. J. 2004. Nomothetic science and idiographic history in twentieth-century americanist anthropology. *Journal of the History of the Behavioral Sciences* **40**, 77–96.

Maynard Smith, J., Burian, R., Kauffman, S. *et al.* 1985. Developmental constraints and evolution. *Quarterly Review of Biology* **60**, 265–287.

Michod, R. E. & Roze, D. 2001. Cooperation and conflict in the evolution of multicellularity. *Heredity* **86**, 1–7.

Minelli, A. 2003. *The Development of Animal Form: Ontogeny, Morphology, and Evolution*. Cambridge: Cambridge University Press.

Minelli, A. & Fusco, G. 2004. Evo-devo perspectives on segmentation: model organisms, and beyond. *Trends in Ecology and Evolution* **19**, 423–429.

Moczek, A. P. 2006. Integrating micro- and macroevolution of development through the study of horned beetles. *Heredity* **97**, 168–178.

Müller, G. B. & Newman, S. A. 2005. The innovation triad: an evodevo agenda. *Journal of Experimental Zoology B (Molecular and Developmental Evolution)* **304**, 487–503.

Peel, A. D., Telford, M. J. & Akam, M. 2006. The evolution of hexapod engrailed-family genes: evidence for conservation and concerted evolution. *Proceedings of the Royal Society of London B* **273**, 1733–1742.

Peterson, K. J., Cameron, R. A. & Davidson, E. H. 1997. Set-aside cells in maximal indirect development: evolutionary and developmental significance. *BioEssays* **19**, 623–631.

Peterson, K. J., McPeek, M. A. & Evans, D. A. D. 2005. Tempo and mode of early animal evolution: inferences from rocks, Hox, and molecular clocks. *Paleobiology* **31**, 36–55.

Pigliucci, M. 2005. Expanding evolution. *Nature* **435**, 565–566.

Raff, R. A. 2000. Evo-devo: the evolution of a new discipline. *Nature Reviews Genetics* **1**, 74–79.

Ransick, A., Cameron, R. A. & Davidson, E. H. 1996. Postembryonic segregation of the germ line in sea urchins in relation to indirect development. *Proceedings of the National Academy of Sciences USA* **93**, 6759–6763.

Reif, W.-E., Junker, T. & Hossfeld, U. 2000. The synthetic theory of evolution: general problems and the German contribution to the synthesis. *Theory in Biosciences* **119**, 41–91.

Richards, E. J. 2006. Inherited epigenetic variation – revisiting soft inheritance. *Nature Reviews Genetics* **7**, 395–400.

Richardson, M. 2003. A naturalist's evo-devo. *Nature Genetics* **34**, 351.

Richardson, M. K., Minelli, A. & Coates, M. I. 1999. Some problems with typological thinking in evolution and development. *Evolution & Development* **1**, 5–7.

Ridley, M. 2002. Natural selection: an overview. In M. Pagel (ed.) *Encyclopedia of Evolution*. New York: Oxford University Press, pp. 797–804.

Rieppel, O. 2006. 'Type' in morphology and phylogeny. *Journal of Morphology* **267**, 528–535.

Robert, J. S., Hall, B. K. & Olson, W. M. 2001. Bridging the gap between developmental systems theory and evolutionary developmental biology. *BioEssays* **23**, 954–962.

Rudel, D. & Sommer, R. J. 2003. The evolution of developmental mechanisms. *Developmental Biology* **264**, 15–37.

Ryan, F. P. 2006. Genomic creativity and natural selection: a modern synthesis. *Biological Journal of the Linnean Society* **88**, 655–672.

Scholtz, G. 2004. Baupläne versus ground patterns, phyla *versus* monophyla: aspects of patterns and processes in evolutionary developmental biology. In G. Scholtz (ed.) *Evolutionary Developmental Biology of Crustacea. Crustacean Issues* 15. Lisse: Balkema, pp. 3–16.

Seo, H.-C., Edvardsen, R. B., Maeland, A. D. *et al.* 2004. *Hox* cluster disintegration with persistent anteroposterior order of expression in *Oikopleura dioica*. *Nature* **431**, 67–71.

Shapiro, M. D., Bell, M. A. & Kingsley, J. S. 2006. Parallel genetic origins of pelvic reduction in vertebrates. *Proceedings of the National Academy of Sciences of the USA* **103**, 13753–13758.

Simpson, P. 2002. Evolution of development in closely related species of flies and worms. *Nature Reviews Genetics* **3**, 1–11.

Sniegowski, P. D. & Murphy, H. A. 2006. Evolvability. *Current Biology* **16**, R831–R834.

Sommer, R. J. 2005. Genomic platforms for "evo-devo". *Current Genomics* **6**, 569–570.

Stoltzfus, A. 2006. Mutationism and the dual causation of evolutionary change. *Evolution & Development* **8**, 304–317.

Tanaka, M., Hale, L. A., Amores, A. *et al.* 2005. Developmental genetic basis for the evolution of pelvic fin loss in the pufferfish *Takifugu rubripes*. *Developmental Biology* **281**, 227–239.

Valentine, J. W. 2004. *On the Origin of Phyla*. Chicago: The University of Chicago Press.

Valentine, J. W. & May, C. L. 1996. Hierarchies in biology and paleontology. *Paleobiology* **22**, 23–33.

Vermeij, G. J. 2004. *Nature: An Economic History*. Princeton: Princeton University Press.

Vrba, E. S. & Gould, S. J. 1986. The hierarchical expansion of sorting and selection: sorting and selection cannot be equated. *Paleobiology* **12**, 217–228.
Wagner, G. P. 2000. What is the promise of developmental evolution? Part I: why is developmental biology necessary to explain evolutionary innovations? *Journal of Experimental Zoology B (Molecular and Developmental Evolution)* **288**, 95–98.
Wagner, G. P. 2001. What is the promise of developmental evolution? Part II: a causal explanation of evolutionary innovations may be impossible. *Journal of Experimental Zoology B (Molecular and Developmental Evolution)* **291**, 305–309.
Wagner, G. P., Chiu, C.-H. & Laublicher, M. 2000. Developmental evolution as a mechanistic science: the inference from developmental mechanisms to evolutionary processes. *American Zoologist* **40**, 819–831.
Wagner, G. P. & Larsson, H. C. E. 2003. What is the promise of developmental evolution? III. The crucible of developmental evolution. *Journal of Experimental Zoology B (Molecular and Developmental Evolution)* **300**, 1–4.
Wilkins, A. S. 2002. *The Evolution of Developmental Pathways*. Sunderland: Sinauer Associates Inc.

Part II   Evo-devo: methods and materials

If evo-devo is a discipline in its own right, is there a distinctive set of biological systems and methods of investigation through which it is currently advancing? Although evo-devo probably does not rely upon specific tools of analysis unknown in other fields of biological research, because of its particular relationships to both evolutionary and developmental biology evo-devo exhibits a specific *combination* of model systems and research tools. In other words, to use a fashionable term in developmental genetics, it has its own toolkit. However, what is most distinctive about evo-devo materials and methods is that, precisely because tools devised in other fields are here used at the borders of their original range of application, investigations in this interdisciplinary territory periodically need a critical evaluation of the sharpness, precision and adequacy of these tools. A survey of this important work is offered in this section.

The model organism approach has become the *lingua franca* in modern biology. However, a good model for medicine, where one searches for conserved features shared with humans, is not necessarily a good model for understanding evolutionary change. Athanasia Tzika and Michel Milinkovitch (Chapter 7) tackle the problem of model organism choice in evo-devo studies. The authors propose a pragmatic optimisation approach that incorporates criteria suggested by evolutionary history such as the phylogenetic position of candidate model species and the presence of ancestral/derived character states, along with practical attributes such as the feasibility of handling, housing and breeding. From this perspective, advantages and disadvantages of some candidate species in amniotes are discussed.

Gerhard Scholtz (Chapter 8) observes that by integrating evolutionary and developmental studies, evo-devo suffers from a tension between the comparative and the experimental approach, both of them with a long tradition in evolutionary and developmental biology respectively. Through a revision of the conceptual foundations of comparison and causation in evolutionary developmental biology with particular focus on the evolution of 'developmental sequences', the author analyses the relationship between experimental and comparative approaches in their contribution to addressing evolutionary questions in evo-devo.

Hans Zauner and Ralf Sommer's contribution (Chapter 9) brings ecology into the picture. This element is missing in most evo-devo studies, but it is sensible to ask how the forms generated through the (evolving) developmental processes confront the demands of an ever-changing environment. The authors attempt to bring these research fields together by illustrating a well-articulated case study: vulva formation in the nematode worms. This study presents a model for combining macro-evolutionary, micro-evolutionary and ecological studies with the aim of contributing to a new synthesis in evolutionary biology.

What is a Hox gene? Discovering that there is no straightforward answer to this apparently easy question can be somehow shocking, especially for those biologists who, having only a general interest in evo-devo, are used to considering Hox genes as a sort of icon for the discipline. David Ferrier (Chapter 10) discusses gene nomenclature critically, with particular reference to the homeobox genes, whose confusing and conflicting names and classifications hamper investigation and understanding of their own evolution and their role in the evolution of development. A more sensible classification of developmental control genes is thus suggested.

The last chapter of this section further illustrates the problem of the categories to which we ascribe biological entities. These categories are useful tools for description and comparison, but only up to a certain point. Beyond an initial, coarser level of description, when we start asking more specific questions, categories can become sly traps that hamper further progress in understanding biological processes in general, and evolution specifically. Rolf Rutishauser, Valentin Grob and Evelin Pfeifer (Chapter 11) accompany us through the 'identity crises' of plant organs as reflected in current morphological nomenclature. Multicellular plants such as angiosperms suffer identity crises on various levels, from cells to meristems and organs and even beyond, revealing the shortcomings issued from inadequate concepts.

7

A pragmatic approach for selecting evo-devo model species in amniotes

ATHANASIA C. TZIKA AND MICHEL C. MILINKOVITCH

One major classical justification of using a model metazoan species for experimentation has been that discoveries of biological phenomena in that species could be extrapolated to other multicellular species. Because the chances that this extrapolation is valid in humans depend on the phylogenetic distance between humans and the model species, many researchers have somewhat sacrificed the major benefits of small size, short generation time and ease of manipulation that characterise some invertebrates in order to use species that humans can more readily relate to, such as the laboratory mouse (*Mus musculus*). However, the community of biologists has continued to use additional model species because each of the selected taxa have specific features that make experimental manipulation easier (e.g. easy-to-score morphological variation and giant polytene chromosomes in *Drosophila melanogaster*, or accurate description of the largely invariant complete cell lineage and full neural connectivity in the roundworm *Caenorhabditis elegans*).

Ever since the molecular genetic revolution, a constant concern has been the possibility of manipulating the genome of model species. For example, generations of *Drosophila* scientists have developed and applied ingenious approaches that allow, in principle, screening for any phenotype at any stage of development (reviewed in St Johnston 2002). Even for the mouse model, multiple techniques, such as homologous recombination, tissue-specific activation/inactivation techniques, cloning and RNA interference (RNAi), have been developed for performing genotype- or phenotype-driven experiments. Furthermore, recent access to full genome sequences makes genome engineering of some

Evolving Pathways: Key Themes in Evolutionary Developmental Biology, ed. Alessandro Minelli and Giuseppe Fusco. Published by Cambridge University Press.

model species easier. However, one of the most important limiting factors to the utility of genome engineering approaches is the difficulty with which phenotypes can be identified. Indeed, many genes exert functions that cannot be investigated by simple examination of a few general morphological and/or physiological parameters. Fortunately, imaging technologies and physiological measuring techniques have recently been miniaturised and adapted for use with small species. Another source of difficulty for phenotyping is due to the epistatic effects among genes. For example, multiple studies have demonstrated that a full knock-out mouse for a gene supposedly essential for a given major process can yield no visible phenotype because another gene has become involved to fulfil the function of the invalidated gene.

MEDICINE-DRIVEN MODELS (SEARCHING FOR EVOLUTIONARY CONSERVATION) VS. EVO-DEVO-DRIVEN MODELS (UNDERSTANDING EVOLUTIONARY CHANGE)

As indicated above, most of the research performed so far with model species has been justified by the potential power of these species for understanding human biology. However, in the context of evo-devo, it is the massive realm of living species that should, ideally, be opened to genome manipulation and phenotypic investigation. Indeed, the interests of evolutionary developmental biologists go well beyond conserved physiologies and developmental processes or patterns as they seek to understand the generative mechanisms underlying biological diversity (Minelli 2003). Uncovering these mechanisms will require the merging of several disciplines (Milinkovitch and Tzika 2007), including molecular developmental biology, evolutionary molecular genetics, palaeontology (Wagner and Larsson 2003) and ecology (Dusheck 2002, Gilbert and Bolker 2003). Hence, one major challenge in evo-devo will be to adapt the tremendous knowledge and sophisticated technologies accumulated on 'classical' model species to model organisms from a wider assortment of lineages on the tree of life in a wider set of environmental conditions. However, as demonstrated in the past, promoting the use of the same set of model species increases the efficiency with which techniques and analytical approaches are developed simply through collaboration, emulation and establishment of public databases. For example, given the necessity to optimise the interplay between in-silico analyses and in-vivo experiments, it is crucial to establish the list of preferred species for which the genome will be sequenced in priority, and this should be done with more input from the evo-devo community (Milinkovitch and Tzika 2007).

CRITERIA FOR CHOOSING NEW MODEL SPECIES

Beside the practical criteria of small size, short generation, abundant progeny, ease of manipulation and of housing/breeding, accessibility of phenotyping and genome manipulation techniques, etc., there are other parameters that should be considered when listing preferred model species.

An intuitive and simple criterion to guide the choice of model species is the evolutionary divergence among these species (Figure 7.1). However, this *phylogenetic distance criterion* is limited by, among others, two parameters: (1) the rate of phenotypic transformation is highly variable among lineages and (2) variation worth investigation exists at multiple phylogenetic levels (one should not focus only on major transformations). Intermingled with the phylogenetic distance criterion, the *ancestrality* of a model species is a decisive factor. For example, the zebrafish is often considered a *canonical vertebrate* (Fishman 2001) because the common ancestor of all vertebrates was fish-like. Although intuitively appealing, this statement is of limited value because, a priori, no extant species is intrinsically more ancestral than any other. So, the real, non-trivial, question is: what is the considered species a model of (i.e. at what hierarchical level(s) of the phylogeny)? The answer depends heavily on what characters one is interested in. For example, as far as anatomy is concerned, the zebrafish might be a better model for teleost fishes than for vertebrates because the species exhibits multiple characters that seem ancestral for the former group, whereas it is very derived (as all teleost fishes are) in respect to the vertebrate ancestor (Metscher and Ahlberg 1999). Furthermore, although we agree that structural and genomic *simplicity* can help in guiding the choice of a model species (as in the case of the cephalochordate amphioxus; Holland *et al.* 2004), it should be carefully investigated whether the observed *simplicity* is ancestral for the group and not a secondary (derived) simplification, as is likely to be the case for flatworms and myzostomes, for example (Bleidorn *et al.* 2007).

The availability of robust and extensive (molecular) phylogenetic hypotheses constitutes an important criterion for choosing evo-devo model species: evolutionary trees constitute the basic framework on which character changes are mapped. For example, mapping of phenotypic characters on robust molecular phylogenies have demonstrated extensive and multiple convergences of ecologically specialised species

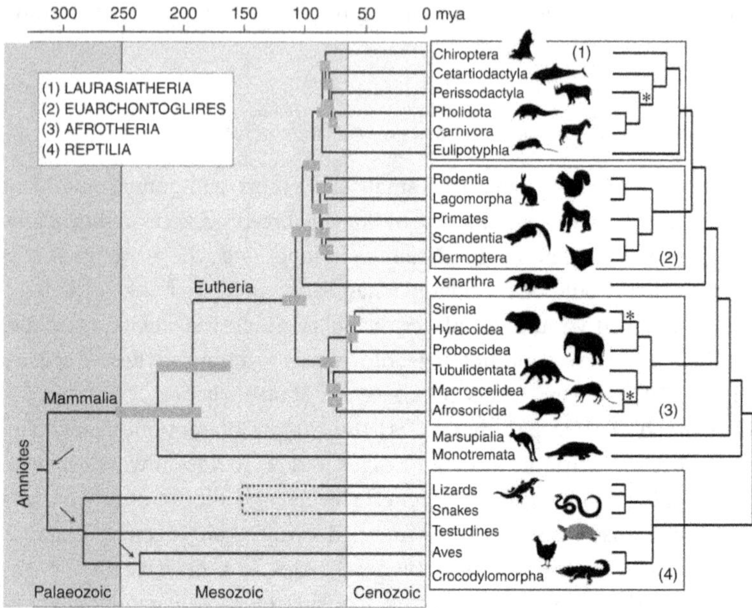

Figure 7.1 Right: cladogram showing the best estimate of the phylogeny of amniotes based on molecular data. The Eutheria cladogram is based on Bayesian and Maximum Likelihood analyses of 16 397 bp of DNA from 19 nuclear and 3 mitochondrial genes from 42 placental and 2 marsupial extant species (Murphy *et al.* 2001b). All nodes are supported by posterior probabilities >95% except the nodes indicated with an asterisk. Note that current analyses of morphological data still disagree with a significant portion of this cladogram. Left: phylogram showing the time of divergence among major amniote lineages. Divergence times (and 95% credibility intervals, grey bars) within Eutheria were obtained (Springer *et al.* 2003) by analysing the 16 397 bp dataset cited above under a Bayesian approach allowing rate variation among lineages and incorporating fossil information in the form of constraints on divergence times (Kishino *et al.* 2001). Some palaeontologists favour a model of radiation of placental mammals after the Mesozoic/Cenozoic boundary. Grey bars on the Marsupials versus Eutherians, and Monotremes versus Theria splits are the lowest and highest estimates of divergence time (i.e. considering standard errors) obtained with two molecular data sets of 2 793 bp for 21 taxa and 10 773 bp from 5 taxa, respectively (van Rheede *et al.* 2006). Divergence time for other nodes on the tree (arrows) are minimum age estimates obtained from palaeontological data (Benton and Donoghue 2007). Dotted lines indicate that no reliable estimate for the divergence of snakes from lizards is available. Two lineages of lizards are indicated to underline the likely paraphyly of that group (with respect to snakes). Geological eras are indicated (Palaeozoic, 570–251 mya; Mesozoic, 251–65 mya; Cenozoic, 65–0 mya).

(ecomorphs), e.g. of cichlid fishes (Kocher *et al.* 1993, Ruber *et al.* 1999), ranid frogs (Bossuyt and Milinkovitch 2000), *Anolis* lizards (Losos *et al.* 1998) and mammals (e.g. between some afrotherian and eulipotyphlan insectivores, see below). Similarly, the snake-like body form has evolved multiple times independently in squamate reptiles (Wiens *et al.* 2006). Investigating the development of such convergent traits in different lineages could form the basis for understanding possible general mechanisms involved in convergence (see Jenner, Chapter 6 of this volume, for a discussion on nomothetic versus idiographic approaches to evo-devo). Such analyses will require a strategy of choosing model organisms based on their traits rather than phylogenetic position per se (Figure 7.1).

A MULTIDISCIPLINARY PRAGMATIC OPTIMISATION APPROACH TO THE SELECTION OF MODEL SPECIES

As further developed elsewhere (Milinkovitch and Tzika 2007), the criteria that are relevant to the choice of a set of model species are multiple and can even be contradictory. This is intrinsic to the highly multidisciplinary nature of evo-devo. For example, a lineage can be characterised by a unique and dramatic set of derived character states that makes it particularly appealing for evo-devo studies, but the group might lack representatives that could reasonably constitute a widespread model species in the laboratory (e.g. take the extreme case of cetaceans). We think that the only possibility is the use of a pragmatic (and partly subjective) optimisation approach, incorporating criteria such as phylogenetic position (Figure 7.1) as well as number and nature of the ancestral/derived character states of the model species, level of diversity within a relevant higher taxon to which the chosen species belongs, ease with which the representative species can be handled, housed and bred, and their protection status. Compromises will have to be made, as it is simply impossible to find species that combine all possible advantageous features. We apply such an approach below for a set of species that could serve as the workhorses for evo-devo research within amniotes. We shortly discuss the advantages and disadvantages of some candidate species and hope this will be used as a starting point for an in-depth analysis with input from morphologists, palaeontologists, animal breeders, physiologists, developmental biologists and molecular phylogeneticists.

REPTILIA

Several lineages within Reptilia (including birds) are more genetically diverse than mammals such that using a single reptilian species from a single lineage seems insufficient, especially given that the Mammalia is already represented by several model species (see below).

Anapsida

Testudines

This clearly monophyletic group (turtles, tortoises and terrapins) includes only about 260 extant species in 13 families and is characterised by several synapomorphies of which the most obvious is the shell (enclosing front and hind limb girdles) comprising a dorsal carapace (made from dermal bones associated with endochondral modified vertebrae and ribs) and a ventral plastron (made of clavicles, interclavicles and abdominal ribs). They have relatively low metabolic rates and are dependent on external sources of body heat, although the leatherback turtle (*Dermochelys coriacea*) seems to show some degree of endothermy. High longevity might also be a character of interest. Long generation time is an obstacle to using testudines as models: in most species, males and females reach sexual maturity not before 3 and 5 years, respectively. Analysis of development is feasible as eggs can be incubated artificially (Gilbert *et al.* 2001).

Two species have been suggested as candidates for full genome sequencing (www.reptilegenome.com): *Trachemys scripta* (red-eared slider) and *Chrysemys picta* (painted turtle). Individual size, and housing and handling techniques, are very similar for the two species but the latter is much more difficult to breed and much less prolific than the former. Shell morphogenesis has been investigated in *Trachemys* (Gilbert *et al.* 2001), whereas a BAC library is available for *Chrysemys* (www.nsf.gov/bio/pubs/awards/bachome.htm), aiding the sequencing of its genome. For sure, only one of these two species should be sequenced as they are redundant phylogenetically (they belong to the same subfamily Deirochelyinae, in the family Emydidae). These species are closely related to the tortoises (family Testudinidae) (Krenz *et al.* 2005), of which several species in the genus *Testudo* are successfully bred around the world and detailed husbandry information is available (Lapid *et al.* 2004).

We think that an additional species, *Pelodiscus sinensis* (Chinese soft-shell turtle, see Figure 7.2A), deserves attention from the evo-devo community. This species is the smallest soft-shell turtle, and belongs to a superfamily (Trionychoidea) phylogenetically distant from the species discussed above. Housing and breeding of *Pelodiscus sinensis* are

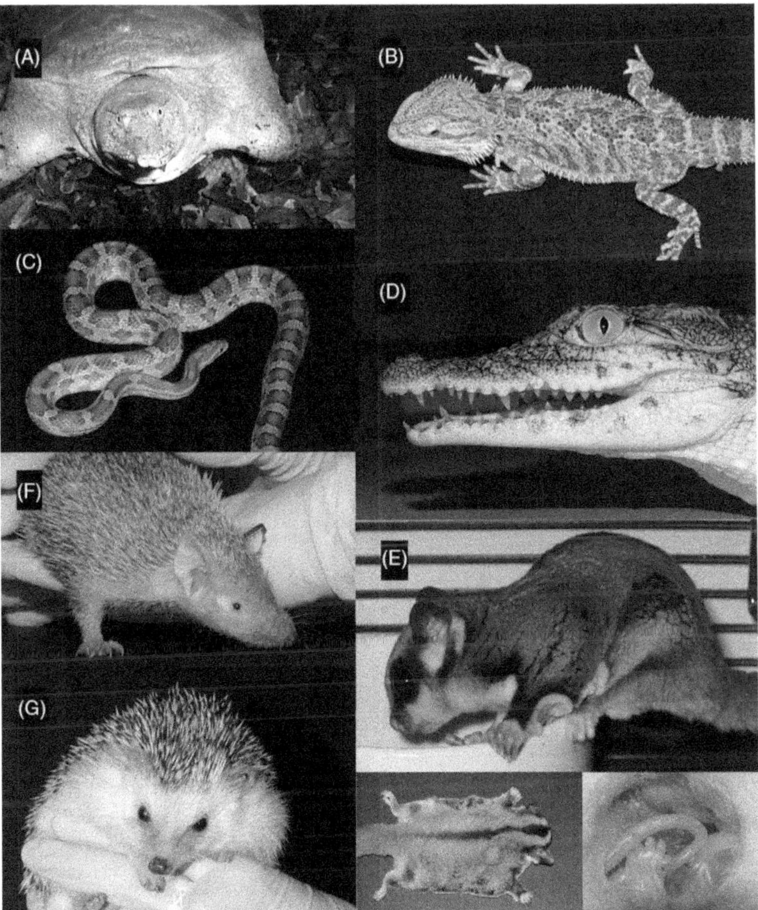

Figure 7.2 Some of the proposed new model species for amniotes: A, *Pelodiscus sinensis*, the Chinese soft-shell turtle; B, *Pogona vitticeps*, the bearded dragon; C, *Elaphe guttata*, the corn snake; D, *Crocodylus niloticus*, the Nile crocodile; E, *Petaurus breviceps*, the sugar glider (the left and right insets show the patagium and the embryo in the mother pouch, respectively); F, *Echinops telfairi*, the lesser hedgehog tenrec, G, *Atelerix albiventris*, the African pygmy hedgehog. A colour version of this figure is available at www.ulb.ac.be/sciences/ueg/modelsp.htm

very well documented because numerous farms exist, mainly in Asia, with a production exceeding six million individuals per year. Furthermore, *Hox* genes have been characterised and their expression patterns investigated (Ohya *et al.* 2005), details of the shell development are known (Kuraku *et al.* 2005) and a turtle–chicken chimera has been generated (Nagashima *et al.* 2005). In addition, several cDNA libraries have been developed (Kuraku *et al.* 2005) and ideal incubation conditions have been identified (Du and Ji 2003).

All species mentioned above belong to the Cryptodira. Less information is available for members of the Pleurodira. Possibly the best candidate as a Pleurodira model species is *Emydura subglobosa*, the pink-bellied sideneck, a mid-sized (about 25 cm) aquatic turtle native to Australia and Papua New Guinea. The species is robust and prolific, and is becoming an increasingly popular pet around the world.

Lepidosauria

Lizards

Sequencing of the full genome of *Anolis carolinensis*, the green anole lizard, has recently been initiated (www.genome.gov/10002154). Individuals are small (average 13 cm for females and 17 cm for males), and their life span is 2–7 years. Their maintenance is, however, not trivial. As most lizards, they are oviparous although their reproduction presents a peculiar feature: females produce a single egg every 1 to 2 weeks with successive alternation of the offspring's sex, a phenomenon that seems hormonally controlled by the female (Lovern and Passek 2002). However, only an average of seven eggs are laid per year (the maximum being 20), limiting the material available for developmental biology studies. Extensive literature on the species (especially behavioural and ecological studies) has accumulated since the 1960s, mainly owing to the animal's small size and abundance of the species in North America (see Lovern *et al.* 2004 for a review).

We consider, however, that an alternative species, *Pogona vitticeps* (bearded dragon; see Figure 7.2B), deserves attention from the evo-devo community. Bearded dragons are larger than anoles as they have an average adult size of about 40 cm, and they have recently become a common pet all around the world. Indeed, these animals are tame, easy to handle and feed, and much information is accessible on their maintenance and reproduction. Female bearded dragons are significantly more prolific than the green anole as the former can lay about 24 eggs up to six times a year.

Snakes

The Reptilian Genomics Working Group has suggested the sequencing of the full genome from *Thamnophis sirtalis* (the common garter snake). This is a relatively small-sized species (46–137 cm) easy to maintain in captivity. Individuals reach sexual maturity at about 55 cm (about 2 years old) and live in captivity up to 10 years. After a three-month gestation, females produce clutches of 4 to 80 offspring, depending on the size of the mother. Given its accessibility in nature in multiple areas of the United States, this species has been widely used for physiology and ecology research programs, but not in developmental biology. A key reproductive feature could prove problematic: *Thamnophis sirtalis* is ovoviviparous. Although ovoviviparity (and associated placental membranes) is by itself a subject worth investigating (Krohmer 2004), this feature seriously limits the accessibility of embryos at any developmental stages unless the mother is sacrificed.

This is why we suggest that *Elaphe guttata* (the corn snake, see Figure 7.2C) constitutes a valid alternative to the common garter snake. It has the great advantage of being oviparous, such that embryos can be incubated artificially for investigation at any stage of the 55–70 days of development after egg laying (i.e. about 50 days after fertilisation) (Kohler 2005).

Archosauria

Crocodylomorpha

The large size and long generation time (sexual maturity depends on size but in no species do individuals reach sexual maturity before the age of 8) is a discouraging factor for their use in evo-devo studies, but we think the task is not impossible because eggs are available in crocodile farms all around the world, mostly for *Alligator mississippiensis* and *Crocodylus niloticus*. Little information is available on the development of crocodiles (Guillette and Gunderson 2001, Tissir *et al.* 2003, Milnes *et al.* 2004), and their sister-group relationship with birds makes this group especially interesting for evo-devo studies. The need to collaborate with crocodile farms is probably the most important constraint on the choice of species. Not surprisingly, the US-based Reptilian Genomics Working Group recommends the sequencing of the *Alligator mississippiensis* (American alligator) genome, but this species is unfortunately common only in the United States, whereas *Crocodylus niloticus* (Nile crocodile, see Figure 7.2D) is much easier to find in European and African farms and zoos and is phylogenetically closely related to *Crocodylus porosus*, the species most studied by Australian laboratories. Furthermore, *A. mississippiensis* could be advantageously replaced by another representative of the Alligatorinae subfamily: the Cuvier's dwarf caiman (*Paleosuchus palpebrosus*), which has the great advantage of reaching a small adult size of about 130 cm. However, its generation time is not significantly shorter than those of larger species and it remains a dangerous animal to handle (Grenard 1991).

Aves

Birds, the sister group to Crocodylomorpha, constitute a highly diverse group comprising about 10 000 extant species in about 30 orders. It is very likely that *Gallus gallus*, the domestic fowl (Galliformes), will remain the focal representative model species of Aves because it exhibits a short generation time (sexual maturity is reached at the age of about 5 months), its reproductive output is staggering (a single hen can lay about 365 eggs a year and a single rooster can produce 100 000 offspring a year), much information is available on the embryonic development, captive holding and breeding techniques incorporating automated feeding and egg collection are very well documented and used all around the world, numerous inbred strains (exhibiting different phenotypes) are available, and regulations associated with captive holding/breeding are well known and much simpler than for other farm animals. The domestic fowl has served as a model system since Aristotle, and new technologies, such as in vivo electroporation and transgenesis, have been continuously incorporated. The development of chicken embryonic stem (ES) cells is an active topic and the genome of the species has been sequenced (Burt 2005, Stern 2005).

Despite the overwhelming advantages discussed above for the domestic fowl, other species have been used as models. For example, multiple species and strains of quails (genus *Coturnix*, Galliformes) have been used in behavioural studies (Mills *et al.* 1997) and reproductive ageing studies (Ottinger *et al.* 2004). The Passeriformes, accounting for about 60 % of the extant bird species (Poole 1999), also includes several model species, especially popular with students of sexual selection and animal communication, such as *Sturnus vulgaris* (starling) (Gentner *et al.* 2006), *Passer domesticus* (house sparrow) (Bonneaud *et al.* 2006) and *Taeniopygia guttata* (zebra finch) (Johnson and Whitney 2005, Birkhead 1995). Captive handling information is available for representatives of a few additional groups, which should aid comparative developmental studies undertaken by evo-devo scientists: *Tyto alba* (barn owl) and *Athene noctua* (little owl) of the Strigiformes are successfully bred in captivity, the widespread *Columba livia* (pigeon) would be an obvious model species for Columbiformes, whereas *Branta bernicla* (brent goose) and numerous pet parrot species could be used as models of Anseriformes and Psittaciformes, respectively (Poole 1999). Finally, *Struthio camelus* (the ostrich, Struthioniformes) is now bred in farms worldwide, making it possible to investigate development in this species (e.g. in relation to the loss of flight abilities).

Monotremata

Egg-laying mammals include only two genera comprising one species of Ornithorhynchidae and two species of Tachyglossidae. *Ornithorhynchus anatinus* (the platypus) is very difficult to maintain and breed in captivity despite efforts to do so starting as early as the 1800s (Temple-Smith and Grant 2001). On the other hand, *Tachyglossus aculeatus* (short-beaked echidna) is a reasonably common animal in zoos throughout the world and can be bred reasonably easily in captivity. Echidnas could serve as model species not only for their peculiar reproductive features but also to investigate other spectacular traits, such as the presence of spines and the characters associated with their ability to feed on ants (Jackson 2003). Obviously, the accessibility to a Monotremata species for evo-devo studies would be of tremendous interest given their phylogenetic position and the retention of ancestral traits in their anatomy in general and their reproductive system in particular. Unfortunately, many zoos are disinclined to provide material given the restricted availability of animals outside Australia. The low-coverage assembly of the platypus genome has recently been made available.

Marsupialia

Monodelphis domestica (the grey short-tailed opossum) is the marsupial species most often used in laboratories because of its small size, large reproductive output and short generation time (males and females reach sexual maturity before 9 months). The species has been used to study the development of some marsupial traits, such as the formation of the placenta (Freyer *et al.* 2002), of the median vagina and the pouch (Regli and Kress 2002, Kress *et al.* 2004), of the neural plate, neural crest and facial region (Smith 2001), and of male morphology (Wilson *et al.* 2002). It is also used as a model for studying cutaneous melanoma induced by ultraviolet radiation (Wang *et al.* 2004). One of the great advantages of using marsupials as model species is the accessibility, from an early developmental stage, of the embryos attached to the mother's nipples. Finally, *Monodelphis domestica* has the great advantage of having had its genome completely sequenced (www.broad.mit.edu/ mammals/opossum). Disadvantages of the species are its aggressive behaviour beyond the age of 4 months, its preference for solitariness (Poole 1999) and the difficulty of obtaining animals outside the United States.

We suggest that *Petaurus breviceps* (the sugar glider, see Figure 7.2E) should also be considered as an additional valid model species. Its size and reproductive output are similar to those of the grey short-tailed opossum but, unlike the latter, it is a social animal so that large groups can be maintained in captivity. Sugar gliders can breed all year around through artificial regulation of the photoperiodicity associated with their reproductive cycle (Jackson 2003). As they are becoming popular pets all around the world, they are much easier to find than short-tailed opossums. Finally, sugar gliders have a skin membrane, called *patagium*, extending from the forelimb to the hind foot and used to glide between trees. This anatomical feature is of interest in the evo-devo context because it evolved independently in three marsupial families, two rodent families and the flying lemurs.

Eutheria

Afrotheria

The representatives of this major clade (Figure 7.1) of mammals exhibit a spectacular diversity of sizes and morphologies, from the tiny *Microgale* species (about 7 grams) to the Asian and African elephants (about 7 tons). Despite the fact that there is, to our knowledge, no known single morphological character that would clearly support that clade, analyses of multiple molecular sequence datasets unambiguously support its existence (Stanhope *et al.* 1998, Murphy *et al.* 2001a, van Dijk *et al.* 2001) as a lineage grouping elephants, sea cows, hyraxes, aardvarks, elephant shrews, golden moles and tenrecs. Literature on afrotherian mammals is scarce, but reliable data (originating mainly from zoos all around the world) are available on their maintenance and reproduction, such that evo-devo studies would almost certainly require collaborations with zoological institutions. Given their size and long generation time, the Tubulidentata (aardvark), Proboscidea (elephants) and Sirenia (dugongs, sea cows and manatees) will obviously be the most difficult to investigate. On the other hand, representatives of the Afrosoricida (tenrecs), Macroscelidea (elephant shrews) and Hyracoidea (hyraxes) are much more amenable to housing and breeding in the laboratory. Tenrecs are particularly interesting as they retain some primitive mammalian features, such as the lack of scrotum in the male or the presence of a cloaca in females (as in marsupials and monotremes). In addition, several species of tenrecs have spines, a character that has convergently evolved in representatives of other mammalian lineages. Finally, tenrecs could constitute useful models for understanding the mechanisms associated with endothermy as their thermoregulation is less efficient than in other eutherians. We suggest *Echinops telfairi* (lesser hedgehog tenrec, see Figure 7.2F) as the model species of choice because (1) it is one of the smallest species for which breeding colonies already exist in research facilities, (2) it has spines macroscopically very similar to those of hedgehogs, and (3) extensive studies have been undertaken on the development of the placenta (Carter *et al.* 2004), as well as the brain structure and physiology (Kunzle 2006). Furthermore, the genome of *E. telfairi* has been recently sequenced.

For elephant shrews (19 living species), only some anatomical and physiological studies are available (e.g. organisation of the somatosensory cortex in *Elephantulus edwardii* [Dengler-Crish *et al.* 2006] and thermoregulation in *E. myurus* [Mzilikazi and Lovegrove 2004, 2006]). Very little information on the reproduction of these species is available, although they have been bred in the past (Rathbun *et al.* 1981). An alternative could be *Macroscelides proboscideus*; less information is available on its biology but established colonies can be found in numerous zoos around the world.

For the Hyracoidea, *Procavia capensis* seems the obvious choice given that it is successfully bred in zoos (although gestation takes about 7 months), it is relatively small, it is a social animal, and its full genome might soon be sequenced (www.genome.gov/10002154).

Xenarthra

Representatives of this mammalian group (anteaters, sloths and armadillos) are confined to South America and are highly specialised animals. Anteaters and sloths are very difficult to study owing to their large size and special requirements for maintenance, whereas armadillos could be bred in laboratory facilities. The *armour shell* of armadillos, a structure consisting of bony scutes covered with keratinous plates, is of interest for evo-devo research, as is an idiosyncrasy of the reproductive strategy of the genus *Dasypus*: litters typically consist of four clones as the offspring develop from a single fertilised egg. Breeding in captivity has proven difficult, although not impossible (Carvalho *et al.* 1997). The nine-banded armadillo (*Dasypus novemcinctus*) genome sequence has recently been made available, although many genes are interrupted or missing because of the low-coverage (2×) assembly: each base in the final sequence is present, on average, in two reads only, as opposed to more than six reads for other sequenced species such as human or mouse.

Euarchontoglires

This large grouping (Figure 7.1) includes the best-studied mammals as it comprises the Rodentia (mice, rats, squirrels, gophers, porcupines, beavers, etc.), Lagomorpha (rabbits, hares and pikas), Primates (lemurs, tarsiers, new and old-world monkeys), Scandentia (tree shrews) and Dermoptera (flying lemurs). Only the latter group has not been extensively studied given the difficulty of keeping these animals in captivity.

Rodentia. Obviously, the laboratory mouse and rat dominate the literature in numerous fields of experimental biology but quite a series of additional species are used as model species for medical and academic research such as, among others: *Microtus agrestis* (short-tailed field vole), *Meriones unguiculatus* (laboratory gerbil), and *Mesocricetus auratus* and *Cricetus grisues* (Syrian and Chinese hamster, respectively) (Poole 1999, Cavanagh *et al.* 2004, Mand *et al.* 2006, Shimozuru *et al.* 2006). Finally, members of the genus *Cavia* (guinea pigs) have been used in sociophysiology studies (Sachser 1998) and are very valuable models for ascorbic acid metabolism because they are the only non-primate laboratory animals that require a dietary source of vitamin C (Burk *et al.* 2006). Guinea pigs are of particular interest in evo-devo studies because they are representatives of the hystricomorph rodents with the largest amount of available information regarding their biology as well as captive holding and breeding. As well as the high-coverage genome sequences of the laboratory mouse and rat, the low-coverage genome sequence of *Cavia porcellus* is now also available.

Lagomorpha. This group is represented by a widely used model species, *Oryctolagus cuniculus* (the laboratory rabbit) for which the full genome sequence is available. Rabbits are widely used for antibody production and have served as a model organism in many physiological and immunological studies. An emerging additional lagomorph model species is *Ochotona curzoniae* (the black-lipped pika) which could be useful for comparative studies given that it belongs to a different family from the laboratory rabbit. Females can breed and produce litters every three months in the summer, with the size of the litter depending on environmental conditions (Dobson *et al.* 1998). Little empirical information is available regarding its captive breeding, but specimens are available in the pet trade. Genome sequencing has been proposed for this species (www.genome.gov/10002154).

Primates. The most frequent and obvious justification for the use of primate species as models is their close phylogenetic relationship with humans. However, beside multiple and significant ethical problems associated with their use as model species (owing to this close relationship with humans), the high demands in maintenance of primates, as well as their vulnerable status in nature, make them difficult and expensive to acquire, breed and rear. Full genome sequences are available for human, the chimpanzee (*Pan troglodytes*), the rhesus macaque (*Macaca mulatta*) and the bushbaby (*Otolemur garnettii*).

Scandentia. Tree shrews form a small group (20 species in five genera) of small-sized mammals that are native to tropical Southeast Asia. The vast majority of experimental work has been done on their visual system because tree shrews are considered good models for understanding the neural organisation of the early primate visual system (Drenhaus *et al.* 2006). The focal tree shrew model species has always been *Tupaia belangeri*, colonies of which have been successfully established in laboratories (Poole 1999). Its value as a medical research model animal and its phylogenetic position as an outgroup of primates are the two main reasons for *Tupaia belangeri* being a genome sequencing target.

Laurasiatheria

This large grouping (Figure 7.1) comprises Carnivora, Pholidota, Perissodactyla, Cetartiodactyla, Chiroptera and Eulipotyphla.

Pholidota. Pangolins are very difficult (if not impossible) to breed in captivity, so that observations can only be based on opportunistic sampling. The situation is particularly unfortunate from an evo-devo perspective because the representatives of this extraordinary group (with only seven living species found in the tropical regions of Africa and Asia) show spectacular features such as, among others, a highly modified skull, an extraordinarily long and muscular tongue, and a set of overlapping scales covering their head, back and tail.

Perissodactyla (horses, tapirs and rhinoceroses) and *Cetartiodactyla* (camels, ruminants, pigs, hippopotamuses, whales, dolphins and porpoises). The large size of most of these mammals complicates their use as evo-devo model species. The situation is, however, easier for families represented by domesticated species (horses, cows, sheep, goats) for which multiple inbred lines, much developmental information, genome engineering techniques and full genome sequences (for pig and cow) are available. Furthermore, a recent investigation (Thewissen *et al.* 2006) of gene expression during early development of hind-limb buds in embryos of the pantropical spotted dolphin, *Stenella attenuata* (providing data on the molecular basis for hind-limb loss during cetacean evolution), demonstrates that a good museum collection of embryo specimens can partially but efficiently compensate for the lack of laboratory animal models.

Carnivora. The diversity of phenotypes among the multiple inbred lines of dogs (*Canis familiaris*) has already been successfully exploited in evo-devo studies (e.g. Fondon and Garner 2004), and representatives of other families (such as mustelids) are becoming increasingly available as possible laboratory model species. Dog and cat full genome sequences are available in public databases.

Eulipotyphla. True insectivores are small, highly mobile animals with long, narrow and often elaborate snouts, adapted to eating insects. Low-coverage assemblies of the European shrew (*Sorex araneus*) and the western European hedgehog (*Erinaceus europaeus*) genomes have recently been made available. The African pygmy hedgehog (*Atelerix albiventris*, Figure 7.2G) would have probably been a better choice of target species for genome sequencing because representatives of this species are small (average size of 20 cm) and tame animals that can be bred all year around in laboratory conditions. They are partially covered with spines, opening the possibility of investigating the development of this trait convergent with the situation observed in some tenrec species (see above). In fact, multiple convergences have evolved between representatives of Eulipotyphla and Afrotheria: Erinaceidae (hedgehogs) versus *Echinops* (Madagascar 'hedgehogs'), Soricidae (shrews) vs. *Microgale* (shrew-like tenrecs), and Talpidae (moles) vs. Chrysochloridae (golden moles). Finally, it would be particularly appealing to identify a model species for the subterranean insectivores. These species exhibit derived traits such as reduced eyes (sometimes covered by skin) and lack of external ears, as well as modified limb morphology. Unfortunately, no information is available on the maintenance of breeding colonies for any of these species, although it is possible to keep them for long periods in captivity (Borroni *et al.* 1999).

Chiroptera. Bats have acquired a number of spectacular evolutionary innovations associated with their adaptation to active flight. Microbats have a number of interesting features including heavily modified limbs (Sears *et al.* 2006), the ability to echolocate (Smotherman and Metzner 2005), precise control over both thermoregulation and gestation (Badwaik and Rasweiler 2001), and an unexpectedly long life span (Dobson 2003). Thus far, the little brown bat (*Myotis lucifugus*) has been proposed as a model species because of its small size and its availability. BAC libraries already exist for the little brown bat, and its full genome sequence has recently been assembled with low coverage. An alternative valid model species for bats would be *Carollia perspicillata* (short-tailed fruit bat) whose maintenance can be readily adjusted to laboratory facilities and whose reproduction takes place at any time of the year (Kiefer 2006). Its embryonic developmental stages have been recently described and illustrated, a fundamental tool for evo-devo studies (Cretekos *et al.* 2005). This basic information allowed further analyses of the development of bat wing digits, providing a plausible scenario as to how bats acquired powered flight soon after their divergence from other mammals about 65 million years ago (Sears *et al.* 2006).

ACKNOWLEDGEMENTS

We thank Alessandro Minelli and Giuseppe Fusco for organising the Evo-Devo workshop in Venice in May 2006 and giving us the opportunity to contribute to the present book. Our manuscript has benefited from discussions with Denis Headon, Olivier Lambert, Christophe Rémy, Adrien Debry and Günter Wagner. Galen B. Rathbun provided information on Macroscelidea. We thank Alessandro Minelli, Giuseppe Fusco and Ronald Jenner for commenting on the first version of the manuscript. This work was supported by grants from the 'Communauté Française de Belgique' (ARC 1164/20022770), and the National Fund for Scientific Research Belgium (FNRS). A.Tzika is a Ph.D. candidate at the *Fonds pour la formation à la Recherche dans l'Industrie et dans l'Agriculture* (FRIA), Belgium.

REFERENCES

Badwaik, N. K. & Rasweiler, J. J. IV 2001. Altered trophoblastic differentiation and increased trophoblastic invasiveness during delayed development in the short-tailed fruit bat, *Carollia perspicillata*. *Placenta* **22**, 124–144.

Benton, M. J. & Donoghue, P. C. J. 2007. Paleontological evidence to date the tree of life. *Molecular Biology and Evolution* **24**, 26–53.

Birkhead, T. R. 1995. Sperm competition: evolutionary causes and consequences. *Reproduction, Fertility and Development* **7**, 755–775.

Bleidorn, C., Eeckhaut, I., Podsiadlowski, L., *et al.* 2007. Mitochondrial genome and nuclear sequnce data support Myzostomida as part of the annelid radiation. *Molecular Biology and Evolution* **24**, 1690–1701.

Bonneaud, C., Chastel, O., Federici, P., Westerdahl, H. & Sorci, G. 2006. Complex Mhc-based mate choice in a wild passerine. *Proceedings of the Royal Society. Biological Sciences* **273**, 1111–1116.

Borroni, A., Loy, A. & Capanna, E. 1999. A flexible arrangement for the study of moles in captivity. *Acta Theriologica* **44**, 207–214.

Bossuyt, F. & Milinkovitch, M. C. 2000. Convergent adaptive radiations in Madagascan and Asian ranid frogs reveal covariation between larval and adult traits. *Proceedings of the National Academy of Sciences of the USA* **97**, 6585–6590.

Burk, R. F., Christensen, J. M., Maguire, M. J. 2006. A combined deficiency of vitamins E and C causes severe central nervous system damage in guinea pigs. *Journal of Nutrition* **136**, 1576–1581.

Burt, D. W. 2005. Chicken genome: current status and future opportunities. *Genome Research* **15**, 1692–1698.

Carter, A. M., Blankenship, T. N., Kunzle, H. & Enders, A. C. 2004. Structure of the definitive placenta of the tenrec, *Echinops telfairi*. *Placenta* **25**, 218–232.

Carvalho, R. A., Lins-Lainson, Z. C. & Lainson, R. 1997. Breeding nine-banded armadillos (*Dasypus novemcinctus*) in captivity. *Contemporary Topics in Laboratory Animal Science* **36**, 66–68.

Cavanagh, R. D., Lambin, X., Ergon, T. 2004. Disease dynamics in cyclic populations of field voles (*Microtus agrestis*): cowpox virus and vole tuberculosis (*Mycobacterium microti*). *Proceedings of the Royal Society of London, Biological Sciences* **271**, 859–867.

Cretekos, C. J., Weatherbee, S. D., Chen, C. H. *et al.* 2005. Embryonic staging system for the short-tailed fruit bat, *Carollia perspicillata*, a model organism for the mammalian order Chiroptera, based upon timed pregnancies in captive-bred animals. *Developmental Dynamics* **233**, 721–738.

Dengler-Crish, C. M., Crish, S. D., O'Riain, M. J. & Catania, K. C. 2006. Organization of the somatosensory cortex in elephant shrews (*E. edwardii*). *Anatomical Record Part A: Discoveries in Molecular, Cellular, and Evolutionary Biology* **288**, 859–866.

Dobson, G. 2003. On being the right size: heart design, mitochondrial efficiency and lifespan potential. *Clinical and Experimental Pharmacology and Physiology* **30**, 590–597.

Dobson, S. E., Smith, A. T. & Gao, W. X. 1998. Social and ecological influences on dispersal and philopatry in the plateau pika (*Ochotona curzoniae*). *Behavioral Ecology* **9**, 622–635.

Drenhaus, U., Rager, G., Eggli, P. & Kretz, R. 2006. On the postnatal development of the striate cortex (V1) in the tree shrew (*Tupaia belangeri*). *European Journal of Neuroscience* **24**, 479–490.

Du, W.-G. & Ji, X. 2003. The effects of incubation thermal environments on size, locomotor performance and early growth of hatchling soft-shelled turtles, *Pelodiscus sinensis*. *Journal of Thermal Biology* **28**, 279–286.

Dusheck, J. 2002. It's the ecology, stupid! *Nature* **418**, 578–579.

Fishman, M. C. 2001. Genomics. Zebrafish: the canonical vertebrate. *Science* **294**, 1290–1291.

Fondon, J. W. III & Garner, H. R. 2004. Molecular origins of rapid and continuous morphological evolution. *Proceedings of the National Academy of Sciences of the USA* **101**, 18058–18063.

Freyer, C., Zeller, U. & Renfree, M. B. 2002. Ultrastructure of the placenta of the tammar wallaby, *Macropus eugenii*: comparison with the grey short-tailed opossum, *Monodelphis domestica*. *Journal of Anatomy* **201**, 101–119.

Gentner, T. Q., Fenn, K. M., Margoliash, D. & Nusbaum, H. C. 2006. Recursive syntactic pattern learning by songbirds. *Nature* **440**, 1204–1207.

Gilbert, S. F. & Bolker, J. A. 2003. Ecological developmental biology: preface to the symposium. *Evolution & Development* **5**, 3–8.

Gilbert, S. F., Loredo, G. A., Brukman, A. & Burke, A. C. 2001. Morphogenesis of the turtle shell: the development of a novel structure in tetrapod evolution. *Evolution & Development* **3**, 47–58.

Grenard, S. 1991. *Handbook of Alligators and Crocodiles*. Malaba: Krieger Publishing Company.

Guillette, L. J. Jr. & Gunderson, M. P. 2001. Alterations in development of reproductive and endocrine systems of wildlife populations exposed to endocrine-disrupting contaminants. *Reproduction* **122**, 857–864.

Holland, L. Z., Laudet, V. & Schubert, M. 2004. The chordate amphioxus: an emerging model organism for developmental biology. *Cellular and Molecular Life Sciences* **61**, 2290–2308.

Jackson, S. 2003. *Australian Mammals: Biology and Captive Management*. Collingwood, Australia: CSIRO Publishing.

Johnson, F. & Whitney, O. 2005. Singing-driven gene expression in the developing songbird brain. *Physiology and Behavior* **86**, 390–398.

Kiefer, J. C. 2006. Emerging developmental model systems. *Developmental Dynamics* **235**, 2895–2899.

Kishino, H., Thorne, J. L. & Bruno, W. J. 2001. Performance of a divergence time estimation method under a probabilistic model of rate evolution. *Molecular Biology and Evolution* **18**, 352–361.

Kocher, T. D., Conroy, J. A., McKaye, K. R. & Stauffer, J. R. 1993. Similar morphologies of cichlid fish in Lakes Tanganyika and Malawi are due to convergence. *Molecular Phylogenetics and Evolution* **2**, 158–165.

Kohler, G. 2005. *Incubation of Reptile Eggs: Basics, Guidelines, Experiences*. Malaba: Krieger Publishing Company.

Krenz, J. G., Naylor, G. J., Shaffer, H. B. & Janzen, F. J. 2005. Molecular phylogenetics and evolution of turtles. *Molecular Phylogenetics and Evolution* **37**, 178–191.

Kress, A., Regli, C., Spornitz, U. M. & Morson, G. 2004. Changes in the epithelium of the vaginal complex during the estrous cycle of the marsupial *Monodelphis domestica*. 2. Scanning electron microscopy study. *Cells Tissues Organs* **178**, 48–59.

Krohmer, R. W. 2004. The male red-sided garter snake (*Thamnophis sirtalis parietalis*): reproductive pattern and behavior. *Ilar Journal* **45**, 54–74.

Kunzle, H. 2006. Thalamo-striatal projections in the hedgehog tenrec. *Brain Research* **1100**, 78–92.

Kuraku, S., Usuda, R. & Kuratani, S. 2005. Comprehensive survey of carapacial ridge-specific genes in turtle implies co-option of some regulatory genes in carapace evolution. *Evolution & Development* **7**, 3–17.

Lapid, R., Nir, I., Snapir, N, & Robinzon, B. 2004. Reproductive traits in the spur-thighed tortoise (*Testudo graeca terrestris*): new tools for the enhancement of reproductive success and survivorship. *Theriogenology* **61**, 1147–1162.

Losos, J. B., Jackman, T. R., Larson, A., de Queiroz, K. & Rodriguez-Schettino, L. 1998. Contingency and determinism in replicated adaptive radiations of island lizards. *Science* **279**, 2115–2118.

Lovern, M. B., Holmes, M. M. & Wade, J. 2004. The green anole (*Anolis carolinensis*): a reptilian model for laboratory studies of reproductive morphology and behavior. *Ilar Journal* **45**, 54–64.

Lovern, M. B. & Passek, K. M. 2002. Sequential alternation of offspring sex from successive eggs by female green anoles, *Anolis carolinensis*. *Canadian Journal of Zoology* **80**, 77–82.

Mand, S., Specht, S., Zahner, H. & Hoerauf, A. 2006. Ultrasonography in filaria-infected rodents: detection of adult *Litomosoides sigmodontis* and *Brugia malayi* filariae. *Tropical Medicine and International Health* **11**, 1382–1387.

Metscher, B. D. & Ahlberg, P. E. 1999. Zebrafish in context: uses of a laboratory model in comparative studies. *Developmental Biology* **210**, 1–14.

Milinkovitch, M. C. & Tzika, A. 2007. Escaping the mouse trap: the selection of new Evo-Devo model species. *Journal of Experimental Zoology Part B: Molecular and Developmental Evolution* **308**, 337–346.

Mills, A. D., Crawford, L. L., Domjan, M. & Faure, J. M. 1997. The behavior of the Japanese or domestic quail *Coturnix japonica*. *Neuroscience and Biobehavioral Reviews* **21**, 261–281.

Milnes, M. R., Allen, D., Bryan, T. A., Sedacca, C. D. & Guillette, L. J. Jr 2004. Developmental effects of embryonic exposure to toxaphene in the American alligator (*Alligator mississippiensis*). *Comparative Biochemistry and Physiolology. C. Toxicology & Pharmacology* **138**, 81–87.

Minelli, A. 2003. *The Development of Animal Form: Ontogeny, Morphology, and Evolution.* Cambridge: Cambridge University Press.

Murphy, W. J., Eizirik, E., Johnson, W. E. 2001a. Molecular phylogenetics and the origins of placental mammals. *Nature* **409**, 614–618.

Murphy, W. J., Eizirik, E., O'Brien, S. J. 2001b. Resolution of the early placental mammal radiation using Bayesian phylogenetics. *Science* **294**, 2348–2351.

Mzilikazi, N. & Lovegrove, B. G. 2004. Daily torpor in free-ranging rock elephant shrews, *Elephantulus myurus*: a year-long study. *Physiological and Biochemical Zoology* **77**, 285–296.

Mzilikazi, N. & Lovegrove, B. G. 2006. Noradrenalin induces thermogenesis in a phylogenetically ancient eutherian mammal, the rock elephant shrew, *Elephantulus myurus*. *Journal of Comparative Physiology B* **176**, 75–84.

Nagashima, H., Uchida, K. 2005. Turtle–chicken chimera: an experimental approach to understanding evolutionary innovation in the turtle. *Developmental Dynamics* **232**, 149–161.

Ohya, Y. K., Kuraku, S. & Kuratani, S. 2005. Hox code in embryos of Chinese soft-shelled turtle *Pelodiscus sinensis* correlates with the evolutionary innovation in the turtle. *Journal of Experimental Zoology Part B (Molecular and Developmental Evolution)* **304**, 107–118.

Ottinger, M. A., Abdelnabi, M., Li, Q. *et al.* 2004. The Japanese quail: a model for studying reproductive aging of hypothalamic systems. *Experimental Gerontology* **39**, 1679–1693.

Poole, T. 1999. *The UFAW Handbook on the Care and Management of Laboratory Animals.* Oxford: Blackwell Science.

Rathbun, G. B., Beaman, P. & Maliniak, E. 1981. Capture, husbandry and breeding of rufous elephant-shrews (*Elephantulus rufescens*). *International Zoo Yearbook* **21**, 176–184.

Regli, C. & Kress, A. 2002. Changes in the epithelium of the vaginal complex during the estrous cycle of the marsupial *Monodelphis domestica*. 1. Transmission electron microscopy study. *Cells Tissues Organs* **172**, 276–296.

Ruber, L., Verheyen, E. & Meyer, A. 1999. Replicated evolution of trophic specializations in an endemic cichlid fish lineage from Lake Tanganyika. *Proceedings of the National Academy of Sciences of the USA* **96**, 10230–10235.

Sachser, N. 1998. Of domestic and wild guinea pigs: studies in sociophysiology, domestication, and social evolution. *Naturwissenschaften* **85**, 307–317.

Sears, K. E., Behringer, R. R., Rasweiler, J. J. IV & Niswander, L. A. 2006. Development of bat flight: morphologic and molecular evolution of bat wing digits. *Proceedings of the National Academy of Science of the USA* **103**, 6581–6586.

Shimozuru, M., Kikusui, T., Takeuchi, Y. & Mori, Y. 2006. Scent-marking and sexual activity may reflect social hierarchy among group-living male Mongolian gerbils (*Meriones unguiculatus*). *Physiology and Behavior* **89**, 644–649.

Smith, K. K. 2001. Early development of the neural plate, neural crest and facial region of marsupials. *Journal of Anatomy* **199**, 121–131.

Smotherman, M. & Metzner, W. 2005. Auditory-feedback control of temporal call patterns in echolocating horseshoe bats. *Journal of Neurophysiology* **93**, 1295–1303.

Springer, M. S., Murphy, W. J., Eizirik, E. & O'Brien, S. J. 2003. Placental mammal diversification and the Cretaceous–Tertiary boundary. *Proceedings of the National Academy of Sciences of the USA* **100**, 1056–1061.

St Johnston, D. 2002. The art and design of genetic screens: *Drosophila melanogaster*. *Nature Reviews Genetics* **3**, 176–188.

Stanhope, M. J., Waddell, V. G., Madsen, O. 1998. Molecular evidence for multiple origins of Insectivora and for a new order of endemic African insectivore mammals. *Proceedings of the National Academy of Sciences of the USA* **95**, 9967–9972.

Stern, C. D. 2005. The chick; a great model system becomes even greater. *Developmental Cell* **8**, 9–17.

Temple-Smith, P. & Grant, T. 2001. Uncertain breeding: a short history of reproduction in monotremes. *Reproduction, Fertility and Development* **13**, 487–497.

Thewissen, J. G., Cohn, M. J., Stevens, L. S. 2006. Developmental basis for hindlimb loss in dolphins and origin of the cetacean bodyplan. *Proceedings of the National Academy of Sciences of the USA* **103**, 8414–8418.

Tissir, F., Lambert De Rouvroit, C., Sire, J. Y., Meyer, G. & Goffinet, A. M. 2003. Reelin expression during embryonic brain development in *Crocodylus niloticus*. *Journal of Comparative Neurology* **457**, 250–262.

van Dijk, M. A., Madsen, O., Catzeflis, F. *et al.* 2001. Protein sequence signatures support the African clade of mammals. *Proceedings of the National Academy of Sciences of the USA* **98**, 188–193.

van Rheede, T., Bastiaans, T., Boone, D. N. 2006. The platypus is in its place: nuclear genes and indels confirm the sister group relation of monotremes and therians. *Molecular Biology and Evolution* **23**, 587–597.

Wagner, G. P. & Larsson, H. C. E. 2003. What is the promise of developmental evolution? III. The crucible of developmental evolution. *Journal of Experimental Zoology B (Molecular and Developmental Evolution)* **300**, 1–4.

Wang, Z., Dooley, T. P., Curto, E. V., Davis, R. L. & VandeBerg, J. L. 2004. Cross-species application of cDNA microarrays to profile gene expression using UV-induced melanoma in *Monodelphis domestica* as the model system. *Genomics* **83**, 588–599.

Wiens, J. J., Brandley, M. C. & Reeder, T. W. 2006. Why does a trait evolve multiple times within a clade? Repeated evolution of snake-like body form in squamate reptiles. *Evolution* **60**, 123–141.

Wilson, J. D., Shaw, G., Leihy, M. L. & Renfree, M. B. 2002. The marsupial model for male phenotypic development. *Trends in Endocrinology and Metabolism* **13**, 78–83.

8

On comparisons and causes in evolutionary developmental biology[1]

GERHARD SCHOLTZ

Denn mit dem Warum der Dinge kommt niemand zu Ende. Die Ursachen alles Geschehens gleichen den Dünenkulissen am Meere: eine ist immer der anderen vorgelagert, und das Weil, bei dem sich ruhen ließe, liegt im Unendlichen.

[For once you begin with the Why you can never get to the end. It is like the dunes by the sea, where behind each dune lies still another and the Because where you might come to final rest lies somewhere in infinity.]

Thomas Mann, Joseph und seine Brüder

Comparison is fundamental to any evolutionary developmental analysis (e.g. Alberch 1985, Rieppel 1988, Dohle 1989, Minelli 2003, Scholtz 2005, Deutsch 2006, Jenner 2006, Breidbach and Ghiselin 2007). However, evo-devo as a discipline evolved from a mix of experimental and descriptive approaches to development. Accordingly, different weight is put on the method of studying development in an evolutionary framework depending on a researcher's scientific background. Here I want to evaluate the different approaches and their contribution to addressing evolutionary questions. I stress that only the comparative approach offers a direct method of studying development with respect to evolutionary changes. Descriptive and comparative approaches are often interpreted as being less 'exact' than experimental studies because they deal with untestable scenarios. Here I want to show

---

[1] This work is dedicated to my teacher and mentor Wolfgang Dohle on the occasion of his 70th birthday.

Evolving Pathways: Key Themes in Evolutionary Developmental Biology, ed. Alessandro Minelli and Giuseppe Fusco. Published by Cambridge University Press.
© Cambridge University Press 2008.

that comparative approaches are a direct means to study evolution if the latter is accepted as the general framework for reasoning about causality and changes and the link between the two. The experimental approach alone, dealing with mechanisms, does not help with respect to evolutionary considerations because developmental mechanisms are not evolutionary mechanisms. In contrast, in comparative approaches to development, ontogenetic mechanisms analysed in terms of independence of developmental steps might reveal the kind of causes operating in evolutionary mechanisms.

COMPARISON AS A GENERAL TOOL FOR THE STUDY OF DEVELOPMENT

Comparisons are always involved in developmental biology and the differences lie only in the theoretical and practical frameworks under which the results are interpreted. Even in an experimental approach using a model organism, comparison of the development of a number of individuals is the prerequisite to conceptualising the normal development of the species and to establish the stages etc. on which the experiments will be carried out. Only against this background can the results be generalised. Focus is not on comparison between species, but rather on comparison between normal development and the results of experimental manipulation or naturally occurring mutants. From these comparisons the mechanisms are inferred within a theoretical framework centred on the question of how we interpret developmental causes. The developmental experimentalist knows what he or she did to achieve the result. Nevertheless, a rest of unknown and unexplained elements remains – that is, the results have to be interpreted!

It is obvious that the role of comparison is even greater in the comparative evolutionary approach to development. Here the comparison is used as a direct means to interpret the observed similarities and differences in the theoretical framework of evolution. There is no contrast of normal versus disturbed development because all observed, existing normal developmental pathways are successful.

CAUSATION

With the advent of experimental developmental biology (Entwicklungsmechanik) at the end of the nineteenth century, the term 'causal morphology' (Kausale Morphologie) was coined (Mocek 1998). Causal morphology claimed to find causal explanations of animal form by performing experiments, in contrast to comparative morphological approaches which, according to the causal morphologists' view, lead only to descriptions of phenomena but not to explanations (Mocek 1998). Since then development has often been conceptualised in terms of causation. Alberch (1985), for instance, discriminates between 'causal developmental sequences' in which every stage is the prerequisite for the next stage, and 'temporal developmental sequences' in which stages lack a causal connection. Mayr (1997) introduced the distinction between functional 'proximate causes' and evolutionary 'ultimate causes' in the explanation of change in biological systems. Because all biological objects are the products of history, only the two levels of causes together explain biological features completely. For instance, the horizontal orientation and the up and down movements of the whale fluke, as opposed to the vertical tail fin of a fish with lateral movements, can be explained in terms of developmental and physiological processes (proximate causes) but to gain a full understanding of these features, the descent of whales from terrestrial placentalian mammals with their characteristic anatomy and movement (ultimate cause) has to be considered as well.

However, causality is a highly problematic issue and a great debate about causality forms a major part of (bio-) philosophy (see e. g. Schopenhauer 1847, Wuketits 1981, Rieppel 1988, Jonas 1997, Mahner and Bunge 1997). The concept of causality has often been used in a broad sense covering most things that produce differences. In contrast to this, Mahner and Bunge (1997) restrict causation to events in a temporal sequence accompanied by energy transfer. According to this narrower concept, other differences in time (e.g. between states or properties) are interpreted in terms of determinants and conditions rather than causes. Furthermore, this view implies that historical (evolutionary) conditions that determine actual biological structures are not causes either. Hence the expression 'ultimate cause' (Mayr 1997) should be abandoned. I acknowledge the merits in these distinctions, but in a developmental biological context the discrimination between causes, determinants and conditions is problematic. Thus, I use developmental causation in a broader sense covering all three of these terms. Furthermore, I know that the occurrence of consecutive or otherwise correlated events does not automatically imply causal relation, i.e. post hoc does not automatically mean propter hoc (Wuketits 1981). Nevertheless, for the sake of clarity in the following I use causality between time-ordered events and states in developmental sequences (as defined by Alberch 1985) as given. Moreover, for the same reason I use formalised linear sequences as examples despite the fact that the relationships between developmental events are often best represented as complex networks.

Biological studies are conceptually divided into those dealing mainly with processes and those dealing with patterns. The distinction between these categories is not always straightforward – this seems evident for the tension between evolutionary process and resulting pattern (e.g. Rieppel 1985, Arthur 2000). Development is almost universally considered as being a continuous process and thus contrasted to things we can categorise as patterns (Cracraft 2005). However, as soon as we deal with development, in particular in a comparative and evolutionary framework, we are forced to use descriptions of discrete steps in time such as stages, instars or phases (Alberch 1985, Rieppel 1988, Scholtz 2004, 2005), and these entities are conceptualised as processes in a theory- and assumption-laden framework as has been discussed by Cracraft (2005). Hence, I do not endorse the distinction made by Alberch (1985) between more static and more dynamic approaches to development, the latter not implying stages. I think it is just a question of perspective and of level of the subdivision of processes – in any case one has to deal with some kind of discrete developmental entity.

There is a long tradition of subdividing development into stages (Richardson *et al.* 2001, Fürst von Lieven 2005, Minelli *et al.* 2006, Hopwood 2007). However, many people associate the term 'stage' with a specific shape taken by an embryo or larva (e.g., nauplius, pharyngula, gastrula). Such stages are often too imprecise to be used for comparison of developmental events and can thus be misleading (Richardson *et al.* 2001, Hopwood 2007). Hence, in contrast to a stage-based approach, I will use in the following the term 'developmental step'. I define a developmental step as a describable and comparable (homologisable) *pattern* at any moment of development. The word step is chosen because of its twofold meaning: as a structure (a step of a ladder) and as an event with a temporal aspect (a walking step). Accordingly, the term developmental step comprises spatial patterns ('frozen' moments of development such as the 16-cell stage, the initial limb bud, a distinct gene expression pattern or early gastrulation) and units within developmental processes and developmental events (e.g. regulation of segmentation genes, gastrulation, cell division sequences) which are also treated as patterns, i.e. patterns in time (Scholtz 2005). A developmental step can correspond to a traditional stage but it also can be just a part of it. Developmental steps can be described and analysed at the morphogenetic, cellular, gene expression or molecular levels. According to this view, development is characterised by complex temporal sequences and interactions of developmental steps.

For the identification of developmental steps a distinct element of comparison is always involved. Furthermore, the recognition of corresponding and similar developmental steps in two or more individuals of one or several species implies their homology. The concept of developmental steps has its basis in empirical observation since it deals with describable patterns. Hence, a developmental step is an ontological entity and not just an artificial construct to subdivide a continuous process into slices. Nevertheless, unavoidable arbitrary elements are involved. These concern, for instance, the temporal and spatial limits of the developmental steps. However, this partial restriction is a problem common to many entities in biology in general (e.g. character, homologue, organ, population, species), and in developmental biology in particular.

It is evident that the concept of developmental steps is related to the character concept in phylogenetics (e.g. Cracraft 2005, Richter 2005). Correspondingly, there is a hierarchy of nested developmental steps with nested homologies: individual steps as well as a sequence of steps can be homologous (see Scholtz 2005). The evolutionary independence of individual developmental steps can be shown in comparative analyses (see below) and, based on this possible independence, a mosaic distribution of homologous developmental steps occurs among taxa. Hence, transformation, insertion, deletion or replacement of developmental steps are the kinds of evolutionary change of developmental processes.

A recent discussion centres on whether developmental processes are modular and whether modules form some sort of functional elements of development (e.g. Wagner 1996, Minelli 2003, Schlosser and Wagner 2003). At first sight this approach seems similar to what I describe here in terms of developmental steps. However, developmental steps in my sense are different from modules insofar as the former are not intended as functional entities. For a critical view of modularity, see Mitchell (2006).

In the following I use letters to represent developmental steps: I am aware of the fact that this is a gross simplification. But if the reader accepts that a letter can stand for a morphological structure, a cell arrangement, a gene expression or a regulatory network, then I think that this simplification is appropriate to clarify what I want to discuss. Each letter represents an observable, comparable and homologisable developmental step.

DESCRIPTIVE STUDIES OF SINGLE SPECIES

The purely descriptive approach to the study of development without any interspecies comparisons leads to an analysis of the temporal sequence of developmental steps. The result is a description of the events of the normal development (normogenesis) of a given species. The descriptive approach leads to a finalistic view because normogenesis is observed to lead to the final and differentiated stage, namely the adult. However, this does not imply any causal relationship between developmental steps. Nevertheless, the regularity of the developmental process observed again and again in every embryo at each generation makes it tempting to infer strict causal connections between subsequent development steps. That this kind of conclusion cannot be legitimately drawn has been already discussed by Roux (1907) and has been shown by classical experiments which revealed the regulatory capacity of development (Müller 1996, Sander 1996).

Nevertheless, exact descriptions of developmental processes are the necessary prerequisite for all following approaches to development as well as for interpreting evidence such as the recently found fossils of Cambrian embryos (e.g. Chen *et al.* 2004, Donoghue *et al.* 2006). This is true in particular of taxa thus far neglected.

THE EXPERIMENTAL MODEL-ORGANISM APPROACH

Experiments are designed to show the independence or dependence of developmental steps or mechanisms, and they sort out what could have an influence on subsequent steps by experimental manipulations (Roux 1907). In other words, this approach is largely an analysis of mal-formations. Given that there is a sequence of developmental steps ABCDE, an experimental developmental biologist would say that A causes B, B causes C etc. if he or she has proven this experimentally, e.g. by deletion experiments or gene silencing. If B is taken away and C does not occur in development as is the case in normal (i.e. not manipulated) development it has been shown that B causes C (Figure 8.1). This is of course a simplification because the causal link between B and C can be quite indirect, but as a general principle this

Observation

$$A \rightarrow B \rightarrow C \rightarrow D \rightarrow E$$

Manipulation

if B^X then C^X

Inference

B is the necessary cause of C

Conclusion

B and C are not independent developmental steps

Proximate causes

physical, chemical properties, sheer existence

Figure 8.1 Reasoning in the framework of the experimental model-organism approach to development. If developmental step B is changed or removed (indicated by B^X), then developmental step C is absent or not properly formed (indicated by C^X).

is justified and is exactly the basis of the Entwicklungsmechanik as founded by Wilhelm Roux (Roux 1894), which is the forerunner of modern molecular developmental biology (see Mocek 1998). If C develops normally despite the turning off or ablation of B, then there is no causal link between the two steps. This experimental analysis allows both positive and negative reasoning. One can conclude that C depends on, or is caused by B if the experiment indicates this, or C does not require B for a proper expression if the ablation of B does not affect C.

The experimental view of development is necessarily finalistic (even more than the descriptive approach) since the result of the normal developmental process is the differentiated adult or reproductive stage, and any severe disturbance of this process causes failure to reach the expected final product. In other words, according to this view development causes the adult stage.

Closely related to the direct experimental manipulation of developmental processes is the use of mutants or otherwise disturbed developmental patterns. In this case those naturally occurring aberrations of normal development are used to infer mechanism and causal relationships between developmental steps. The difference with respect to experimental manipulation is that the reason for the aberration remains largely unknown. The best-known example for this approach is the use of mutations in the fruit fly *Drosophila melanogaster* in understanding segmentation (Nüsslein-Volhard and Wieschaus 1980).

How do these approaches relate to evolutionary change and evolutionary mechanisms? First of all one has to stress that there is no a-priori reason to generalise the outcome of these experiments. This is a fundamental problem of the model-organism centred approach. A single counterexample falsifies any generalisation. Furthermore, the experimental model-organism based approach is an indirect method for evolutionary inference. The experimental changes are just an analogy to evolutionary changes – developmental mechanisms are not evolutionary mechanisms. The new experimentally created phenotype is compared with naturally occurring differences between taxa and it is deduced that a similar developmental change has happened in the course of evolution. The problem is that the deduced evolutionary scenario might be correct or might be incorrect, since we simply do not know whether evolution carried out a corresponding experiment which led to evolutionary change. One has to discriminate between the functional proximate causes and the evolutionary ultimate causes or conditions for change (Mayr 1997, Sudhaus 2007). Experiments deal with proximate functional causes and not with ultimate evolutionary causes (Mayr 1997). Accordingly, there is an unbridgeable gap between experimental results and evolutionary changes.

The other aspect that has to be considered in the experimental framework is the fitness approach. According to this approach the performance of developmentally manipulated organisms or naturally occurring mutants is compared with that of normal embryos. This approach is often used to explain evolutionarily conservative characters as constraints caused by selective forces. An example is the investigation of Galis (1999) of the problem of why most mammals exhibit seven cervical vertebrae. Here malformed human and mouse embryos are studied showing that the occurrence of cervical ribs (i.e. a different number of true cervical vertebrae) is associated with a dramatic reduction in health and survival rate and fitness in general (Galis 1999). However, this view faces the problem that all living systems are functionally balanced and that mutations and experiments artificially disturb this functional balance. Accordingly, a reduced fitness may not be surprising and might not explain the evolutionary stability of characters in larger groups. Furthermore, the fact that there are mammals such as sloths and manatees that possess a different number of cervical vertebrae reveals that even in mammals this character has the freedom to evolve. Again, the proximate causes of functional disadvantages of onto-genetic change offer no direct explanation for the ultimate causes of evolutionary change. Obviously, the latter has a degree of freedom not realised by today's ontogenies.

In summary, experimental approaches cannot directly tell us about evolutionary change. They only show the possibility of what might have happened. They produce analogies to evolutionary processes but not direct evidence on them.

THE DESCRIPTIVE COMPARATIVE APPROACH

In the descriptive comparative approach the developmental sequence ABCDE of one or more species is compared with the corresponding (homologous) development of other species which, for instance, exhibit an ABXDE sequence. This comparison allows the conclusion that C (or X) is not the necessary prerequisite of D and the following stages and that the third stage is independent of the other stages, i.e. not causally related to them (Figure 8.2). This is true in both developmental directions, namely the relationship to B and to D because C (or X) is not caused by B and C (or X) does not cause D. Furthermore, this conclusion is independent of the direction of the evolutionary transformation. It is

<table>
<tr><td colspan="2">Observation</td></tr>
<tr><td>species</td><td>sequence</td></tr>
<tr><td>1</td><td>$A{\to}B{\to}C{\to}D{\to}E$</td></tr>
<tr><td>2</td><td>$A{\to}B{\to}C{\to}D{\to}E$</td></tr>
<tr><td>3</td><td>$A{\to}B{\to}X{\to}D{\to}E$</td></tr>
<tr><td colspan="2">Inference</td></tr>
<tr><td colspan="2">B does not necessarily cause C(X), D is not necessarily caused by C(X)</td></tr>
<tr><td colspan="2">Conclusion</td></tr>
<tr><td colspan="2">C(X) is an independent evolutionary step</td></tr>
<tr><td colspan="2">Ultimate causes</td></tr>
<tr><td colspan="2">evolutionary mechanisms</td></tr>
</table>

Figure 8.2 Reasoning in the framework of the descriptive comparative approach to development.

true in any instance independent of whether C or X is the plesiomorphic developmental character.

The comparative approach falsifies generalisations that two or more developmental steps are necessarily causally connected. At the same time this approach reveals that the developmental steps that differ are 'evolutionary developmental steps', i.e. developmental steps that have had the freedom to evolve independently of other developmental steps. This does not automatically imply an evolutionary direction or polarisation of the change but deals with the general possibility of change. Only a phylogenetic analysis allows assessment of the direction of change. Moreover, from comparison alone one cannot show that there was a causal nexus between developmental steps in the first place. If we assume that, based on a phylogenetic analysis, we 'know' that evolution went from C to X, then it can be concluded that the causal relationship between C and D has been broken. However, this can be concluded only if experiments have shown that C does indeed cause D in the normal development of that particular species or group of species (see above). It means that at some stage in evolution C has become an evolutionary developmental step which gained the potential for independent evolutionary alteration. In evolutionary terms, C is no longer the cause of D etc. and is not caused by B. If we concluded from experiments or based on comparisons that C is the necessary cause of D, this is falsified because the evolutionary transformation of C to X under conservation of the other stages reflects the independence of the subsequent stages from the occurrence of C.

The literature is full of examples for this phenomenon at all levels from genes to morphogenesis (e.g. Sewertzoff 1931, Remane 1952, Dohle 1989, Raff 1999, Richardson *et al.* 2001, Scholtz 2005). Hence, these examples will not be repeated here.

This result provides a different perspective on development. According to this view, development is not finalistic since every developmental step including the adult stage can be altered or lost in the course of evolution (see Scholtz 2005, and for arguments against the current general 'adultocentric' view see Minelli 2003). Accordingly, in evolutionary terms development does not necessarily cause the final stage. Furthermore, only negative causal conclusions ('falsifications') can be drawn, i.e. it can be concluded that C is independent of B and D or that C is not caused by B and does not cause D. In no case is it possible to make a positive inference that B depends on A even if this is the case in all observed examples.

The comparative approach can show that a developmental step is an independent evolutionary step. In other words, the developmental step is free to evolve independently of other steps of developmental sequence of which it is a part. The comparative approach shows, so to speak, experiments carried out by evolution (Dohle and Scholtz 1988, Scholtz and Dohle 1996). The absence of a necessary causal relation can be directly shown if the outcome of a developmental sequence is different.

In summary, comparison is a direct means to study changes in evolution because it deals directly with the differences and causes in developmental sequences.

THE EXPERIMENTAL COMPARATIVE APPROACH

As stated above, the descriptive comparative approach allows the conclusion that a developmental step is not the necessary condition or cause for subsequent developmental steps. If this is the case, an experimental comparative analysis does not necessarily add further support to this conclusion. An example for this is the role of the pair rule gene *even-skipped* (*eve*) in hexapods. From experiments in *Drosophila* it has been deduced that *eve* is a necessary prerequisite for proper segmentation because it regulates the segment polarity gene *engrailed* (*en*) which in turn establishes segmental boundaries (Fujioka *et al.* 1995). Comparative analyses have shown that in the grasshopper *Schistocerca americana*, *eve* is not expressed as a pair rule gene, but proper segmental *en* stripes are formed nevertheless (Patel *et al.* 1992). Accordingly, this result falsified the generalisation drawn from the *Drosophila* condition that *en* expression and thus segmentation is necessarily caused by *eve*. Liu and Kaufman (2005) tested the functional role of *eve* with RNAi experiments in the bug *Oncopeltus fasciatus*. This study revealed that indeed *eve* does not play a role as a pair rule gene in *Oncopeltus*, where nevertheless, as in *Schistocerca*, normal *en* stripes are generated (Liu and Kaufman 2005).

Comparative experiments add useful information if they reveal that in corresponding sequences ABCDE in two different species, in one case B is the cause of C and C is the cause of D, and in the other case B is not the cause of C and C is not the cause of D. This indicates that despite similar sequences of developmental stages, the underlying causal chains have been altered during evolution. C is an independent developmental step (Figure 8.3). Again, together with a phylogenetic analysis this allows the stepwise analysis of changes, including those that are not visible with a descriptive comparative approach alone

Observation

species	*sequence*
1	$A{\rightarrow}B{\rightarrow}C{\rightarrow}D{\rightarrow}E$
2	$A{\rightarrow}B{\rightarrow}C{\rightarrow}D{\rightarrow}E$

Manipulation

    species 1: if B^X then C^X then D^X

    species 2: if B^X then C, if C^X then D

Inference

    species 1: B is the necessary cause of C etc.

    species 2: B is not the cause of C, C is not the cause of D, C has the
        potential to evolve independently

Proximate causes

    set the starting point for ultimate causes

Figure 8.3 Reasoning in the framework of the experimental comparative approach to development. Experimentally manipulated steps and the resulting alterations of subsequent steps are indicated by an [x].

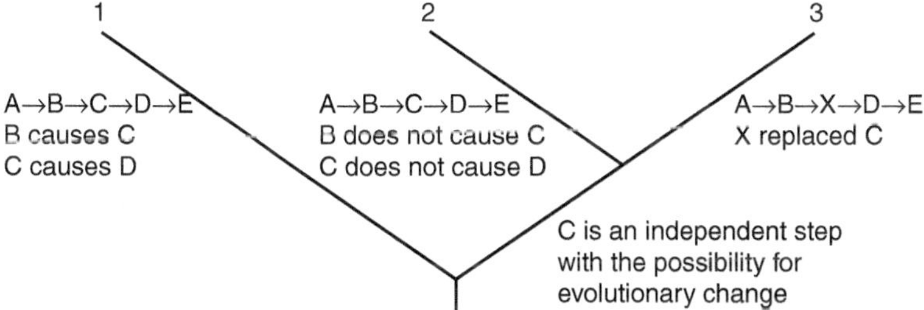

Figure 8.4 Putative evolutionary sequence showing the stepwise independence of developmental step C. In species 1 and 2 there is the same sequence of developmental steps. In species 1 there are causal relationships between all developmental steps in the sequence. In species 2 the developmental step C is not caused by the preceding developmental step B and does not cause the next step D. It can be concluded that C has the potential to evolve independently of B and D. In species 3 the developmental step C is replaced by X. The comparative analysis reveals that C is an independent evolutionary developmental step.

(Figure 8.4). That this is not just an academic exercise is shown by the comparative studies in nematodes by Schierenberg and co-workers. In *Caenorhabditis elegans* and representatives of the Plectidae, the EMS blastomere cannot be replaced by other blastomeres when it is artificially eliminated (Lahl *et al.* 2003); as a consequence, a gut is not formed. In another nematode *Acrobeloides nanus*, however, the ablated EMS cell is replaced by its neighbour cell C which instead gives rise to gut structures (Wiegner and Schierenberg 1999). This shows that the causal connection between the EMS blastomere and the gut is not in every case as strict as in *Caenorhabditis elegans* despite a great similarity in the initial cleavage patterns of these nematode species.

Through a comparative experimental approach combined with a phylogenetic analysis and reasoning it is possible to reconstruct the temporal sequence of the evolutionary independence (Figure 8.4). Depending on the evolutionary sequence (polarisation) this could mean either the establishment of a new causal relationship or the interruption of a pre-existing one. Clearly, this interpretation is only possible in a phylogenetic context.

DIFFERENCES AND CONTRADICTIONS BETWEEN EXPERIMENTAL AND COMPARATIVE APPROACHES TO DEVELOPMENT

According to what has been discussed above, experiments and comparisons can lead to contradictory conclusions about causal links between developmental steps. Contradiction occurs, for instance, if in the experimental approach C is the cause of D while, in contrast, the comparative analysis suggests that C is not necessarily the cause of D (see above). Both statements are correct but each one holds true only within its specific reference systems. The experimental approach deals with the normal development of living species or group of species; specifically, it deals with what have been called proximate or functional causes (Mayr 1997). As said above, one can interpret existing organisms as balanced, but a disturbance of this balance is only buffered to a certain degree and then the system collapses. This need not be the case in an evolutionary perspective under which seemingly closely interacting and dependent developmental aspects of one species can be decoupled in another species. The comparative approach is clearly evolutionary because it relates causal relationships to changes in time. Hence it deals with the evolutionary or ultimate causes (Mayr 1997). However, the evolutionary causes in this case are twofold: they are the reason for the transformation of the developmental step itself but, in addition, the changes in the causal connection which allows the transformation in the first place are caused by evolution. In particular, the latter aspect touches the domain of experimental developmental biology since it relates to causal connections between developmental steps. Hence, the experimental model organism approach leads to functional statements whereas the comparative approach leads to evolutionary as well as functional statements. The general principle of comparison is the potential falsification of a causal nexus – claiming that a step is not necessarily caused by another step. Furthermore, this falsification of causal relationships explains (as a precondition) the possibility of evolutionary change or transformation of developmental steps.

CONCLUSIONS

The comparative approach to developmental biology is a direct means to study evolution because it deals with evolutionary change. The comparison of developmental sequences reveals differences in developmental steps which falsify a generalisation of causal relationships between subsequent developmental steps. With this approach independent evolutionary developmental steps are identified. They also show what is allowed within the developmental system without destroying it. Accordingly, the evolutionary framework is set to address further questions. Comparative experiments may reveal a functional dissociation of subsequent developmental steps which could be a starting point for evolutionary dissociation. The phylogenetic framework allows the polarisation of the stated differences, thus indicating the direction of evolutionary changes, but it does not address the general notion of evolutionary independence. However, all this increases the precision of the analysis of what has happened and how evolutionary changes have occurred. This is the prerequisite for addressing important questions: why development changes, and whether the inclusion of development adds new aspects to evolutionary mechanisms.

ACKNOWLEDGEMENTS

The author thanks Alessandro Minelli and Giuseppe Fusco for the invitation to the Venice workshop on evolutionary developmental biology. Greg Edgecombe and two anonymous reviewers inspired important changes in the manuscript.

REFERENCES

Alberch, P. 1985. Problems with the interpretation of developmental sequences. *Systematic Zoology* **34**, 46–58.

Arthur, W. 2000. The concept of developmental reprogramming and the quest for an inclusive theory of evolutionary mechanisms. *Evolution & Development* **2**, 49–57.

Breidbach, O. & Ghiselin, M. T. 2007. Evolution and development: Past, present, and future. *Theory in Biosciences* **125**, 157–171.

Chen, J., Braun, A., Waloszek, D., Peng, Q.-Q. & Maas, A. 2004. Lower Cambrian yolk-pyramid embryos from southern Shaanxi, China. *Progress in Natural Science* **14**, 167–172.

Cracraft, J. 2005. Phylogeny and evo-devo: characters, phylogeny and historical analysis of the evolution of development. *Zoology* **108**, 345–356.

Deutsch, J. S. 2006. Introduction. Development and phylogeny of the arthropods: Darwin's legacy. *Development Genes & Evolution* **216**, 357–362.

Dohle, W. 1989. Zur Frage der Homologie ontogenetischer Muster. *Zoologische Beiträge (N.F.)* **32**, 355–389.

Dohle, W. & Scholtz, G. 1988. Clonal analysis of the crustacean segment: the discordance between genealogical and segmental borders. *Development* **104** (supplement), 147–160.

Donoghue, P. C. J., Bengtson, S., Dong, X. *et al.* 2006. Synchrotron X-ray tomographic microscopy of fossil embryos. *Nature* **442**, 680–683.

Fürst von Lieven, A. 2005. The embryonic moult in diplogastrids (Nematoda): homology of developmental stages and heterochrony as a prerequisite for morphological diversity. *Zoologischer Anzeiger* **244**, 79–91.

Fujioka, M., Jaynes, J. B. & Goto, T. 1995. Early *even-skipped* stripes act as morphogenetic gradients at the single cell level to establish *engrailed* expression. *Development* **121**, 4371–4382.

Galis, F. 1999. Why do almost all mammals have seven cervical vertebrae? Developmental constraints, *Hox* genes, and cancer. *Journal of Experimental Zoology B (Molecular and Developmental Evolution)* **285**, 19–26.

Hopwood, N. 2007. A history of normal plates, tables and stages in vertebrate embryology. *International Journal of Developmental Biology* **51**, 1–26.

Jenner, R. A. 2006. Unburdening evo-devo: ancestral attractions, model organisms, and basal baloney. *Development Genes & Evolution* **216**, 385–394.

Jonas, H. 1997. *Das Prinzip Leben – Ansätze zu einer philosophischen Biologie.* Frankfurt am Main: Suhrkamp.

Lahl, V., Halama, C. & Schierenberg, E. 2003. Comparative and experimental embryogenesis of Plectidae (Nematoda). *Development Genes & Evolution* **213**, 18–27.

Liu, P. Z. & Kaufman, T. C. 2005. *Even-skipped* is not a pair-rule gene but has segmental and gap-like functions in *Oncopeltus fasciatus*, an intermediate germband insect. *Development* **132**, 2081–2092.

Mahner, M. & Bunge, M. 1997. *Foundations of Biophilosophy.* Berlin: Springer.

Mayr, E. 1997. *This is Biology.* Cambridge, MA: Harvard University Press.

Minelli, A. 2003. *The Development of Animal Form.* Cambridge: Cambridge University Press.

Minelli, A., Brena, C., Deflorian, G., Maruzzo, D. & Fusco, G. 2006. From embryo to adult: beyond the conventional periodization of arthropod development. *Development, Genes & Evolution* **216**, 373–383.

Mitchell, S. D. 2006. Essay Review: Modularity – more than a buzzword? *Biological Theory* **1**, 98–101.

Mocek, R. 1998. *Die werdende Form.* Marburg: Basilisken-Presse.

Müller, W. A. 1996. From the Aristotelian soul to genetic and epigenetic information: the evolution of the modern concepts in developmental biology at the turn of the century. *International Journal of Developmental Biology* **40**, 21–26.

Nüsslein-Volhard, C. & Wieschaus, E. 1980. Mutations affecting segment number and polarity in *Drosophila*. *Nature* **287**, 795–801.

Patel, N. H., Ball, E. E. & Goodman, C. S. 1992. Changing role of *even-skipped* during the evolution of insect pattern formation. *Nature* **357**, 339–342.

Raff, R. A. 1999. Larval homologies and radical evolutionary changes in early development. In G. R. Bock & C. Cardew (eds.) *Homology (Novartis Foundation Symposium 222)*. Chichester: Wiley, pp. 110–121.

Remane, A. 1952. *Die Grundlagen des natürlichen Systems, der vergleichenden Anatomie und der Phylogenetik*. Leipzig: Geest und Portig.

Richardson, M. K., Jeffery, J. E., Coates, M. I. & Bininda-Emonds, O. R. P. 2001. Comparative methods in developmental biology. *Zoology* **104**, 278–283.

Richter, S. 2005. Homologies in phylogenetic analyses – concepts and tests. *Theory in Biosciences* **124**, 105–120.

Rieppel, O. C. 1985. Muster und Prozeß: Komplementarität im biologischen Denken. *Naturwissenschaften* **72**, 337–342.

Rieppel, O. C. 1988. *Fundamentals of Comparative Biology*. Basel: Birkhäuser.

Roux, W. 1894. Einleitung. *Archiv für Entwicklungsmechanik* **1**, 1–42.

Roux, W. 1907. Über die Verschiedenheit der Leistungen der deskriptiven und der experimentellen Forschungsmethode. *Archiv für Entwicklungsmechanik* **23**, 344–356.

Sander, K. 1996. On the causation of animal morphogenesis: concepts of German-speaking authors from Theodor Schwann (1839) to Richard Goldschmidt (1927). *International Journal of Developmental Biology* **40**, 7–20.

Schlosser, G. & Wagner, G. P. (eds) 2003. *Modularity in Development and Evolution*. Chicago: University of Chicago Press.

Scholtz, G. 2004. Baupläne versus ground patterns, phyla versus monophyla: aspects of patterns and processes in evolutionary developmental biology. In G. Scholtz (ed.) *Evolutionary Developmental Biology of Crustacea. Crustacean Issues 15*. Lisse: A.A. Balkema, pp. 3–16.

Scholtz, G. 2005. Homology and ontogeny: pattern and process in comparative developmental biology. *Theory in Biosciences* **124**, 121–143.

Scholtz, G. & Dohle, W. 1996. Cell lineage and cell fate in crustacean embryos: a comparative approach. *International Journal of Developmental Biology* **40**, 211–220.

Schopenhauer, A. 1847. *Ueber die vierfache Wurzel des Satzes vom zureichenden Grunde, 2. Auflage*. Frankfurt a. M.: J. Chr. Hermann.

Sewertzoff, A. N. 1931. *Morphologische Gesetzmäßigkeiten der Evolution*. Jena: Fischer.

Sudhaus, W. 2007. Die Notwendigkeit morphologischer Analysen zur Rekonstruktion der Stammesgeschichte. *Species, Phylogeny and Evolution* **1**, 17–32.

Wagner, G. P. 1996. Homologues, natural kinds and the evolution of modularity. *American Zoologist* **36**, 36–43.

Wiegner, O. & Schierenberg, E. 1999. Regulative development in a nematode embryo: a hierarchy of cell fate transformations. *Developmental Biology* **215**, 1–12.

Wuketits, F. M. 1981. *Biologie und Kausalität: Biologische Ansätze zur Kausalität, Determination und Freiheit*. Berlin: Paul Parey.

9

Evolution and development: towards a synthesis of macro- and micro-evolution with ecology

HANS ZAUNER AND RALF J. SOMMER

Until our population-based evolutionary theory can be reconciled with our homology-based evolutionary theory, we live without a true synthesis of evolutionary thought.

Amundson 2005: 249–250

Evolutionary theory is the philosophical backbone of biology. Interestingly, contemporary research in evolutionary biology involves several parallel lines of investigations that build on different philosophies and aim for different kinds of explanations and mechanisms. At its extreme, at least three independent research activities are actively promoted in evolutionary biology: neo-Darwinism with a population genetics research agenda analyses the evolution of populations by natural selection (Amundson 2005). Molecular phylogeny tries to reconstruct historical patterns and the phylogenetic relationship of organisms using cladistic approaches. And finally, comparative morphology, and more recent 'evo-devo' research, build on the evolution of ontogeny and try to show how modifications of development (ontogeny) result in evolutionary novelties (Valentine 2004, Kirschner and Gerhard 2005).

All of these agenda are actively propagated and they all consider themselves to follow the Darwinian logic. Surprisingly, however, there is hardly any cross-talk between these disciplines and even worse, these research fields ignore each other to a certain extent. Several authors have emphasised the different research strategies and philosophies in contemporary evolutionary biology, i.e. neo-Darwinism and

Evolving Pathways: Key Themes in Evolutionary Developmental Biology, ed. Alessandro Minelli and Giuseppe Fusco. Published by Cambridge University Press.
© Cambridge University Press 2008.

evolutionary developmental biology (Wilkins 2002, Amundson 2005). Despite these obvious problems and lack of interactions, we are in need of a true synthesis of evolutionary thought. And such a synthesis must include both population genetic and developmental thinking. In this context, homology could be an important concept. Homology has been a central aspect of comparative morphology and evolutionary developmental biology, but was long considered to be of limited importance in population genetics (Mayr 1966).

Here, we will summarise our attempts to bring these research fields together. Building on and believing in the 'case study' philosophy of experimental biology, we summarise our research on nematode macro- and micro-evolution of developmental processes, population genetics and nematode ecology. We hope that such studies will help – in the long run – to bridge barriers and to bring together different research strategies from evolutionary developmental biology over population genetics all the way to ecology.

THE SYSTEM: VULVA FORMATION IN THE NEMATODE
CAENORHABDITIS ELEGANS

The nematode *Caenorhabditis elegans* was introduced as a model system in genetics and developmental biology more than 30 years ago (Brenner 1974). Several features made this organism attractive to developmental biologists. For example, the adult organism consists of roughly 1000 somatic cells that develop through a fixed cell lineage, which is identical among the individuals of the species. Building on the complete description of post-embryonic development, genetic and molecular studies have helped to identify the molecular principles of ontogenetic processes in *C. elegans* (Sulston and Horvitz 1977). The formation of several organs and tissues during post-embryogenesis has been studied in great depth, providing a detailed understanding of developmental mechanisms.

One well-studied developmental process is the formation of the vulva, the egg-laying structure of nematode females and hermaphrodites (Sternberg 2005). The vulva is formed during the third larval stage by a subset of ventral epidermal precursor cells. Like all other nematodes studied to date, *C. elegans* have a total of 12 ventral epidermal cells, called P1.p to P12.p from anterior to posterior (Figure 9.1). Three of these cells, P(5–7).p are selected in wild-type animals to form vulval tissue. These cells adopt one of two alternative cell fates. P6.p generates eight progeny, which form the central part of the vulva, and this cell has been designated as the cell with the 1° (primary) fate (see below). P(5,7).p generate seven progeny each and form the anterior and posterior part

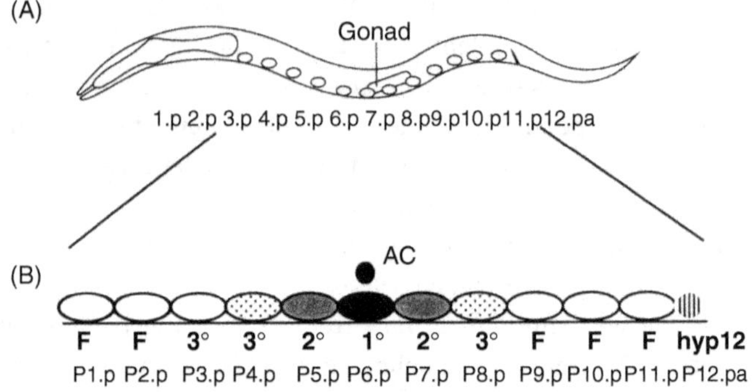

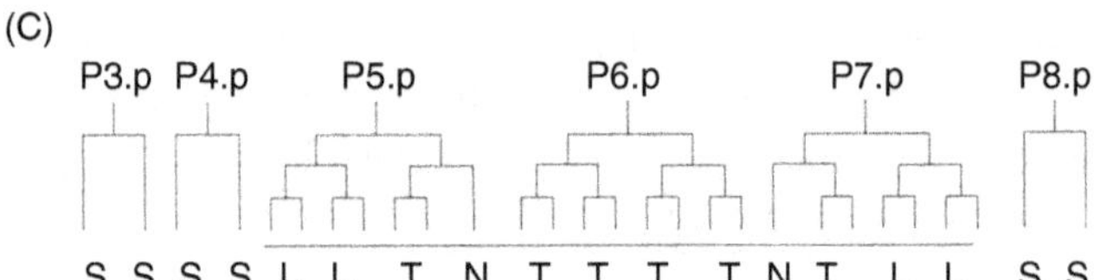

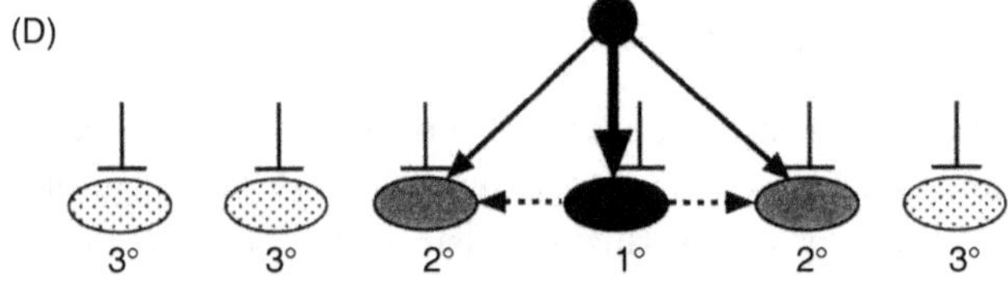

(E)

Manipulation	P3.p	P4.p	P5.p	P6.p	P7.p	P8.p
Wild-type	3°	3°	2°	1°	2°	3°
P5.p ablated	3°	2°	-	1°	2°	3°
P(5-7).p ablated	2°	1°	-	-	-	2°
AC ablated	3°	3°	3°	3°	3°	3°
Vulvaless mutant	3°	3°	3°	3°	3°	3°

Figure 9.1 Schematic summary of vulva formation in *C. elegans*. A,
During the L1 stage, the 12 ventral epidermal cells P(1–12).p are equally
distributed between pharynx and rectum. B, P(1,2,9–11).p fuse with the
hypodermal syncytium *hyp7* (F, white ovals). P(3–8).p form the vulva
equivalence group and adopt one of three alternative cell fates. P6.p has a
1° fate (black oval), and P(5,7).p have a 2° fate (grey ovals). P(3,4,8).p have a
3° fate and remain epidermal (dotted ovals). The anchor cell (AC, black
circle) provides an inductive signal for vulva formation. C, Cell lineage
pattern of the vulval precursor cells. P(3,4,8).p divide once and then fuse

of the vulva, respectively. The fate of these two cells has been designated as 2° (secondary). Two features made *C. elegans* vulva formation of special interest to developmental biologists. First, there are three more cells in the ventral epidermis that have the developmental competence to participate in vulva formation (Figure 9.1E). P(3,4,8).p can form vulval tissue after P(5–7).p, or a subset of these cells, have been ablated by laser microbeam irradiation early in development. In wild-type animals, P(3,4,8).p remain epidermal and adopt the 3° (tertiary) fate. Second, vulva formation requires inductive input by the gonadal anchor cell (AC). If the AC is ablated at birth, P(5–7).p remain epidermal and have a 3° fate, like their neighbouring cells P(3,4,8).p (Figure 9.1E).

Genetic and molecular studies in the past two decades have revealed the mechanistic basis of vulva formation in *C. elegans* (for review see Sternberg 2005, Eisenmann 2005). The inductive signal is provided by the AC and encodes an EGF-like molecule. Within the vulval precursor cells (VPCs), this signal is transmitted by EGF/RAS/MAPK signalling. A Wnt signalling pathway plays a redundant role in vulva formation. Cross-talk among the VPCs requires a Notch-like signalling system. After more than 20 years of active research, *C. elegans* vulva formation represents a paradigm for the molecular understanding of the interaction of signalling pathways in animal development and provides a platform for evolutionary developmental biology and comparative studies of vulva formation between different nematode species.

A SATELLITE ORGANISM IN EVOLUTIONARY DEVELOPMENTAL BIOLOGY: THE NEMATODE *PRISTIONCHUS PACIFICUS*

Comparative studies on vulva formation were initiated more than a decade ago and evolutionary changes of nearly all aspects of vulva

Figure 9.1 (cont.) with *hyp7* (S). P(5,7).p generate seven progeny each. The first two cell divisions occur along the antero-posterior axis, the third division can be longitudinal (L), or transversal (T), or can be absent (N). P6.p generates eight progeny. D. Schematic summary of signalling interactions during vulva formation in *C. elegans*. An inductive EGF-like signal originates from the AC (black arrows). P6.p signals its neighbours to adopt a 2° fate via 'lateral signalling' (dotted arrows). Negative signalling (bars) prevents inappropriate vulva differentiation. E, Summary of cell ablation experiments. After ablation of P5.p, P4.p adopts a 2° fate and forms part of the vulva. After ablation of P(5–7).p, P(3,4,8).p can form a functional vulva. After ablation of the AC, all precursor cells adopt a 3° fate. Vulvaless mutants have a similar phenotype as AC-ablated animals.

formation have been identified (Sommer and Sternberg 1994, 1995, 1996, Félix and Sternberg 1998, Félix *et al.* 2000). Basically, changes involve (1) the size of the vulva equivalence group, (2) the number of VPCs participating in vulva formation, (3) involvement of the gonad/AC in vulva induction, (4) novel signalling activities and (5) cell lineage alterations (for review see Sommer 1997). For example, in species with a posterior vulva, such as *Teratorhabditis palmarum* and *Mesorhabditis* sp. PS1179, the vulva is made from P(5–7).p, as in *C. elegans*. However, these cells migrate towards the posterior prior to differentiation and form a vulva even in the absence of the gonad (Sommer and Sternberg 1994). Further ablation experiments revealed that the VPCs in these species have strong autonomous properties to form vulva tissue, but the genetic program involved in vulva formation has not been investigated.

As a result of comparative studies of more than 50 nematode species, a few of them were selected for more detailed comparisons. We have developed the diplogastrid *Pristionchus pacificus* as a satellite organism in evolutionary developmental biology (Sommer *et al.* 1996, Sommer and Sternberg 1996, Hong and Sommer 2006). *P. pacificus* shows many substantial differences from *C. elegans* in the development of the vulva and at the same time fulfils many requirements for a model organism: *P. pacificus* is a hermaphroditic species that can feed on *E. coli* and has a 3–4 day generation time (20 ⏃). Large-scale mutagenesis screens have been performed in *P. pacificus* and many mutants defective in sex determination, vulva and gonad formation have been isolated (Eizinger and Sommer 1997, Pires-daSilva and Sommer 2004, Rudel *et al.* 2005). More recently, a genomic initiative including the generation of a genetic linkage map and a physical map has complemented the developmental and genetic studies (Srinivasan *et al.* 2002, 2003). A whole genome-sequencing project is currently ongoing and should result in a draft of the complete genome sequence in 2007 (www.pristionchus.org).

MACRO-EVOLUTION: VULVA FORMATION DIFFERS STRONGLY
BETWEEN *P. PACIFICUS* AND *C. ELEGANS*

Detailed experimental, genetic and molecular studies identified at least four major differences between vulva development in *P. pacificus* and *C. elegans*. First, non-vulval cells in the anterior and posterior body region fuse with the surrounding hypodermis in *C. elegans*, but die of programmed cell death in *P. pacificus* (Figure 9.2) (Eizinger and Sommer 1997). Second, vulva induction relies on a continuous interaction

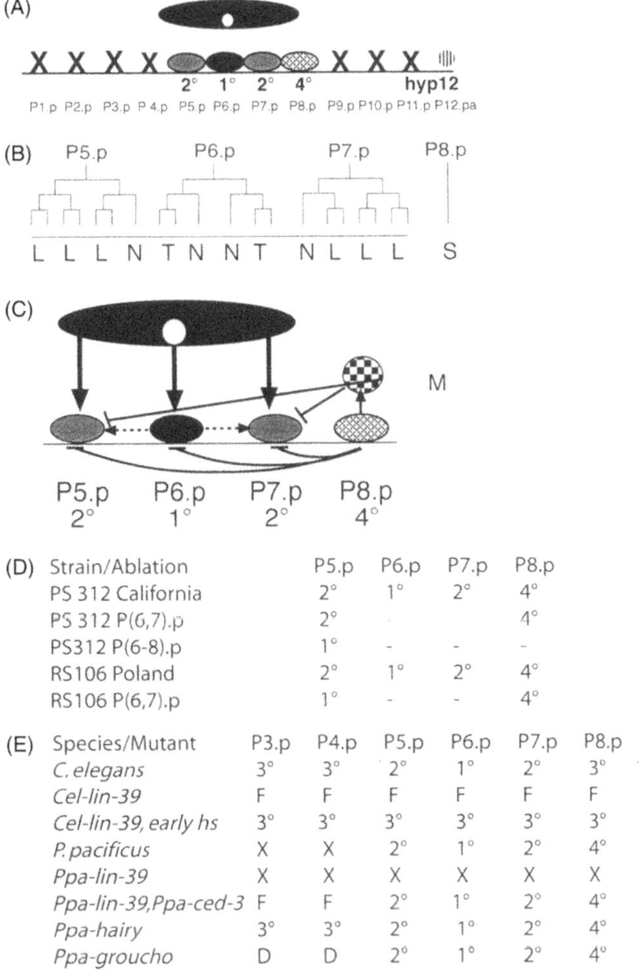

(D) Strain/Ablation	P5.p	P6.p	P7.p	P8.p
PS 312 California	2°	1°	2°	4°
PS 312 P(6,7).p	2°			4°
PS312 P(6-8).p	1°	–	–	–
RS106 Poland	2°	1°	2°	4°
RS106 P(6,7).p	1°	–	–	4°

(E) Species/Mutant	P3.p	P4.p	P5.p	P6.p	P7.p	P8.p
C. elegans	3°	3°	2°	1°	2°	3°
Cel-lin-39	F	F	F	F	F	F
Cel-lin-39, early hs	3°	3°	3°	3°	3°	3°
P. pacificus	X	X	2°	1°	2°	4°
Ppa-lin-39	X	X	X	X	X	X
Ppa-lin-39, Ppa-ced-3	F	F	2°	1°	2°	4°
Ppa-hairy	3°	3°	2°	1°	2°	4°
Ppa-groucho	D	D	2°	1°	2°	4°

Figure 9.2 Schematic summary of vulva formation in *P. pacificus*. A, Cell fate specification of the 12 ventral epidermal cells. P(1–4,9–11).p die of programmed cell death during late embryogenesis. P(5–7).p form the vulva with a 2°-1°-2° pattern. P8.p (shaded oval) has a special fate designated as 4°. B, Cell lineage pattern of the vulval precursor cells. P(5,7).p generate seven progeny each, whereas P6.p generates six progeny. P8.p does not divide and finally fuses with *hyp7*. C, Model for cell–cell interactions during vulva development in *P. pacificus*. P8.p provides lateral inhibition to P(5,7).p, mediated by the mesoblast M (chequered circle). Lateral inhibition influences the 1° vs. 2° cell fate decision of P(5,7).p. P8.p also provides a negative signal (black bars), which influences the vulva vs. non-vulva cell fate decision. For clarity, negative signalling is shown here as an interaction between P8.p and P(5–7).p. It is possible that indirect interactions involving

(Cont.)

between several cells of the somatic gonad and the vulval precursor cells (VPCs) rather than an interaction between the VPCs and the AC as in *C. elegans* (Sigrist and Sommer 1999). Third, P8.p represents a novel cell type in *P. pacificus* and is involved in multiple cell–cell interactions during vulva formation, not known from *C. elegans* or other nematodes (Jungblut and Sommer 2000). For instance, P8.p inhibits P5.p and P7.p to adopt the 1° cell fate, a process called 'lateral inhibition'. Additional experiments indicated that the mesoblast M is also involved in lateral inhibition and that P8.p and M interact to inhibit P(5,7).p from taking a 1° fate (Jungblut and Sommer 2000). No interaction between P8.p and the M cell has been observed in *C. elegans*. Finally, a negative signalling system provided by the VPCs themselves prohibits precocious vulva differentiation and counteracts vulva induction by the somatic gonad (Jungblut and Sommer 2000, Zheng *et al.* 2005).

P. *pacificus* vulva defective mutants have been isolated in large-scale mutagenesis screens and the phenotypes of mutations have

---

Figure 9.2   (cont.) other cells could exist. Inductive signalling from the somatic gonad is a continuous process (black arrows). Lateral signalling occurs between P6.p and P8.p (not indicated) and perhaps also between P6.p and P(5,7).p (dotted arrows). D, Summary of cell ablation experiments. After ablation of P(6,7).p, P5.p adopts a 2° fate in the presence of P8.p. After ablation of P(6–8).p, P5.p predominantly has a 1° fate, indicating that the presence of P8.p influences the cell fate decision of P5.p. E, Comparison of the function of the homeotic gene *lin-39* between *C. elegans* and *P. pacificus*. In *C. elegans lin-39* mutant animals, the central body region shows a homeotic transformation, and P(3–8).p fuse with the surrounding hypodermis like their anterior and posterior lineage homologues. *P. pacificus lin-39* mutant animals also show a homeotic transformation and P(5–8).p die of programmed cell death. If the first function of *Cel-lin-39* is rescued by providing *lin-39* under the control of a heat-shock promoter, P(3–8).p have a 3° fate because *Cel-lin-39* is required during vulva induction. The first function of *Ppa-lin-39* can be overcome by generating a *Ppa-lin-39 Ppa-ced-3* double mutant. *Ppa*-CED-3 is a general regulator of programmed cell death and mutations in *Ppa-ced-3* result in animals unable to undergo apoptosis. Such double mutants form a normal vulva indicating that in contrast to *Cel-lin-39*, *Ppa-lin-39* is not required during vulva induction. The size of the vulva equivalence group in *P. pacificus* is regulated by two genes, *Ppa-hairy* and *Ppa-groucho*. Mutations in these genes result in the survival of P(3,4).p, but not P(1,2,9–11).p. Biochemical studies have shown that *Ppa*-HAIRY and *Ppa*-GROUCHO form a heterodimer that regulates the activity of the Hox gene *Ppa-lin-39*. See text for details. X, programmed cell death; F, cell fusion; D, ectopic vulva differentiation.

helped in elucidating the molecular mechanisms of evolutionary change. So far, major differences involve the function of the Hox genes *Ppa-lin-39* and *Ppa-mab-5*, the basic helix-loop-helix (bHLH) gene *Ppa-hairy* and the Wnt signalling genes *Ppa-lin-17/Frizzled* and *Ppa-groucho* (Eizinger and Sommer 1997, Jungblut and Sommer 1998, Sommer *et al.* 1998, Zheng *et al.* 2005, Schlager *et al.* 2006). In the following, we highlight three of these functional differences.

Vulva formation is initiated in *P. pacificus* and *C. elegans* by the formation of the vulva equivalence group (VEG). Genetic studies in the 1990s have shown that both organisms use the Hox gene *lin-39*, which is most similar to the *Drosophila* genes *Scr* and *Dfd*, for the establishment of the VEG (Clark *et al.* 1993, Wang *et al.* 1993, Eizinger and Sommer 1997). Mutations in *Cel-lin-39* and *Ppa-lin-39* result in the VPCs adopting the fate of their anterior and posterior lineage homologues. That is, in *C. elegans* these cells fuse with the surrounding hypodermis in the L1 stage and in *P. pacificus* they undergo programmed cell death (Figure 9.2E). However, later studies indicated substantial differences with regard to an additional function of LIN-39 in vulva formation in *C. elegans*. It has been shown that *Cel-lin-39* is required during the L3 stage in response to EGF/RAS signalling, and mutants in which the first function of *Cel-lin-39* has been rescued are induction vulvaless (Maloof and Kenyon 1998). Thus, *Cel*-LIN-39 is actively involved in vulva induction. In contrast, *Ppa-lin-39; Ppa-ced-3* double mutants, in which cell death no longer occurs, form a normal vulva indicating that *Ppa*-LIN-39 is dispensable for vulva induction (Figure 9.2E) (Sommer *et al.* 1998). This observation represented the first fundamental difference in a function of a homologous developmental control gene in nematode development between different species (Eizinger *et al.* 1999).

More recent studies have revealed major differences in the signalling systems acting during vulva formation of the two species. The first mutation with a multivulva phenotype in *P. pacificus* was originally isolated as *ped-7*. Positional cloning identified *ped-7* to encode *Ppa-lin-17*, one of the Frizzled-type Wnt receptors in nematodes (Zheng *et al.* 2005). *Ppa-lin-17/Frizzled* shows conserved functions during vulva formation (i.e. the regulation of cell lineage symmetry of the posterior cell P7.p) as well as divergent functions: in *P. pacificus*, *Ppa-lin-17/Fz* is part of the negative signalling system that counteracts vulva formation. In contrast, no such negative signalling function is known from *C. elegans*.

The most substantial differences so far have been identified by the molecular cloning of two genes that regulate the size of the VEG in *P.*

pacificus. Mutations in two genes result in the survival of P(3,4).p and expand the size of the VEG in the anterior region. These mutations have an atavistic phenotype, resembling nearly exactly the pattern known from *C. elegans* (Figure 9.2E). Molecular cloning revealed that one of these two genes encodes for a bHLH-like protein of the HAIRY-type (Schlager *et al.* 2006). Further studies revealed that *Ppa*-HAIRY forms a heterodimer with the product of the second gene, *Ppa-groucho*, which when mutated also causes the survival of P(3,4).p. *Ppa*-HAIRY and *Ppa*-GROUCHO form a heterodimer and *Ppa*-HAIRY binds to HES-binding sites in the *Ppa-lin-39* promoter. *Ppa-lin-39* is up-regulated in *Ppa-hairy* mutants, further indicating that *Ppa*-HAIRY and *Ppa*-GROUCHO restrict the size of the VEG in *P. pacificus* by repressing the Hox gene *Ppa-lin-39*. Interestingly, the *Ppa-hairy* gene does not have a 1:1 orthologue in *C. elegans*. Thus, a *Ppa*-HAIRY and *Ppa*-GROUCHO module that is absent in *C. elegans* regulates the size of the VEG in *P. pacificus* (Schlager *et al.* 2006). Phylogenetic considerations strongly suggest that the pattern observed in *P. pacificus* and all other diplogasterid nematodes represents a derived character and that the HAIRY/GROUCHO module was involved in the evolutionary restriction of the VEG.

ECOLOGY

Ultimately, the development of organisms has to be studied in the context of the natural environment, and several recent studies have pointed towards the importance of ecology in developmental biology (Gilbert 2001, Dusheck 2002, Herrmann *et al.* 2006a). In general, little is known about the ecology of model organisms in developmental biology. This holds true also for the nematode laboratory organisms, *P. pacificus* and *C. elegans*. For example, the environmental niche of the model organism *C. elegans* is largely unknown and it is only recently that studies have indicated that *C. elegans* occurs predominantly in compost heaps (Barrière and Félix 2005, Kiontke and Sudhaus 2005). We have recently shown that nematodes of the genus *Pristionchus* live in close association with scarab beetles and the Colorado potato beetle in Western Europe and the United States (Herrmann *et al.* 2006a,b). Intensive samplings in Europe in 2004 and 2005 generated 371 isolates that fell into six species, most of which are morphologically indistinguishable from one another. The two hermaphroditic species *P. entomophagus* and *P. maupasi* accounted for 226 of these 371 (60%) isolates and occurred on dung beetles and cockchafers, respectively.

A relationship between nematodes and their hosts, such as seen in the case of *Pristionchus* and scarab beetles, has long been known as 'necromeny' (Sudhaus 1976). In general, existing types of nematode–insect associations can be divided into three types (Sudhaus 1976). In phoresy, nematodes attach themselves to passing insects/invertebrates for transportation, an interaction that is usually not species-specific (Kiontke and Sudhaus 2005). For example, *C. elegans* can occasionally exhibit phoretic behaviour through unspecific associations with invertebrates. In necromeny, nematodes not only use insects for transport, but also subsequently as food. This association is much more specific than a phoretic relationship. Finally, entomopathogenic nematodes are obligate parasites preying on insects. It has been suggested that necromeny is an intermediate step preceding true parasitism. In this context it will be interesting to study the association of *Pristionchus* with beetles in more detail to learn more about the pre-adaptations towards parasitism (Herrmann *et al.* 2006a).

The *Pristionchus*–beetle association represents a unique opportunity to couple research in evolutionary developmental biology with ecology. Some scarab beetles and the Colorado potato beetles can be cultured in the laboratory, and thus the nematode–beetle interaction can be studied under laboratory conditions. Such studies will be of importance for understanding many aspects of the biology of the nematode, e.g. the genetic regulation of dauer formation and olfaction. Ultimately, knowledge of the natural environment of *Pristionchus* and the development of nematode cultures under similar conditions will also be of importance for the analysis of the evolution of developmental processes, because any adaptation (like the developmental adaptations discussed above) results from environmental conditions to which organisms are exposed.

MICRO-EVOLUTION

In the long run, the comparison of developmental processes between *P. pacificus* and *C. elegans* will benefit from knowledge of the natural environment of these nematodes. However, micro-evolutionary studies are necessary to bridge macro-evolution of development (as represented by the comparison between *P. pacificus* and *C. elegans*) and ecology. To this end we have initiated micro-evolutionary studies to better understand, first, the population structure of *P. pacificus* and second, how developmental characters change among wild isolates.

Sampling efforts by various nematologists throughout the world and detailed field studies as described above have revealed many strains of *P. pacificus*. However, in contrast to many other *Pristionchus* species for which we could identify specific beetle hosts, no strong beetle association has yet been observed for *P. pacificus*. Instead, sampling efforts during the past 10 years have identified more than 15 strains of *P. pacificus* from several countries in Africa, Asia, Europe and North America (Srinivasan *et al.* 2001, Zauner *et al.* 2007). However, *P. pacificus* collections were rare in all samples from Europe, North America and South Africa. At the same time, the available strains indicate that *P. pacificus* is a cosmopolitan species and interestingly, *P. pacificus* is currently the only species in the genus *Pristionchus* with such a worldwide distribution (Figure 9.3).

The availability of multiple strains of *P. pacificus* from different countries and continents allows for the initiation of population genetic investigations. Micro-evolutionary studies can help address many questions of importance for a comprehensive understanding of the biology of this organism. Are the strains of *P. pacificus* in permanent genetic contact with one another? Do males occur in nature, and is out-crossing observed in the natural environment? We have started to use mitochondrial sequence data for various *P. pacificus* strains to address these and related questions. Mitochondrial sequence analysis of more than 3300 bp has shown that the molecular diversity between *P. pacificus* strains is much higher than between *C. elegans* strains (Zauner *et al.* 2007). At the same time, single stranded conformational polymorphism analysis and amplified fragment length polymorphism (AFLP) studies of *P. pacificus* strains reveal genome-wide linkage disequilibrium indicating

Figure 9.3 *Pristionchus pacificus* is a cosmopolitan species.

low out-crossing rates. The highly diverse molecular signatures of *P. pacificus* strains hint at a long-lasting colonisation of new habitats, likely to be in association with a beetle host.

Do the different *P. pacificus* strains also show differences in their development, i.e. the formation of the vulva? Some of the *P. pacificus* strains have been compared with one another and several differences in developmental properties have been observed. The strain of *P. pacificus* that serves as the basis for genetic and molecular studies is PS312 from Pasadena (California, USA). Mitochondrial sequence data revealed that this strain is untypical for American isolates of the species. While all other North American strains fall into a distinct mitochondrial clade, PS312 from California is more closely related to strains from Poland (RS106) and Hawaii (JU138). Surprisingly, these molecular signatures do not correlate with vulva developmental characters. The strongest difference in vulva development was observed in the strain from Poland RS106 (Srinivasan *et al.* 2001). Specifically, the developmental competence of individual VPCs differs between the strain from Poland and most of the other investigated strains. For example, if P(6,7).p are ablated, P5.p will adopt a 2° fate as long as P8.p is present and prevents P5.p from taking a 1° fate (Figure 9.2D). This result is observed in the majority of P(6,7).p-ablated animals for the laboratory strain PS312 and most other strains of *P. pacificus*. However, after similar ablations in RS106 from Poland, P5.p can have a 1° fate in the majority of the treated worms (Srinivasan *et al.* 2001). Considering that AFLP analysis and mitochondrial haplotype analysis have revealed that PS312 and RS106 are nearly identical at the molecular level, these developmental differences suggest that small molecular alterations (mutations), which have yet to be identified, can account for developmental novelties. Building on these promising observations, we hope to create a micro-evolutionary analysis of vulva development between *P. pacificus* strains and closely related *Pristionchus* species.

CONCLUSIONS

We have summarised here three different lines of investigation that try to use the nematode *P. pacificus* and related species of the same genus for studies in different areas of evolutionary biology. Genetic and developmental studies of vulva development in *P. pacificus*, when compared with the knowledge available from *C. elegans*, show how developmental processes change at a macro-evolutionary level and how genetic alterations can create developmental novelty. Studies on the ecology of

Pristionchus nematodes provide the first indication of what the natural environment of these organisms looks like. In the long run, such investigations can help to provide a basis for an understanding of those adaptations that in response to the environment have shaped morphological and developmental patterns that differ between species. And finally, micro-evolutionary comparisons of different *P. pacificus* strains might reveal the actual molecular nature of developmental 'mutations' that ultimately result in novel patterns and structures. We envisage coupling genetic analysis to environmental studies by testing, for example, vulval patterning mutants for their effect in the beetle environment rather than under agar-plate (laboratory) conditions. Also, we plan to test different isolates from the wild for their genetic and developmental variations. We hope that by combining macroevolutionary, micro-evolutionary and ecological studies, we can contribute to a much-needed synthesis in evolutionary biology.

REFERENCES

Amundson, R. 2005. *The Changing Role of the Embryo in Evolutionary Thought.* Cambridge: Cambridge University Press.

Barrière, A. & Félix, M. A. 2005. High local genetic diversity and low outcrossing rate in *Caenorhabditis elegans* natural populations. *Current Biology* **15**, 1176–1184.

Brenner, S. 1974. The genetics of *Caenorhabditis elegans*. *Genetics* **77**, 71–94.

Clark, S. G., Chisholm, A. D. & Horvitz, H. R. 1993. Control of cell fates in the central body region of *C. elegans* by the homeobox gene *lin-39*. *Cell* **74**, 43–55.

Dusheck, J. 2002. It's ecology, stupid! *Nature* **418**, 578–579.

Eisenmann, D. 2005. Wnt signaling. In *WormBook*; The *C. elegans* Research Community, WormBook, doi/10.1895/wormbook.1.37.1, www.wormbook.org.

Eizinger, A., Jungblut, B. & Sommer, R. J. 1999. Evolutionary change in the functional specificity of genes. *Trends in Genetics* **15**, 197–202.

Eizinger, A. & Sommer, R. J. 1997. The homeotic gene *lin-39* and the evolution of nematode epidermal cell fates. *Science* **278**, 452–455.

Félix, M. A., De Ley, P., Sommer, R. J. *et al.* 2000. Evolution of vulva development in the Cephalobina (Nematoda). *Developmental Biology* **221**, 68–86.

Félix, M. A. & Sternberg, P. W. 1998. A gonad-derived survival signal for vulval precursor cells in two nematode species. *Current Biology* **8**, 287–290.

Gilbert, S. F. 2001. Ecological developmental biology: developmental biology meets the real world. *Developmental Biology* **233**, 1–12.

Herrmann, M., Mayer, E. W. & Sommer, R. J. 2006a. Nematodes of the genus *Pristionchus* are closely associated with scarab beetles and the Colorado potato beetle in western Europe. *Zoology* **109**, 96–108.

Herrmann, M., Mayer, E. W. & Sommer, R. J. 2006b. Sex, bugs and Haldane's rule: the nematodes genus *Pristionchus* in the United States. *Frontiers in Zoology* **3**, 14.

Hong, R. L. & Sommer, R. J. 2006. *Pristionchus pacificus*: a well-rounded nematode. *BioEssays* **28**, 651–659.

Jungblut, B. & Sommer, R. J. 1998. The *Pristionchus pacificus mab-5* gene is involved in the regulation of ventral epidermal cell fates. *Current Biology* **8**, 775–778.

Jungblut, B. & Sommer, R. J. 2000. Novel cell-cell interactions during vulva development in *Pristionchus pacificus*. *Development* **127**, 3295–3303.

Kiontke, K. & Sudhaus, W. 2005. Ecology of *Caenorhabditis* species. In *WormBook*; The *C. elegans* Research Community, WormBook, doi/10.1895/wormbook.1.37.1, www.wormbook.org.

Kirschner, M. W. & Gerhart, J. C. 2005. *The Plausibility of Life*. New Haven and London: Yale University Press.

Maloof, J. N. & Kenyon, C. 1998. The HOX gene *lin-39* is required during *C. elegans* vulval induction to select the outcome of Ras signaling. *Development* **125**, 181–190.

Mayr, E. 1966. *Animal Species and Evolution*. Cambridge: Harvard University Press.

Pires-daSilva, A. & Sommer, R. J. 2004. Conservation of the global sex determination gene *tra-1* in distantly related nematodes. *Genes & Development* **18**, 1198–1208.

Rudel, D., Riebesell, M. & Sommer, R. J. 2005. Gonadogenesis in *Pristionchus pacificus* and organ evolution: development, adult morphology and cell/cell interactions in the hermaphrodite gonad. *Developmental Biology* **277**, 200–221.

Schlager, B., Röseler, W., Zheng, M., Gutierrez, A. & Sommer, R. J. 2006. HAIRY-like transcription factors and the evolution of the nematode vulva equivalence group. *Current Biology* **16**, 1386–1394.

Sigrist, C. B. & Sommer, R. J. 1999. Vulva formation in *Pristionchus pacificus* relies on continuous gonadal induction. *Development, Genes & Evolution* **209**, 451–459.

Sommer, R. J. 1997. Evolution and development: the nematode vulva as a case study. *BioEssays* **19**, 225–231.

Sommer, R. J., Carta, L. K., Kim, S. Y. & Sternberg, P. W. 1996. Morphological, genetic and molecular description of *Pristionchus pacificus* sp. n. (Nematoda, Diplogastridae). *Fundamental and Applied Nematology* **19**, 511–521.

Sommer, R. J., Eizinger, A., Lee, K. Z. *et al.* 1998. The *Pristionchus* HOX gene *Ppa-lin-39* inhibits programmed cell death to specify the vulva equivalence group and is not required during vulva induction. *Development* **125**, 3865–3873.

Sommer, R. J. & Sternberg, P. W. 1994. Changes of induction and competence during the evolution of vulva development in nematodes. *Science* **265**, 114–118.

Sommer, R. J. & Sternberg, P. W. 1995. Evolution of cell lineage and pattern formation in the vulva equivalence group of rhabditid nematodes. *Developmental Biology* **167**, 61–74.

Sommer, R. J. & Sternberg, P. W. 1996. Apoptosis limits the size of the vulval equivalence group in *Pristionchus pacificus*: a genetic analysis. *Current Biology* **6**, 52–59.

Srinivasan, J., Pires-daSilva, A., Gutierrez, A. *et al.* 2001. Microevolutionary analysis of the nematode genus *Pristionchus* suggests a recent evolution of redundant developmental mechanisms during vulva formation. *Evolution & Development* **3**, 229–240.

Srinivasan, J., Sinz, W., Jesse, T. *et al.* 2003. An integrated physical and genetic map of the nematode *Pristionchus pacificus*. *Molecular Genetics and Genomics* **269**, 715–722.

Srinivasan, J., Sinz, W., Lanz, C. *et al.* 2002. A BAC-based genetic linkage map of the nematode *Pristionchus pacificus*. *Genetics* **162**, 129–134.

Sternberg, P. W. 2005. Vulval development. In *WormBook*; The *C. elegans* Research Community, WormBook, doi/10.1895/wormbook.1.37.1, www.wormbook.org.

Sudhaus, W. 1976. Vergleichende Untersuchungen zur Phylogenie, Systematik, Ökologie, Biologie und Ethologie der Rhabditidae (Nematoda). *Zoologica* **43**, 1–229.

Sulston, J. E. & Horvitz, H. R. 1977. Postembryonic cell lineages of the nematode *Caenorhabditis elegans*. *Developmental Biology* **56**, 110–156.

Valentine, J. W. 2004. *On the Origin of Phyla*. Chicago and London: Chicago University Press.

Wang, B. B., Müller-Immerglück, M. M., Austin, J. *et al.* 1993. A homeotic gene cluster patterns the anteroposterior body axis of *C. elegans*. *Cell* **74**, 29–42.

Wilkins, A. S. 2002. *The Evolution of Developmental Pathways*. Sunderland: Sinauer Associates Inc.

Zauner, H., Mayer, W., Herrmann, M. *et al.* 2007. Distinct patterns of genetic variation in *Pristionchus pacificus* and *Caenorhabditis elegans*, two partially selfing nematodes with cosmopolitan distribution. *Molecular Ecology* **16**, 1267–1280.

Zheng, M., Messerschmidt, D., Jungblut, B. & Sommer, R. J. 2005. Conservation and diversification of Wnt signaling function during the evolution of nematode vulva development. *Nature Genetics* **37**, 300–304.

10

When is a Hox gene not a Hox gene? The importance of gene nomenclature

DAVID E. K. FERRIER

GENE CLASSIFICATION IS AN ESSENTIAL PRECURSOR
TO EVO-DEVO

A sensible classification of developmental control genes and an understanding of their phylogeny are essential to any endeavour of molecular evolutionary developmental biology (evo-devo) or comparative genomics, since it is crucial that the structure, expression and function of orthologous genes are being compared between taxa. This is particularly true for the homeobox genes, for which there are confusing and conflicting names and classifications that hinder our investigation and understanding of their evolution and their role in animal evo-devo (I will restrict myself here to consideration of animal homeobox genes). Since these genes are central components of most developmental processes, are important indicators of major transitions in animal genome evolution, and are often found to be targets and/or agents of the evolution of development, then we must continue to improve and coordinate our classifications of these genes as more data become available from a greater array of taxa in this age of genomics.

CONVENTIONS

Animal homeobox genes can be divided, on the basis of their sequence similarities, into two major classes (ANTP and PRD) along with several minor classes (TALE, LIM, POU, ZF, cut, prox, HNF and SIX; Bürglin 2005, Edvardsen *et al.* 2005, Holland and Takahashi 2005). It is in the

Table 10.1 *Homeobox gene families in the ANTP-class.*

Human names are taken from the HUGO nomenclature website (www.gene.ucl.ac.uk/nomenclature/), with old names and synonyms given in brackets. *Drosophila* gene names are taken from Flybase (http://flybase.bio.indiana.edu/), along with synonyms. The 'Other animals' column provides the names of orthologues from other species that are relevant to the naming of the family. The prefix 'Amphi' denotes genes from amphioxus (*Branchiostoma floridae*), CapI = *Capitella I* (polychaete worm), Csa = *Cupiennius salei* (spider), Dti = *Discocelis tigrina* (flatworm), Gga = *Gallus gallus* (chicken), Odi = *Oikopleura dioica* (appendicularian urochordate), Nve or Nv = *Nematostella vectensis* (sea anemone), Nvi = *Nereis virens* (polychaete worm), Pdu = *Platynereis dumerilii* (polychaete worm), Pst = *Phascolion strombus* (sipunculan worm), Sko = *Saccoglossus kowalevskii* (hemichordate), Sp = *Strongylocentrotus purpuratus* (sea urchin), Tricho (and Ta of NotTa gene) = *Trichoplax adhaerens* (placozoan), X or Xl = *Xenopus laevis* (frog). Double-headed arrows demarcate groups and subclasses of gene families. Black represents the Hox and ParaHox clusters, white is the NK cluster genes (see Luke *et al.* (2003) and Bürglin

	Human (synonyms)	Drosophila (synonyms)	Other animals
Hox1	HOXA1 (HOX1F) HOXB1 (HOX2I) HOXD1 (HOX4G)	labial (lab)	AmphiHox1, LsHox1 (Lineus)
Hox2	HOXA2 (HOX1K) HOXB2 (HOX2H)	proboscipedia (pb)	AmphiHox2
Gsx	GSH1 GSH2	intermediate neuroblasts defective (ind)	AmphiGsx, PstGsx
Hox3	HOXA3 (HOX1E) HOXB3 (HOX2G) HOXD3 (HOX4A)	zerknüllt 1 & 2 (zen1 & 2)	AmphiHox3, CsaHox3
Xlox	IPF1 (PDX1, IDX1, STF1, MODY4)	-	AmphiXlox, Xlhbox8, Lox3 (leech)
Central Hox (Hox4-8, Dfd-abdA)	A4(1D), A5(1C), A6(1B), A7 (1A) B4(2F), B5(2A), B6(2B), B7(2C), B8(2D) C4(3E), C5(3D), C6(3C), C8(3A) D4(4B), D8(4E)	Deformed(Dfd), Sex combs reduced (Scr), Antennapedia (Antp), Ultrabithorax (Ubx), abdominalA (abdA)	AmphiHox4 – 8
Posterior Hox (Hox9+, AbdB, Post1 & Post2)	A9(1G),A10(1H), A11(1I), A13(1J) B9(2E), B13 C9(3B), C10(3I), C11(3H), C12(3F), C13(3G), D9(4C), D10(4D), D11(4F), D12(4H),D13(4I)	AbdominalB (AbdB)	AmphiHox9 – 14, NviPost-1 & -2
Cdx	CDX1 CDX2 (CDX3) CDX4	caudal (cad)	AmphiCdx, CapI-Cdx
Mox	MEOX1 (MOX1) MEOX2 (MOX2, GAX)	buttonless (btn)	AmphiMox, NveMOXA-D
Evx	EVX1, EVX2	even-skipped (eve)	AmphiEvxA & B, Pdu-eve
Gbx	GBX1, GBX2	unplugged (unpg)	AmphiGbx, Pdu-Gbx
Mnx	HLXB9 (HB9)	dHB9	AmphiMnx, NveMNX, TrichoMnx
En	EN1, EN2	engrailed (en), invected (inv)	AmphiEn,
Rough	-	rough (ro)	NveROUGH
Dlx	DLX1, DLX2 (TES-1), DLX3, DLX4 (DLX7, DLX8, DLX9, BP1), DLX5, DLX6	Distal-less (Dll)	AmphiDlx, NveDLX, TrichoDlx
Msx	MSX1 (HOX7, HYD1), MSX2 (HOX8, MSH, PFM, CRS2, FPP, PFM1)	muscle segment homeobox (msh)	AmphiMsx, NveMSXC & B
NK4/tin	NKX2-3 (NKX2.3, NKX2C), NKX2-5 (NKX2.5, CSX1, CSX, NKX2E), NKX2-6 (NKX2.6)	tinman (tin) (msh2, NK-4)	AmphiNK4, NveNK2-TIN
NK3/bap	NKX3-1 (NKX3A, NKX3.1), BAPX1 (NKX3B, NKX3.2)	bagpipe (bap) (NK-3)	AmphiNK3, NveNK3
Lbx	LBX1 (HPX-6, LBX1H), LBX2	ladybird early (lbe), ladybird late (lbl)	AmphiLbx
Tlx	TLX1 (HOX11, TCL3), TLX2 (HOX11L1, NCX, Enx, Tlx2), TLX3 (HOX11L2, RNX)	C15 (93Bal, clawless, Hox11-311, Ect5)	AmphiTlx
NK1/slou	NKX1-1 (HSPX153, SAX1, NKX1.1), NKX1-2 (NKX1.2, SAX2)	slouch (slou) (paired-like9, S59, NK-1)	AmphiNK1a & b, NveNK1/NveSLOU PduNK1

(Cont.)

Table 10.1 (cont.)

	Human (synonyms)	Drosophila (synonyms)	Other animals
Hex	HHEX (PRHX, HEX, HOX11L-PEN)	CG705 6/RT01131p	OdiHex, NveHEX
NK5/Hmx	HMX1 (H6), HMX2 (Nkx5-2, H6L), HMX3 (Nkx5-1)	Hmx (H6-like homeobox, CG5832)	SpHmx, TrichoHmx
NK6	NKX6-1 (Nnk6.1, NKX6A), NKX6-2 (NKX6.1, NKX6B, GTX), NKX6-3	HGTX (Nk6, Nkx6, CG13475, CG4745)	NveNK6
NK7	-	N K7.1 (CG8524)	DtiNK7 NveNK7
Hlx	HLX1 (HB24)	H2.0 (CG11607)	NveHLXA-G
Emx	EMX1, EMX2, EMX2OS (=opposite strand)	empty spiracles (ems), E5 (CG9930)	SkoEmx, AmphiEmxA & B, NveEMXA & B
Dbx	DBX1, DBX2 (FLJ16139)	CG12361	SkoDbx
BarH	BARHL1, BARHL2	B-H1 (BarH1), B -H2 (BarH2)	SkoBarH
BarX	BARX1, BARX2	-	OdiBarX
Bsx	LOC390259	brain -specific homeobox (bsh) (Bashed (Bsh))	GgaBsx
Vent	VENTX (VENTX2, HPX42B), (pseudogenes VENTXP1-7 include VENTX1/HPX42)	-	Xvent2(Vox, Xbr-1a), AmphiVent1 & 2
Vax	VAX1, VAX2 (DRES93)	-	SkoVax
NK2	NKX2 2 (NKX2.2, NKX2B), NKX2-4 (NKX2.4, NKX2D), NKX2-8 (NKX2.8, NKX2H, NKX2-9), TITF1 (NKX2A, Nkx2.1, TTF-1, BCH, BHC, TEBP)	ventral nervous system defective (vnd) (NK -2)	AmphiNK2-2 & 2-1,
Not	NOTO	CG18599	Xnot, NotTa
CG13424	-	CG13424	Nve13424A & B
CG15696	-	CG15696	Nv046 & Nv066
CG34031	-	CG34031	SpCG34031
CG11085	-	CG11085	SpCG11085

(2005) for mosquito Msx inclusion); the cross-hatched arrow is the EHGbox cluster
(Pollard and Holland 2000), and the grey arrows are the different versions of the
'Extended Hox' subclass: Mox + Evx (Pollard and Holland 2000),
Mox + Evx + Gbx + Mnx + En (Castro and Holland 2003, and
Mox + Evx + Gbx + Mnx + En + Rough + Dlx (proposed here on the basis of
established gene linkages rather than poorly resolved phylogenetic trees, as
described in the text, which now becomes the 'Hox-linked' subclass). Lineage
specific genes, such as *Nanog* (Booth and Holland 2004) and *bicoid*, are omitted for
simplicity. The name MSXLX has been suggested for the CG15696 family based on
weak phylogenetic support for a sister group relationship with the Msx family in a
Bayesian tree (Ryan *et al.* 2006). This level of support is low, and so following the
reasoning presented here the MSXLX name is at present not adopted for the
CG15696 family. The subclasses of the ANTP-class that are identified here are thus
(1) Hox + ParaHox, (2) Hox-linked, (3) NK cluster, including NK4/tin, NK3/bap, Tlx,
Lbx, NK1/slou and Msx, with NK5/Hmx potentially joining following closer
examination (García-Fernández 2005), and (4) NK-linked. There will be one or two
families of the ANTP-class that fall outside these subclass definitions that will be
clarified by analysis of further genomes. Such families would then also be
categorised as subclasses.

ANTP-class that most confusion and discrepancy exists, and so I shall concentrate on this class and attempt to resolve at least some of the confusion. The classification of the groups of homeobox genes that I will use here broadly follows those of Wada *et al.* (2003), Edvardsen *et al.* (2005), Booth and Holland (2007) and Ryan *et al.* (2006), with a group of orthologous genes being a family, e.g. the Msx family, and the major distinctive groups of animal homeobox genes being classes, e.g. PRD class, and then the term 'subclass' being reserved for an intermediate grouping of several families within a class, e.g. the NKL subclass.

This classification of homeobox genes, which is based upon molecular phylogenies and sequence similarities, is very robust at the family level for the majority of gene families (e.g. Hox1/lab, Dlx or Evx), across the animal kingdom (Gauchat *et al.* 2000, Bürglin 2005, Holland and Takahashi 2005). Orthologous genes can usually be recognised even between cnidarians and humans or flies, despite their lineages having diverged over 550 million years ago. There are of course exceptions due to lineage-specific mutation rate elevations, and lineage-specific duplications or losses, which I will come to later. On the whole, however, each homeobox family is united by high support values in phylogenetic trees (e.g. Kamm and Schierwater 2006, Ryan *et al.* 2006). The families of the ANTP-class and their members from humans and flies, with a few other selected representatives, are given in Table 10.1.

With this clarity of the family-level phylogeny, and the increasing availability of whole genome sequences from around the animal kingdom, our understanding of complete homeobox complements from several animals, and hence lineage-specific gene gains and losses, is improving. Importantly, the possibility of having every homeobox gene from a particular animal means that the uncertainty of whether a particular orthologue has been missed or not is largely eliminated, provided the genome sequencing has been done to sufficient coverage, the assembly is sound, and enough care has been taken with thoroughly searching for all homeobox genes so that problems with inadequate computational gene prediction algorithms are overcome. This last point is clearly illustrated by the different analyses of one of the most carefully sequenced and analysed genomes of all, that of humans. The initial predictions of numbers of homeobox genes were 160 and 267 (Venter *et al.* 2001, Lander *et al.* 2001, respectively). Both are wrong, and careful searching and analysis actually reveals 235 homeobox-containing genes in the euchromatin of the human genome, of which 100 are in the ANTP-class (Anne Booth, Peter Holland and Elspeth Bruford, personal communication).

Armed with these large, comprehensive datasets we are now in a position to agree on the nomenclature of homeobox families on a sound, phylogenetically driven basis. This should overcome the inevitable problems generated by orthologous genes, or even the same gene, being given a multitude of different names because of different laboratory traditions or biases (Table 10.1). What cannot be overcome is the problem of different names caused by taxon-specific conventions, e.g. mutant-based names in *Drosophila*, three-letter and number conventions in nematodes, and acronyms in mammals. But these different taxon conventions can easily be accommodated once a gene from any particular animal is given just a single recognised name.

Ideally this nomenclature convention should be extended throughout a phylum. A convention that is rapidly gaining acceptance, at least in animals for which an alternative convention has not already been established, is naming genes with three letters that denote the species (the first capital letter being the first letter of the genus name, and the following two lower-case letters being the first two letters of the species name) followed by the gene name, which is deduced from phylogenetic analysis and sequence similarity (e.g. de Rosa *et al.* 1999). The *Drosophila melanogaster labial* gene thus becomes *Dmelab* whilst the *Tribolium castaneum labial* gene is *Tcalab*.

Difficulty arises when the gene phylogeny is not well resolved, as can occur when genes are isolated from a new taxon that is phylogenetically divergent from previously sampled animals, or the newly sampled taxon is 'long-branch' with rapidly evolving sequences. This problem is prevalent in the nematode *Caenorhabditis elegans*, and is illustrated by the changing views on the affinity of the *C. elegans* Hox genes. The gene *egl-5* was originally designated as the Posterior Hox gene of *C. elegans*, and hence orthologous to the *AbdB* gene of flies and the Hox9+ genes of chordates (Bürglin 1994). The classification of this gene, based on phylogenies, was never particularly robust, but was accepted because *egl-5* was the closest thing to a Posterior Hox gene known from nematodes at that time. Two further Hox genes were subsequently found during the whole genome sequencing of *C. elegans*, *php-3* and *nob-1*, which had greater similarity to Posterior Hox genes (Van Auken *et al.* 2000). Now views are appearing that have usurped *egl-5* from its designation as a Posterior Hox gene, and pushed it towards a Central Hox gene. However, such 'shoe-horning' of the gene is not done with great conviction, owing to the poorly resolved phylogenies of the genes (de Rosa *et al.* 1999, Bürglin 2005).

To avoid gene-naming difficulties such as two species with the same three-letter abbreviation, the homeobox community could adopt some form of hierarchical system, with precedence being given to the first species code used. Also, when phylogenies are poorly resolved, or interpreted in different ways by different authors, the community would need to adopt a set of 'rules' in which a particular level of phylogenetic resolution (with agreed forms of phylogeny building) is taken as warranting classification as a family member unless further lines of evidence (see below) justify otherwise. Incidentally these regions of phylogenetic ambiguity are often where some of the most interesting biology exists, which we can more easily focus on if clearer gene nomenclature reduces ambiguities and confusion.

A further gene classification problem can arise when there have been lineage-specific duplications, divergences and even gene losses. Further taxon sampling can often help to resolve such classification problems, effectively breaking the long branches in the molecular phylogenies of difficult-to-classify genes. This has proven to be the case for the *zerknüllt* (*zen*) and *bicoid* (*bcd*) genes of flies. These genes are in the Hox cluster of drosophilids between the group 2 Hox gene, *proboscipedia*, and the group 4 Hox gene, *Deformed*. On the basis of their genomic location they were thought to have probably been derived from a Hox3 gene. However, in phylogenetic trees the *zen* and *bcd* genes did not group robustly with the Hox3 genes available at the time. Only once other arthropod taxa were sampled, such as beetles and grasshoppers, and more basal lineages of flies than *Drosophila*, did stronger phylogenetic signals start to appear (Falciani *et al.* 1996, Stauber *et al.* 1999). The conclusion that insect *zen* and *bcd* genes are derived from Hox3, *zen* by sequence divergence and *bcd* via subsequent duplication and further divergence, is further supported when combined with the observations that the expression of the arthropod Hox3 genes can be seen to have evolved from a typical Hox-like expression (restricted along the anterior–posterior axis) into the derived role of the *zen* genes in extra-embryonic membrane and dorsal–ventral patterning, followed by the origin of *bcd* and its role as a maternal morphogen in fly anterior–posterior axis patterning (Damen and Tautz 1998, Telford and Thomas 1998, Stauber *et al.* 1999).

The *zen/bcd* example illustrates another important source of information that can be used to supplement phylogenetic information when attempting to classify homeobox genes, namely the genomic location. Regions of synteny can be analysed, and are particularly useful if the syntenic regions do not merely contain orthologous gene neighbours, but also conserve these genes in the same order along the chromosome. In the case of the *zen/bcd* example these genes are in the location of the Hox3 gene, in between the insect versions of Hox2 and Hox4. The resolution of homeobox evolutionary patterns can also be aided by synteny analysis of neighbouring genes other than homeobox genes as well. Examples of this have been the resolution of vertebrate ParaHox gene clusters evolving by whole cluster duplication, followed by some gene losses (Brooke *et al.* 1998, Coulier *et al.* 2000, Ferrier *et al.* 2005, Mulley *et al.* 2006), or the cryptic orthology of the *TPRX1* and *TPRX2P* genes of humans to the Obox genes of mice (Booth and Holland 2007). So, although homeobox phylogenies are immensely valuable in understanding gene orthologies, they sometimes need to be supplemented. Other useful sources of information come from conservation of other domains outside of the homeodomain (Bürglin 2005), or with further taxon sampling, information on genomic location and expression data.

There are one or two other families from the list in Table 10.1 that also suffer from some ambiguities at present, which are confounded by poorly resolved phylogenies and inconsistent or confusing nomenclature. The NK2 and NK4/tin genes are one important example, which can clearly be resolved once gene expression and genomic location are also taken into account. Unfortunately the names of the NK4/tin genes of humans are *NKX2-3*, *NKX2-5* and *NKX2-6*, whilst the NK2 genes proper are *NKX2-2*, *NKX2-4* and *NKX2-8* (along with *TITF1-2* which has bizarrely escaped the NKX2-X nomenclature theme). These NK4/tin and NK2 genes do indeed group together on phylogenies, suggesting that the two groups are sister groups, or a single multi-gene family if phylogeny alone is used (e.g. Ryan *et al.* 2006). However, the genomic location of the NK4/tin genes in the NK cluster of flies and mosquitoes and their tight linkage with the NK3/bap orthologues in chordates as well (Luke *et al.* 2003) (including their conserved role in 'heart' development; Harvey 1996), indicates that the NK4/tin and NK2 groups are distinct families. Following the logic that the genomic location is of prime importance in the classification of these two particular groups of genes, the genomic position of a sea anemone (*Nematostella vectensis*) NK4/NK2-like gene next to an NK3/bap gene permitted its identification as the *Nematostella* NK4/tin orthologue (Chourrout *et al.* 2006).

HOX-LIKE AND NK-LIKE SUBCLASSES: ARE THE NAMES MISLEADING?

Despite the lack of phylogenetic resolution between the families of the ANTP-class, the families are usually divided into the Hox-like subclass and NK-like (or NKL) subclass (Bürglin 2005, Holland and Takahashi 2005). At the broad scale this can be useful, but there is a serious problem with regards to reconstructing the relationships between these different homeobox families. The support values at the nodes that might define the divergence patterns between different homeobox families are almost always very poorly resolved, and so will usually be collapsed if standard molecular phylogenetic 'rules' are applied (such as collapsing or ignoring Neighbour-Joining nodes with less than 70% bootstrap support). When drawing homeobox trees, however, most workers tend to be more liberal, and avoid collapsing poorly supported nodes. Whilst this may be said to be invalid on strict molecular phylogenetic grounds, there is nevertheless some justification for taking this liberal approach when genomic organisation of the genes is also considered. For example, Mox is a 'Hox-like' gene that also is relatively closely linked to the Hox cluster genes in chordate genomes (Minguillón and Garcia-Fernàndez 2003).

Other 'Hox-like' genes have been designated by several authors as Evx, Mnx, Gbx and possibly En, and have been distinguished as the 'Extended Hox' genes by Pollard and Holland (2000) (who distinguish En, Gbx and Mnx as the EHGbox genes), and Castro and Holland (2003) (who dispense with the EHGbox classification) (Table 10.1), although Bürglin (2005) instead classifies Gbx and Mnx as NKL genes. However, the phylogenetic support for such a distinction is weak, or even absent, and Evx genes sometimes relocate away from the Hox cluster genes into the so-called NKL genes in phylogenetic trees (Gauchat *et al.* 2000, Kamm and Schierwater 2006), and En remains so ambiguously placed in the phylogenetic trees that Bürglin (2005) groups it with neither the 'Hox-like' genes nor the NKL genes. It is the genomic locations of these genes (Evx, Mox, En, Mnx and Gbx) that has warranted their classification as Hox-like or Extended-Hox genes. This logic, however, creates a problem when we consider the Mega-homeobox cluster hypothesis.

THE ANCESTRAL MEGA-HOMEOBOX CLUSTER: FACT OR FICTION, AND HOW WILL WE KNOW?

One of the most pervasive and appealing ideas with regards to animal homeobox evolution is the hypothesis that ancestrally the Hox-like and NKL genes were clustered in a Mega-homeobox cluster, which has subsequently been largely dispersed by inversions and translocations in extant lineages (Pollard and Holland 2000; reviewed in Garcia-Fernàndez 2005, and summarised in Figure 10.1). This hypothesis arises from the important deduction that these homeobox genes predominantly originate by tandem duplications. Whilst non-tandem forms of gene duplication clearly are possible, they are much rarer than tandem duplications. Such evolution by tandem duplication has led to genes which group together in phylogenetic trees, such as the ANTP-class, tending to be genomically linked. This linkage does not have to be tight any more, owing to the prevalence of inversions that dissociate genes along a chromosome unless there is a functional reason for them to remain clustered, e.g. the Hox genes. Nevertheless, pooling linkage data from several taxa, and following the reasoning that a tight association of two homeobox genes is more likely to reflect an ancestral tight linkage rather than a chance coming together along a particular lineage, allowed the reconstruction of larger arrays of homeobox genes and a hypothetical ancestral Mega-homeobox cluster (Pollard and Holland 2000).

The key gene in this hypothesis is Dlx. Dlx is classified as an NKL gene because of its location (with weak support) in phylogenetic trees, but it is genomically linked to the Hox cluster of chordates. However, as is the case for all of the 'NKL' genes, the divergence patterns for the families cannot be robustly resolved (Ryan *et al.* 2006, Kamm and Schierwater 2006). So can we really exclude the possibility that Dlx is not an NKL gene after all, but a Hox-like gene, whose sequence divergence is comparable to the Evx, Mnx or Gbx families, so that it cannot be placed reliably in molecular phylogenies?

The NKL gene Msx was originally thought to also provide evidence for Hox-NKL clustering, but this would have required breakage of the linkage after the genome duplications at the origin of the vertebrates (to accommodate two different break-points, between Dlx-Msx and

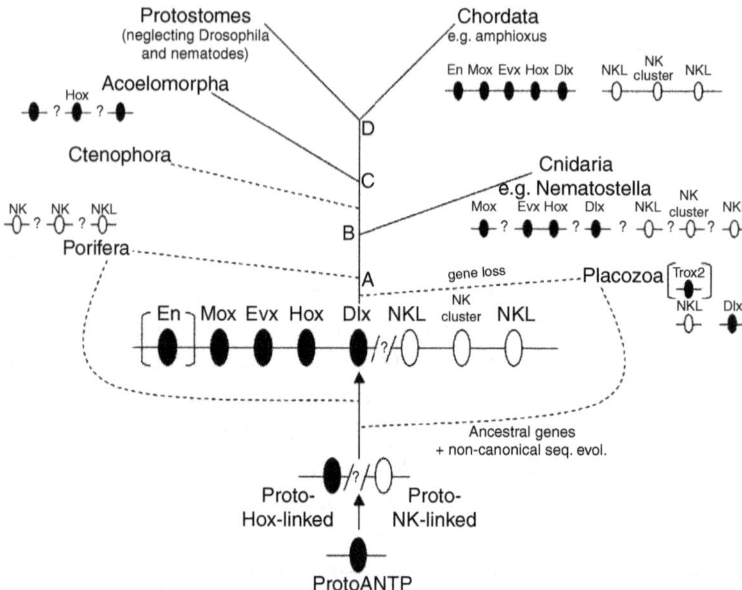

Figure 10.1 The diversity of the Hox-linked and NK-linked genes was already established early in animal evolution, but were the genes linked in a Mega-homeobox cluster or not? Conventionally a distinction has been made between Hox-like and NK-like (NKL) genes in the ANTP-class. However, the boundary between these two groups is very fuzzy, with poorly resolved molecular phylogenetic trees of homeodomain sequences. The coloration used here instead emphasises the gene linkage patterns discernable in chordates (particularly amphioxus), rather than the traditional Hox-like/NKL division (see text and Table 10.1 for further discussion). Black ovals represent genes linked to the Hox genes ('Hox-linked') and the Hox genes themselves, and white denotes genes linked to the NK cluster ('NK-linked') and the NK cluster itself (or its broken remains in chordates). *Trox2* is bracketed because its orthology with bilaterian genes is at present debated (Jakob et al. 2004, Ferrier 2006). *Engrailed* (*En*) is bracketed because it has not yet been found in a non-bilaterian animal, and so its presence on the baseline of the mega-homeobox array is purely conjectural (Ryan *et al.* 2006). En may instead have been a bilaterian innovation. Horizontal lines denote established genomic linkage, apart from in the hypothetical Mega-homeobox array and the Proto-Hox-linked and Proto-NK-linked genes, which remain to be established by further comparative genomics from the animal lineages outlined here. The dotted lines connecting the Placozoa and Porifera to the trunk of the phylogeny reflect the uncertainty over their origin in relation to the full array of the Mega cluster genes. Gene order and the numbers of Hox-linked and

Msx-NKL) (Pollard and Holland 2000). However, the *Msx4* gene (now called *MSX2P*; Genbank accession NR_002307) that is linked to the HOXB cluster in humans is a processed pseudogene, and has been dismissed as useless for reconstructing ancestral linkage patterns (Castro and Holland 2003). It is still intriguing, however, that of all of the places in the genome that an Msx processed pseudogene could jump into, it should happen to land right next to a Hox cluster, within 10 Mb.

The Mega-homeobox cluster hypothesis thus hangs in the balance, on some poorly resolved molecular phylogenies. How could the veracity (or not) of the hypothesis be resolved? The discovery of Dlx linkage with other NKL genes, or preferably with an NK cluster, would be one source of evidence. Similarly the clustering or linkage of other supposed NKL genes with the Extended Hox genes would provide another major line of evidence in support of the Mega-homeobox cluster hypothesis.

We naturally assume that when we find genes tightly linked together over large phylogenetic distances (between phyla) then it must be significant. Surely after 550 million years on two or even three lineages, if the genes did not have to be together in order to function, then they would have been broken apart by now by inversions and other genomic rearrangements. This is indeed reasonable when inversion rates are considered, although there is the caveat that inversion

---

Fig. 10.1 (cont.) NK-linked genes shown in the hypothetical ancestral array (s) are not of significance in this figure. There are more Hox-linked and NK-linked genes than those shown. Clear gaps along horizontal lines denote lack of linkage, whilst gaps with question marks denote unresolved linkage. The *Nematostella* 'NK cluster' consists of at least NK4/tin and NK3/bap (Chourrout *et al.* 2006). The protostome gene linkage has been left blank because the linkages in *Drosophila* are of dubious comparative relevance in the present context, owing to the low number of chromosomes in the fly, and hence the greatly increased likelihood that two genes can be linked 'by chance' rather than by the deep ancestral linkages being discussed here. The gene linkages from nematodes are not used either, owing to the extreme shuffling of the nematode genome (Coghlan and Wolfe 2002), and small genome size coupled with low chromosome number. The ancestral ProtoANTP gene duplicated to give rise to the ancestors of the Hox-linked and NK-linked genes. Whether this duplication event was tandem or not is crucial to whether the Mega-homeobox cluster existed, or was divided between the Hox-linked and NK-linked genes right from this earliest stage by a rare non-tandem duplication.

rates are known for only a few animal taxa, and those are often from lineages that are known, or suspected, to have experienced elevated levels of genomic rearrangements (e.g. *Drosophila*, nematodes) (Ranz *et al.* 2001, Coghlan and Wolfe 2002). An understanding of inversion rates from a wider diversity of animal taxa is desirable, and may be obtainable with further genomic sequencing from around the animal kingdom rather than from just within groups close to already established model systems such as vertebrates, nematodes and insects.

An important form of genomic rearrangement, in addition to inversion, is translocation. This is important in the context of homeobox gene evolution in view of the Mega-homeobox cluster hypothesis. Chordates possess a 'Hox' chromosome, a 'ParaHox' chromosome and a 'NKL' chromosome, typified by the cephalochordate amphioxus (*Branchiostoma floridae*) and neglecting the paralogy in vertebrates due to whole genome duplication events (Castro and Holland 2003). Focusing on the Hox-like/ NKL split, and ignoring the dispersion of the constituent genes along the respective chromosomes by the supposed relatively high rates of inversions, there has only been one major translocation that has broken the ancestral Mega-homeobox chromosome in the lineage leading to chordates (the moot point of whether the ProtoHox duplication into the Hox and ParaHox clusters was tandem or not is not discussed here; Minguillón and Garcia-Fernàndez 2003). Does this mean that the translocation and break of the Mega cluster was close to the origin of the chordates? Cnidarians have most of the Mega cluster genes, and sponges also have NKL genes, so expansion by tandem duplications was well under way and may well have largely happened right at the base of the animal kingdom (for reasoning see Garcia-Fernàndez 2005). Are the rates of translocations so low that the Mega cluster could have arisen at the base of the animals, been scattered along the particular parent chromosome and simply not been split apart owing to genomic evolutionary inertia until the origin of the chordates? Therefore what is the genomic arrangement of these Mega cluster genes in other lineages, the protostomes and the diploblasts? Also, how frequently do translocations occur? Can we find evidence of the Mega cluster being split in a different position? Would we expect Dlx to be the closest NKL gene to the Hox genes before translocation in poriferan lineages (if sponges have/had Hox genes), and the cnidarian and ctenophore lineages, or the acoelomorph and protostome lineages?

The flip-side to breakage of gene arrays by translocations is the coming together of genes by chromosome fusions. How frequently have chromosome fusions occurred during animal evolution? This is an essential consideration when examining gene linkage patterns in the context of the Mega cluster hypothesis, and is most clearly highlighted by the fact that most of the NKL and Hox-like genes are on the same chromosome in the fruit fly *Drosophila* (Castro and Holland 2003). This linkage cannot be used as support for the hypothesis, however, since the fruit fly has only four chromosomes, and so it is highly likely that gene linkage patterns can have arisen just by chance (note we are talking about mere linkage in this context and not tight linkage or clustering). The linkage of *Drosophila* NKL and Hox-like genes is thus quite possibly secondary. This *Drosophila* situation of course highlights a further complication when assessing the Mega-homeobox cluster hypothesis. How prevalent was chromosome fusion (followed by inversions) and then fission in different lineages, thus clouding our views of how likely it is that gene linkages can have evolved by chance? Synteny comparisons (not just restricted to homeobox genes) across phyla will reveal these parameters as more taxa are sequenced, and it is hoped that genome projects will proliferate in lineages that have 'normal' genomes, rather than specialised, derived, shrunken, rearranged organisations, so that we can reconstruct genomes that were in existence over 550 mya.

Consequently, in order to put the Mega-homeobox cluster hypothesis into context, we need a better understanding of the rates of genome rearrangements, including inversions, translocations, chromosome fusions and fissions, all in the context of the ancestry of different animal lineages. Did the ancestors at the points A–D of Figure 10.1 have very few chromosomes, or many? How has chromosome number changed along the branches diverging from these ancestors, and how often do lineages include drastic decreases in chromosome number and then secondary increases due to chromosome break-up? What are the rates of translocation, and therefore, given the dispersal of a gene cluster along a chromosome by the inversions that are known to occur relatively frequently, how frequently might we expect this ancestral homeobox chromosome to be broken apart in different ways or even secondarily linked together again?

AN ALTERNATIVE: HOX-LINKED AND NK-LINKED

What is the alternative to the Mega-homeobox cluster? The genes clearly evolved predominantly by tandem duplication, but could have been scattered around the genome by translocations whilst these duplications were in progress. Or there may have been one or two basal non-tandem duplications, such as the types associated with transposable elements or repetitive sequences leading to non-homologous recombination during meiosis. Both of these modes are much rarer than tandem duplications, and if the origin of animal homeobox genes was via an explosive radiation of duplications, then there was simply not enough time for many non-tandem rearrangements to occur before the full(ish) complement of ANTP-class genes arose. Kamm and Schierwater (2006) and Ryan *et al.* (2006) show that all of the major homeobox families are already there by the origin of the Cnidaria. Therefore sponge data will be interesting, to gauge just how basally all of these families existed and gauge the extent of the homeobox explosion at the origin of the animals (Gauchat *et al.* 2000, Garcia-Fernàndez 2005).

An important first step in this endeavour has been made by the analysis of the ANTP-class homeobox complement in the whole genome sequence of the sponge *Amphimedon queenslandica*, which contains clustered NKL genes but seems to lack Hox and Hox-related genes (Larroux *et al.* 2007).

Trichoplax might be another interesting basal animal genome, but early indications of homeobox content are of a low number of genes that still classify robustly with established homeobox families (Monteiro *et al.* 2006). Correlated with a low density of genes, this is suggestive of extensive gene loss, which corresponds to the simple morphology of *Trichoplax*. The alternative possibility, that the *Trichoplax* homeobox genes are in fact descended from the ancestral genes of multiple homeobox families, requires that the evolution of these genes is occurring via an unusual mode, which will confound attempts to distinguish gene loss from an ancestral gene-poor condition; see Ferrier's (2006) 'Trox-2' model of gene evolution, and Jakob *et al.* (2004).

Alternatively, Dlx is not an NKL gene but another Extended Hox gene, and the Hox-like and NKL subclasses were never together, the founders of each subclass having arisen by one rare non-tandem duplication event (Figure 10.1). Larger-scale synteny analyses than are currently available in diploblasts such as *Nematostella* may help (Chourrout *et al.* 2006), as long as the chromosome number of the animal is higher than that of *Drosophila*, so that non-relevant, chance linkages are not prevalent.

Is the Hox-like and NKL terminology valid and useful, or misleading? It is useful if it alerts us to possibilities of classifying NKL genes with Hox-like genes, e.g. Dlx, but misleading if Dlx is not an NKL gene after all. Therefore we need to examine linkage and clustering patterns with detachment from the Hox-like and NKL terminology, and perhaps base our terminology more on genomic organisation rather than poorly resolved phylogenies. A better terminology than the Hox-like and NK-like designations would thus be Hox-linked and NK-linked (Hox and NK here denoting the genes of the Hox and NK clusters) (Figure 10.1 and Table 10.1).

SO WHAT IS A HOX GENE?

From the preceding discussion, along with Table 10.1, we can now return to the original question: when is a Hox gene not a Hox gene? There are several relatively straightforward answers to this question, such as when it is instead a ParaHox gene, or a ProtoHox gene (Brooke *et al.* 1998, Ferrier and Holland 2001), or even a Hox-like or Hox-linked gene (as discussed above), and when it is derived from a Hox gene but has become something else such as a segmentation gene like *Drosophila fushi-tarazu (ftz)* (Dawes *et al.* 1994, Damen 2002), an anterior–posterior determinant such as *bcd* (Stauber *et al.* 1999), or a dorso-ventral pattern-ing gene such as *zen* (Falciani *et al.* 1996). The Hox genes are thus a para-phyletic group of genes, with the 'derived Hox genes' evolving from within the Hox genes proper, just as reptiles are paraphyletic when the birds are classified as a distinct group. Such distinctions (such as saying *Drosophila ftz* is not a Hox gene, or a bird is not a reptile), whilst not strictly following the phylogeny of the genes or organisms, are nevertheless extremely useful in an evolutionary context since they highlight a significant evolutionary transition. Finally, a Hox gene is not a Hox gene when it is an NK gene such as Tlx or Hex, whose old names of Hox11 and Hox11L-PEN really must be retired from use.

The other side to the question is what is a Hox gene? Clustering and collinearity are often implicit assumptions within Hox gene research, and form part of a Hox gene definition for many (Scott 1993, Kamm *et al.* 2006). There are now, however, an increasing number of examples of genes which people would still want to call Hox, but which are neither collinear nor clustered (reviewed in Monteiro and Ferrier 2006). It is thus clear that we must be more cautious when assuming clustering (and collinearity) in investigations of Hox genes without explicitly testing for it.

Given this relatively frequent occurrence of non-clustering of sup-posed Hox genes, along with the occurrence of some genes that are derived from Hox genes remaining within the cluster, but are no longer true Hox genes (e.g. *ftz, zen, bcd*), then it is clear that sequence similarity must be combined with genomic data as well as gene expression data before a gene can reliably be called a Hox gene. The hierarchy in importance of these different sources of data is sequence similarity, which can reveal where a gene evolved from even if it is no longer a true Hox gene, followed by expression data, which should reveal a potential role in axial patterning, followed by genomic linkage data to resolve ambiguous assignments and test for gene order and collinearity. The contentious issue of whether cnidarians possess Hox and ParaHox genes or not (Kamm *et al.* 2006) provides a recent example of the importance of genomic linkage data. Gene linkage data has unambiguously shown the existence of two gene clusters in the sea anemone *Nematostella*, one of which has greater sequence similarities to the bilaterian Hox genes, whilst the second cluster has genes with greater similarities to the ParaHox genes of bilaterians, even though the specific orthologies of some of the individual genes remains ambiguous (Chourrout *et al.* 2006, Ryan *et al.* 2007).

Hox genes then do not have to be clustered, but they are certainly derived from clustered genes in all cases, and they must be expressed in a coordinated axially restricted or staggered fashion to distinguish them from Hox-derived non-Hox genes. The genomic context in which this system evolved remains to be resolved. The clear classification of homeobox genes, such as the Hox genes, is essential in order to accurately formulate hypotheses on how the genes evolved (e.g. within a Mega-homeobox cluster or not) and how they have contributed to the evolution of the animal kingdom.

ACKNOWLEDGEMENTS

The author would like to thank Alessandro Minelli and Giuseppe Fusco for the opportunity to attend the Venice Evo-Devo workshop and contribute to this book. The ideas outlined in this chapter have developed from helpful discussions with Peter Holland, Ana Sara Monteiro, Thomas Butts and Jerome Hui. The author's research is funded by the BBSRC.

REFERENCES

Booth, H. A. F. & Holland, P. W. H. 2004. Eleven daughters of *NANOG. Genomics* **84**, 229–238.
Booth, H. A. F. & Holland, P. W. H. 2007. Annotation, nomenclature and evolution of four novel homeobox genes expressed in the human germ line. *Gene* **7**, 7–14.

Brooke, N. M., Garcia-Fernàndez, J. & Holland, P. W. H. 1998. The ParaHox gene cluster is an evolutionary sister of the Hox gene cluster. *Nature* **392**, 920–922.

Bürglin, T. R. 1994. A comprehensive classification of homeobox genes. In D. Duboule (ed.) *Guidebook to the Homeobox Genes*. Oxford: Oxford University Press, pp. 25–71.

Bürglin, T. R. 2005. Homeodomain proteins. In R. A. Meyers (ed.) *Encyclopedia of Molecular Cell Biology and Molecular Medicine*, Weinheim: Wiley-VCH Verlag GmbH & Co., pp. 179–222.

Castro, L. F. C. & Holland, P. W. H. 2003. Chromosomal mapping of ANTP class homeobox genes in amphioxus: piecing together ancestral genomes. *Evolution & Development* **5**, 459–465.

Chourrout, D., Delsuc, F., Chourrout, P. *et al.* 2006. Minimal ProtoHox cluster inferred from bilaterian and cnidarian Hox complements. *Nature* **442**, 684–687.

Coghlan, A. & Wolfe, K. H. 2002. Fourfold faster rate of genome rearrangement in nematodes than in *Drosophila*. *Genome Research* **12**, 857–867.

Coulier, F., Burtey, S., Chaffanet, M., Birg, F. & Birnbaum, D. 2000. Ancestrally duplicated paraHOX gene clusters in humans. *International Journal of Oncology* **17**, 439–444.

Damen, W. G. 2002. *Fushi tarazu*: a Hox gene changes its role. *BioEssays* **24**, 992–995.

Damen, W. G. & Tautz, D. 1998. A Hox class 3 orthologue from the spider *Cupiennius salei* is expressed in a Hox-like fashion. *Development Genes & Evolution* **208**, 586–590.

Dawes, R. E., Dawson, I., Falciani, F., Tear, G. & Akam, M. E. 1994. *Dax*, a locust Hox gene related to *fushi-tarazu* but showing no pair-rule expression. *Development* **120**, 1561–1572.

de Rosa, R., Grenier, J. K., Andreeva, T. *et al.* 1999. Hox genes in brachiopods and priapulids and protostome evolution. *Nature* **399**, 772–776.

Edvardsen, R. B., Seo, H., Jensen, M. *et al.* 2005. Remodelling of the homeobox gene complement in the tunicate *Oikopleura dioica*. *Current Biology* **15**, R12–R13.

Falciani, F., Hausdorf, B., Schröder, R. *et al.* 1996. Class 3 Hox genes in insects and the origin of *zen*. *Proceedings of the National Academy of Sciences of the USA* **93**, 8479–8484.

Ferrier, D. E. K. 2006. Evolution of Hox gene clusters. In S. Papageorgiou (ed.) *HOX Gene Expression*. Landes Bioscience, pp. 53–67.

Ferrier, D. E. K., Dewar, K., Cook, A. *et al.* 2005. The chordate ParaHox cluster. *Current Biology* **15**, R820–R822.

Ferrier, D. E. K. & Holland, P. W. H. 2001. Ancient origin of the Hox gene cluster. *Nature Reviews Genetics* **2**, 33–38.

Garcia-Fernàndez, J. 2005. The genesis and evolution of homeobox gene clusters. *Nature Reviews Genetics* **6**, 881–892.

Gauchat, D., Mazet, F., Berney, C. *et al.* 2000. Evolution of Antp-class genes and differential expression of Hydra Hox/ParaHox genes in anterior patterning. *Proceedings of the National Academy of Sciences of the USA* **97**, 4493–4498.

Harvey, R. P. 1996. NK-2 homeobox genes and heart development. *Developmental Biology* **178**, 203–216.

Holland, P. W. H. & Takahashi, T. 2005. The evolution of homeobox genes: implications for the study of brain development. *Brain Research Bulletin* **66**, 484–490.

Jakob, W., Sagasser, S., Dellaporta, S. L. *et al.* 2004. The Trox-2 Hox/ParaHox gene of *Trichoplax* (Placozoa) marks an epithelial boundary. *Development Genes & Evolution* **214**, 170–175.

Kamm, K. & Schierwater, B. 2006. Ancient complexity of the non-Hox ANTP gene complement in the anthozoan *Nematostella vectensis*: implications for the evolution of the ANTP superclass. *Journal of Experimental Zoology B (Molecular and Developmental Evolution)* **306**, 589–596.

Kamm, K., Schierwater, B., Jakob, W., Dellaporta, S. L. & Miller, D. J. 2006. Axial patterning and diversification in the Cnidaria predate the Hox system. *Current Biology* **16**, 920–926.

Lander, E. S., Linton, L. M., Birren, B. *et al.* 2001. Initial sequencing and analysis of the human genome. *Nature* **409**, 860–921.

Luke, G. N., Castro, L. F., McLay, K. *et al.* 2003. Dispersal of NK homeobox gene clusters in amphioxus and humans. *Proceedings of the National Academy of Sciences of the USA* **100**, 5292–5295.

Larroux, C., Fahey, B., Degnan, S. M. *et al.* 2007. The NK homeobox gene cluster predates the origin of Hox genes. *Current Biology* **17**, 706–710.

Minguillón, C. & Garcia-Fernàndez, J. 2003. Genesis and evolution of the Evx and Mox genes and the extended Hox and ParaHox gene clusters. *Genomic Biology* **4**, R12.

Monteiro, A. S. & Ferrier, D. E. K. 2006. Hox genes are not always colinear. *International Journal of Biological Sciences* **2**, 95–103.

Monteiro, A. S., Schierwater, B., Dellaporta, S. L. & Holland, P. W. H. 2006. A low diversity of ANTP class homeobox genes in Placozoa. *Evolution & Development* **8**, 174–182.

Mulley, J. F., Chiu, C. H. & Holland, P. W. H. 2006. Breakup of a homeobox cluster after genome duplication in teleosts. *Proceedings of the National Academy of Sciences of the USA* **103**, 10369–10372.

Pollard, S. L. & Holland, P. W. H. 2000. Evidence for 14 homeobox gene clusters in human genome ancestry. *Current Biology* **10**, 1059–1062.

Ranz, J. M., Casals, F. & Ruiz, A. 2001. How malleable is the eukaryotic genome? Extreme rate of chromosomal rearrangement in the genus *Drosophila*. *Genome Research* **11**, 230–239.

Ryan, J. F., Burton, P. M., Mazza, M. E. *et al.* 2006. The cnidarian-bilaterian ancestor possessed at least 56 homeoboxes: evidence from the starlet sea anemone, *Nematostella vectensis*. *Genome Biology* **7**, R64.

Ryan, J. F., Mazza, M. E., Pang, K. *et al.* 2007. Pre-bilaterian origins of the Hox cluster and the Hox code: evidence from the sea anemone, *Nematostella vectensis*. *PLoS ONE* **2**(1), e153.

Scott, M. P. 1993. A rational nomenclature for vertebrate homeobox (HOX) genes. *Nucleic Acids Research* **21**, 1687–1688.

Stauber, M., Jäckle, H. & Schmidt-Ott, U. 1999. The anterior determinant *bicoid* of *Drosophila* is a derived Hox class 3 gene. *Proceedings of the National Academy of Sciences of the USA* **96**, 3786–3789.

Telford, M. J. & Thomas, R. H. 1998. Of mites and *zen*: expression studies in a chelicerate arthropod confirm *zen* is a divergent Hox gene. *Development Genes & Evolution* **208**, 591–594.

Van Auken, K., Weaver, D. C., Edgar, L. G. & Wood, W. B. 2000. *Caenorhabditis elegans* embryonic axial patterning requires two recently discovered posterior-group Hox genes. *Proceedings of the National Academy of Sciences of the USA* **97**, 4499–4503.

Venter, J. C., Adams, M. D., Myers, E. W. *et al.* 2001. The sequence of the human genome. *Science* **291**, 1304–1351.

Wada, S., Tokuoka, M., Shoguchi, E. *et al.* 2003. A genomewide survey of developmentally relevant genes in *Ciona intestinalis* II. Genes for homeobox transcription factors. *Development Genes & Evolution* **213**, 222–235.

11

Plants are used to having identity crises

ROLF RUTISHAUSER, VALENTIN GROB AND EVELIN PFEIFER

Macroscopic nature is never really anomalous. Abnormalities, like other exceptional cases, at least show incontestably, what the plants *can* do.

Arber 1950: 6

However, regardless of how much faith one has in anatomical definitions, they should not be taken as more than a means of communication prior to subsequent genetic analysis.

Scheres *et al.* 1996: 963

Truth, except as a figure of speech, does not exist in empirical science.

Brower 2000: 18

INTRODUCTION

Our green and living world is a continuum in space and time. This view is well expressed in the 'continuum model' proposed by botanists and biophilosophers such as Arber (1950) and Sattler (1996). As an opposite view we may accept the green world around us as consisting of discrete units on several hierarchical levels. This view is called here the 'discontinuum model' or the 'classical model' because it has been the predominant view in biological textbooks for decades. Branching and repetition of developmental units (e.g. cells, meristems, modules, leaves, phytomers) are omnipresent as developmental processes in multicellular plants. These processes resemble the process of segmentation in various metazoan phyla, also occasionally leading to fuzzy borderlines between consecutive developmental units (Minelli and Fusco 2004, Prusinkiewicz 2004, Rutishauser and Moline 2005). Perspectivists studying plants accept structural and developmen-

Evolving Pathways: Key Themes in Evolutionary Developmental Biology, ed. Alessandro Minelli and Giuseppe Fusco. Published by Cambridge University Press.
© Cambridge University Press 2008.

tal categories such as cells, meristems, modules, leaves and phytomers as mind-born, simplified concepts reflecting certain aspects of the structural diversity (Sattler and Rutishauser 1990, Hay and Mabberley 1994). The best choice is to combine the heuristic advantages of both the continuum and the classical model by accepting the living world (especially the green one) as a heterogeneous continuum where typical forms are more frequent than intermediate ones. The present chapter focuses on unusual plant forms outside or beyond the geneticist's lab.

Classical model

A striking feature of vascular plants (i.e. seed-plants and ferns) is their apparent morphological simplicity. During the life cycle of a typical flowering plant, only three vegetative organ systems (leaves, stems and roots) and four reproductive organ systems (sepals, petals, stamens and carpels) are formed (Sattler 1996, Soltis *et al.* 2005, Endress 2006). According to the classical model the various structural categories are crisp sets, perfectly excluding each other. For example, for a clear leaf–stem distinction the relative position of an organ is taken as the most useful criterion (Rutishauser and Moline 2005).

Continuum model

Developmental geneticists (e.g. Tsukaya 1995, Jackson 1996, Sinha 1999, Hofer *et al.* 2001) have pointed to the fact that some vascular plants transcend the classical model. The continuum model accepts the same organ systems as the classical model, but allows them to have fuzzy (blurred) borderlines and intermediates, as described by, for example, Sattler and Jeune (1992) and Lacroix *et al.* (2003). This approach coincides with the fractal paradigm in that the whole is repeated in the parts to some extent, or the holographic structure where the whole can be retrieved from a part of it (Sattler 2001). For example, the continuum model accepts developmental mosaics of plant organs and mixed homologies between root, shoot (including stem), leaf and their parts (Rutishauser 1995, 1999, Sattler 1996, Baum and Donoghue 2002, Hawkins 2002). Developmental mosaics can be defined by giving equal weight to both the position criterion and the criterion of linkage through intermediate forms (Rutishauser and Moline 2005).

Organ identity

An organ in multicellular animals and plants is a part of a living organism with a certain set of functions besides its positional and constructional characters. In the context of plant developmental genetics 'organ identity' means the developmental fate of an uncommitted primordium. This term is used in zoology (e.g. Blochlinger *et al.* 1991) and botany as well. 'Organ identities' can be defined by morphological criteria *and* by their gene expression pattern, including organ identity genes that sculpt, for instance, the structure of angiospermous flowers (Yu *et al.* 1999, Soltis *et al.* 2005, Theissen 2005, Endress 2006, Jaramillo and Kramer 2007). The 'organ identity' concept is closely related to the concept of 'homology'; both have multiple and sometimes conflicting meanings, as reviewed by Sattler (1994), Bock and Cardew (1999), Brower (2000), Rutishauser and Moline (2005), and Theissen (2005). 'Organ identity' as a concept is also used outside the reproductive zone. The vegetative body of vascular plants thus shows primordia that are committed during early development to take over organ identities such as 'leaf identity' or 'root identity'. Acquisition of organ identity often happens progressively rather than at once (Sylvester *et al.* 1996, Bey *et al.* 2004). In some ferns and aberrant flowering plants ('morphological misfits') such as *Utricularia* the commitment of a primordium (meristematic area) to become a 'leaf' or 'shoot' (including 'stem') can be considerably delayed (Steeves *et al.* 1993, Rutishauser and Isler 2001).

MULTICELLULAR PLANTS HAVING IDENTITY CRISES ON VARIOUS LEVELS

The concept of an identity crisis is better known in psychology and sociology where it describes a condition of disorientation and role confusion as a result of conflicting pressures and expectations. Within the past few years the term 'identity crisis' has been adopted by developmental geneticists (Elledge 1996, Geuten *et al.* 2006, Lugassi *et al.* 2006). They have started to understand pattern repetition (reiteration) better at different levels of structural complexity. Multicellular plants such as angiosperms are used to having identity crises on various levels, from cells to meristems and organs and even beyond. Identity crises, however, are not the problem of the plants, but of our inadequate thinking and concepts.

Cell identity crisis

Cells in multicellular organisms acquire different identities in an ordered spatial arrangement: 'How do cells learn about their identity?' is a question asked by Scheres (2001: 112). Cell theory identifies the cell as the elementary unit from which all living organisms are constructed. In contradiction of cell theory, cells of higher plants are neither physically separated nor structurally independent. Plasmodesmata (i.e. cell-to-cell channels) connect each plant cell to its neighbours, aiding the exchange of proteins, RNAs and other molecules. Thus, higher plants are nowadays accepted as being both multicellular and supracellular. This view coincides with Tsukaya's (2002: 33) 'neo-cell theory': 'Each cell is also controlled by factors that govern the morphogenesis of the organ of which the cells are a part.' Baluska *et al.* (2004) admit: 'This identity crisis of the "cell" is not simply a problem confined to plants, as nanotubular intercellular bridges are also generated "de novo" between animal cells.'

Meristem identity crisis

Meristems in vascular plants are tissues primarily concerned with the formation of new cells by division. They are responsible for making roots, shoots (including stems and leaves) and (in seed plants) also flowers. The term 'meristem identity' is used to characterise the growth phases of a shoot apical meristem (SAM), with vegetative meristem, inflorescence meristem and floral meristem as three possible 'identities'. Involved in the change of meristem identities in flowering plants (such as *Antirrhinum* and *Arabidopsis*) are *FLO*-like genes such as *FLORICAULA* and *LEAFY* (inflorescence meristem identity). Also responsible for floral meristem identity are MADS-box genes such as *SQUAMOSA* (*Antirrhinum*) and *APETALA1* (*Arabidopsis*), whereas MADS-box genes such as *AGAMOUS* and *PLENA* are involved in floral meristem determination (Theissen 2000, 2005). In *Arabidopsis* the switch from one meristem identity to the next means a change of identities of lateral appendages. For example, Parcy *et al.* (1998) described the conversion of vegetative to floral meristems in *Arabidopsis* as follows: 'After floral induction in wild-type *Arabidopsis*, primordia that would otherwise have become leaves develop into flowers instead.' A similar situation is found in *Nymphaea* and *Nuphar* (Nymphaeaceae) which share an extra-axillary mode of floral inception in the shoot (i.e. rhizome) apical meristem. Some leaf sites along the ontogenetic spiral are occupied by floral primordia lacking a subtending leaf or bract. This pattern of flower initiation in leaf sites is repeated inside the 'branching flowers' of *Nymphaea prolifera*, a species occurring in Central and South America (Figures 11.1B, 11.2A, 11.2B). Instead of producing a single flower, an individual floral meristem of certain angiosperms (wild-type plants or mutants) can branch continuously giving rise to a complex inflorescence with many flowers. This process is called floral reversion and is found in, for example, *Arabidopsis* when floral meristems switch back to an indeterminate state and restart as a shoot apical meristem (Stahl and Simon 2005). Floral reversion also occurs in *Nymphaea prolifera*: each 'branching flower' first produces some perianth-like leaves, then it switches back to SAM identity forming the rhizome tip (Figure 11.1B, 11.2A, 11.2B). This switch is repeated up to three times giving rise to a branched complex of more than 100 sterile flowers serving as vegetative propagules (Grob *et al.* 2006).

There are developmental parallels between the flower level (as a subsystem) and the inflorescence level (as a system) in various groups of angiosperms. Baum and Donoghue (2002: 64) used the double-term 'inflorescence-flowers' when developmental programs of both flower and inflorescence are mixed. By examining MADS-box genes Yu *et al.* (1999) have shown that gene activities found during flower initiation are also found during early development of the head-like inflorescences of Asteraceae. This could explain why simple flowers and compound 'flowers' (i.e. inflorescences) are in certain taxa quite similar to each other – 'a resemblance sometimes carried into the minutest details of form and coloration' (Arber 1947: 233).

Floral organ identity crises

Developmental geneticists began to use the term 'organ identity' in botany while studying the genes that determine the developmental fate of flowers and their appendages (Coen 1999, Theissen 2000, 2005). For example, 'complete' bisexual flowers are observable in the eudicot *Clerodendrum minahassae*, a member of the eudicot family Verbenaceae (Figure 11.1D). Four whorls of organs are distinguishable: sepals, petals, stamens and carpels (the latter fused into a superior ovary). *Clerodendrum minahassae* adds 'petalness' to its sepals *after* anthesis in order to attract birds for fruit dispersal. In *Mussaenda* and *Warszewiczia* (both Rubiaceae) one out of five sepals is transformed into a showy, bract-like organ, serving as an optical signal for the whole inflorescence (Figure 11.1E).

Figure 11.1 Flowers and their parts having identity crises. A, *Nymphaea tuberosa* (Nymphaeaceae). Continuum of identities in floral organs. Upper row seen from ventral (inner) side, lower row seen from dorsal (outer) side. Outermost four perianth members as sepals green outside, followed by several completely white petals. Innermost petals turning into yellowish petaloid stamens, inner stamens yellow with narrow filaments. B, *Nymphaea prolifera* (Nymphaeaceae). Mother flower (F) branching into daughter flowers (F') and grand-daughter flowers (F"), acting as vegetative propagules. (For anatomical and developmental details see Figure 11.2.) C, *Jacquinia pungens* (Theophrastaceae). Two flowers, with male stage on the left, female stage (ovary visible) on the right. There are two pentamerous whorls of red 'petals': the outer whorl equals the 'true' corolla, the inner whorl is equivalent to five petaloid staminodes. D, *Clerodendrum minahassae* (Verbenaceae). Temporal continuum of flower shape from anthesis to fruit maturation. Anthetic flowers white, with long corolla tube and five petal lobes. After anthesis the perianth is dropped. The five sepals (S, white during anthesis) become firm and turn into a spreading red star surrounding the nearly black fleshy fruit. Thus, the calyx is adding 'petalness'

(Cont.)

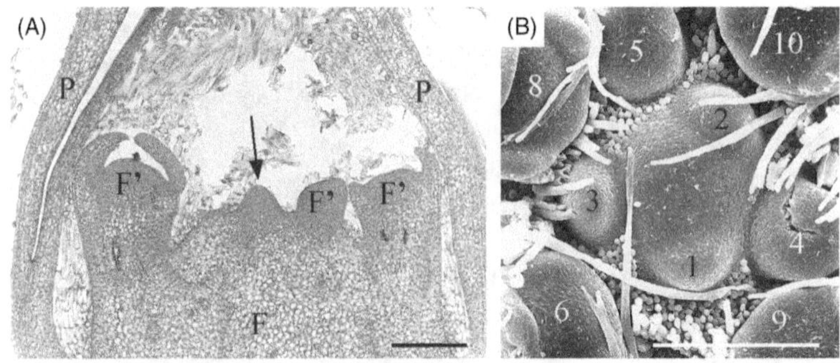

Figure 11.2 *Nymphaea prolifera* with meristem identity crisis of branching flowers (see overview Figure 11.1B). A, Longitudinal section of mother flower (F) with daughter flowers (F'). Note perianth (P) of mother flower. Arrow points to obliquely cut young leaf primordium next to apical meristem. Scale bar = 200 μm (reproduced from Grob *et al.* 2006). B, Top view on central portion of mother flower (perianth removed). The flower centre behaves like a shoot apical meristem, showing lateral primordia (1–10) along an ontogenetic spiral (site 7 is out of the frame). Most sites (4–6, 8–10) are occupied by floral primordia without any subtending leaves. Dorsiventral shape of primordia 1 and 3 (next to apical meristem) reveals their leaf identity. Hemispherical primordium 2 is the first stage of another young flower. Scale bar = 200 μm (reproduced from Grob *et al.* 2006).

Geuten *et al.* (2006) used the term 'floral organ identity crisis' or, more specifically, 'petal identity crisis' while focusing on the developmental genetics of petaloid sepals and the evolution of petaloidy in *Impatiens* (Balsaminaceae) and related eudicots. In basal angiosperms such as *Nymphaea* there is a continuum of forms (a 'morphocline') bridging sepals, petals and stamens. In, for example, *N. tuberosa* (Figure 11.1A) there are intermediates between typical sepals (green outside), typical petals (completely white) and typical stamens (yellow, with anthers fixed to thread-like filaments). Do these sepals and petals suffer from an identity crisis? Endress (2006: 7) seems to agree for *Nymphaea*: 'Therefore, it makes more sense to speak of tepals and to use the modifiers sepaloid tepals and petaloid tepals.' Fuzziness of floral organ identity gene action in various angiosperms is described by Soltis *et al.* (2005: 190). They

Caption for Figure 11.1 (Cont.)   after anthesis. E, *Warszewiczia coccinea* (Rubiaceae). The flower clusters along the twig are provided with flag-like leaves (red, 6 cm long) which are modified sepals (S). Photograph by P. Peisl (Zurich).

concluded: 'Morphology, developmental data and genetic data may provide conflicting evidence of homology (organ identity) and yet ultimately a more complete, and complex, view of a structure.' This is in agreement with a combinatorial notion of homology, such as suggested, for example, by Minelli (1998). Developmental mosaics between sepals, petals and stamens are acceptable because these kinds of floral appendages are usually taken as modified leaves. Arber (1950: 55) summarised the history of this idea: 'We may indeed agree with Goethe and deCandolle that petals and stamens show so much affinity that it is evidently reasonable to group them together. The petals will then be regarded as transition members between the vegetative and the actively reproductive parts of the floral shoot.'

Many angiosperms produce petaloid stamens, i.e. showy floral organs which according to their relative position in the developing flower are stamens and not petals. Jaramillo and Kramer (2007) call this phenomenon 'a decoupling of position from the morphological similarity of the structures in question.' For example, flowers of *Jacquinia pungens* (Figure 11.1C) have two pentamerous whorls of petaloid structures, with 'true' petals forming the outer whorl and petaloid staminodes forming the inner one (Walker-Larsen and Harder 2000). In *Costus* (related to ginger), 'true' sepals and petals are rather inconspicuous whereas the showy and broad lip (attractive for pollinators) is formed by the fusion of five staminodes (i.e. sterile stamens) (Figure 11.3A–D; Kirchoff 1991).

Although stamens and carpels are often said to be homologous structures, both being sporophylls, they are basically different in some respects. According to Endress (2006: 9) 'an ovule can also be compared with a stamen in some way, and the carpel is then more complex.' Nagasawa *et al.* (2003) described double mutants of rice (*Oryza*) having 'organs with unknown identity'. They are 'neither stamens nor carpels, but have partial floral identity.'

Leaf identity crisis: leaf–shoot indistinction

Some vascular plants transcend the classical model with respect to leaves and shoots (Cronk 2001, Hofer *et al.* 2001, Bharathan *et al.* 2002). For example, compound leaves of *Chisocheton* (Meliaceae) with indeterminate apical growth and three-dimensional branching due to epiphyllous shoots are developmental mosaics sharing growth processes with whole shoots. 'Indeterminate leaves' in *Chisocheton* and *Guarea* (Meliaceae) and examples of leaf–shoot indistinction in bladderworts

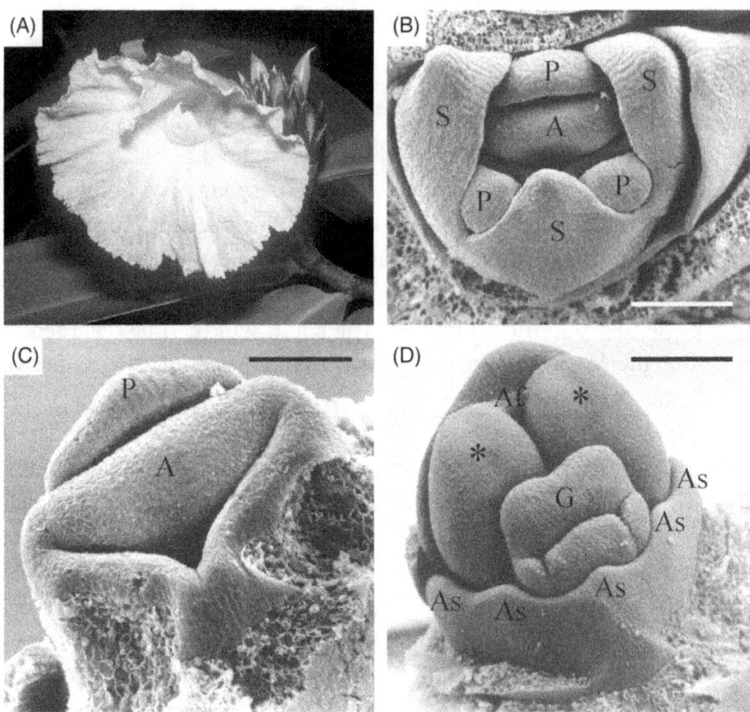

Figure 11.3 *Costus* (Costaceae) with petaloid androecium having an identity crisis. A, *Costus* sp.: frontal view of flower in anthesis, showing prominent white petaloid lip resulting from fusion of five staminodes. There is only one fertile stamen per flower. The 'true' sepals and petals are inconspicuous and hidden. Photograph by P. Peisl (Zurich). B–D, *Costus cuspidatus*. Flower development as shown by scanning electron micrographs before and after removal of sepals (S) and petals (P). The androecium arises as a collar-like girdling primordium (labelled A). Its prominent dorsal portion develops into the only fertile stamen (Af) with two anther halves (asterisks). The less prominent lateral and ventral portions of the girdling primordium have teeth equalling the five staminodal teeth (As) that will form the showy androecial lip. G = stigma (gynoecium). Scale bars = 200 μm.

(*Utricularia*, Lentibulariaceae) were presented by Rutishauser (1999), Rutishauser and Isler (2001), Fisher (2002) and Fukuda *et al.* (2003). Epiphyllous flowers are also known from African Podostemaceae (riverweeds) which are angiosperms adapted to waterfalls (Figure 11.5A; Rutishauser and Moline 2005).

Induction of leaves directly from leaves occurs in maize mutants (Schichnes *et al.* 1997), and induction of ectopic shoot meristems on leaves is known from *Arabidopsis* mutants (Byrne *et al.* 2003).

A genetically well-understood example illustrating the switch from leaf identity to shoot (or stem) identity is the *Hooded* mutant of *Hordeum vulgare*, also called hooded barley (Yagil and Stebbins 1969): the subtending leaf ('lemma' = bract) produces one or more ectopic spikelets on the awn. Molecular studies (e.g. Reiser *et al.* 2000) have elucidated that the hooded phenotype of barley is caused by misexpression of a *knox* gene. Williams-Carrier *et al.* (1997) suggested that the inverse polarity of the ectopic spikelets seen in the *Hooded* mutant of barley and transgenic *KNOTTED1* plants of maize results from the homeotic transformation (conversion) of the lemma awn into a reiterative inflorescence axis. Hooded barley is therefore an example of conversion of organ identity. In a phylogenetic context the above-mentioned cases of leaf–shoot indistinction are consistent with the hypothesis that leaves are derived from stem-like (or shoot-like) organs, at least in most ferns and seed-plants (Cronk 2001, Friedman *et al.* 2004).

We should also consider the identity question for tiny, scale-like leaves (1–2 mm long) of aquatic angiosperms such as Podostemaceae. For example, the moss-like leaves of *Tristicha* and allies lack vascular tissue completely (Rutishauser 1995). Thus, there are good reasons for doubt about their leaf identity: where is the borderline between scale-like hairs (trichomes) and rudimentary leaves?

Stipule identity crisis: stipule–leaf indistinction

Typical stipules are two lateral appendages of the leaf base. Leaf blade and stipules usually arise from a common primordial bulge at the shoot apex. Stipules occur in many dicot families (e.g. Fabaceae, Rosaceae) whereas they are absent in most monocots. Charlton (1991) found in *Azara microphylla* (Flacourtiaceae) a homeotic replacement of the stipule by a leaf. The same phenomenon is observable in pea (*Pisum sativum*) mutants such as *cochleata* (Yaxley *et al.* 2001). There are tendrils forming the blades as well as tendrils arising from stipular positions, a situation not known from any wild-type member of Fabaceae (Marx 1987, Hofer *et al.* 2001). In the *afila* (*af*) mutant all primary pinnae are replaced by a bunch of tendrils, whereas the stipules are not altered. A gene known to interact with the *af* gene is *sinuate leaf* (*sil*), which results in undulating margins of both leaflets and stipules. When combined with *af*, *sil* plants have adventitious tendrils arising from clefts in the distal portion of the stipule (Figure 11.4A, B; Marx 1987). Thus, pertinent characters of the leaf blade can be ectopically expressed in stipular sites. Or, as said by Yaxley *et al.* (2001), these

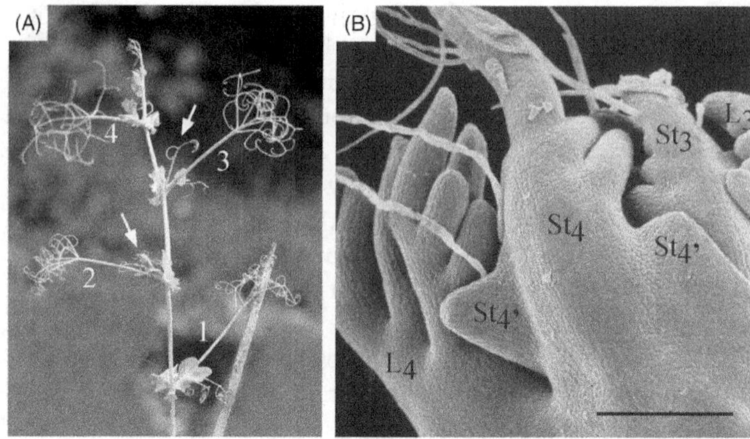

Figure 11.4 *Pisum sativum* (Fabaceae) 'afila'/'sinuate leaf ' (*af/sil*) double mutant, cultivated at the Botanical Garden Zürich from seeds received from G. A. Marx B777-188-(4) (fixed 3.10.1985). Stipules are changed into a more compound leaf-like identity. A, climbing shoot with compound leaves (1–4), each one associated with two leafy stipules. All leaflets of the blade are replaced by a bunch of tendrils (as typical for the *af* mutant). Moreover, *af/sil* plants have adventitious tendrils (arrows) arising from a cleft in the distal portion of each stipule. B, growing shoot tip of *af/sil* plant, showing two young leaves (including stipules). The blade of leaf 4 (L4) consists of primordial tendrils. The associated stipule (St4) gives rise to an upper portion again forming tendrils and two lateral stipular lobes (St4'). Scale bar = 200 µm.

mutants 'change stipules into a more "compound leaf-like" identity'. More examples of ectopic expression of leaf identity in stipular position are mentioned in Rutishauser and Sattler (1986), and Rutishauser (1999). Stipules in flowering plants are, by definition, restricted to the leaf base. However, a few mutants in *Arabidopsis* and pea are known to have supernumerary stipules which are ectopically expressed as part of the leaf blade or rachis (Tattersall *et al.* 2005). The so-called stipels at the base of the lateral leaflets in the compound leaves of the garden bean (*Phaseolus vulgaris*) may be understood as ectopically expressed supernumerary stipules (Arber 1950, Rutishauser and Isler 2001).

Stem identity crisis

Stems are the carriers of leaves and flowers in seed plants. Stems and roots are usually cylindrical and provided with apical meristems

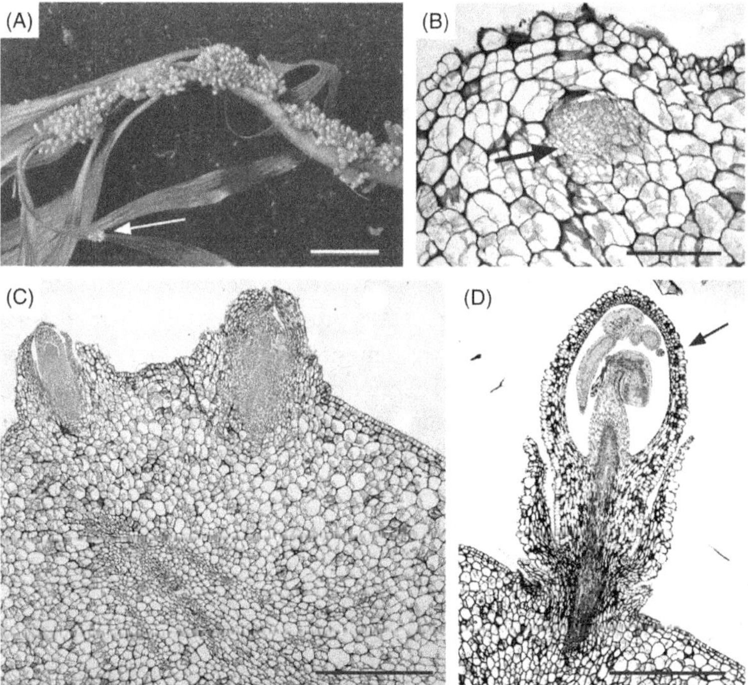

Figure 11.5 Endogenous flower formation along leafy stems of *Lederman-niella letouzeyi* (Podostemaceae = river-weeds). Do they suffer from an identity crisis? A, Elongate flowering shoot with forked leaves and many flower buds arising along the stem. Arrow points to epiphyllous flowers in cleft of forked leaf blade. Scale bar = 2 cm. B, Cross-section of stem cortex. Arrow points to endogenous shoot bud, still surrounded by cortex of mother stem. Several parenchyma cells of stem cortex start to divide up into meristematic cells (as part of dedifferentiation). Scale bar = 200 μm. C–D, Later developmental stages with endogenous flower buds penetrating the stem periphery. Each flower is protected by a sac-like cover = spathella (arrow), as typical for most Podostemaceae. Scale bars = 1 mm.

whereas typical leaves are dorsiventrally flattened and lacking an apical meristem. Stems and leaves result from the production of a chain of phytomers as reiterative units, each unit consisting of the node with the leaf attached, its axillary bud (if present) and a stem unit (internode) (Sylvester *et al.* 1996, Jaramillo and Kramer 2007). Thus, a stem may be understood as a composite structure (i. e. having 'compound identity'), consisting of the basal portions of the phytomers as developmental units. Stems and roots are usually distinguished by the position of outgrowing lateral shoots. In stems, lateral shoots (twigs) arise from exogenous buds along the stems, subtended by a leaf or bract in seed-plants. Endogenous shoot buds (without a subtending leaf) usually arise from roots and not from stems. There are, however, examples of endogenous flower bud origin along the stems in African Podostemaceae such as *Ledermanniella letouzeyi* (Figure 11.5A–D). Most flowers are initiated inside the stem cortex (owing to dedifferentiation) and penetrate the stem periphery prior to anthesis. This is a peculiar solution to increase flower numbers along stems *ad infinitum*. The flowers in these aquatic angiosperms can be viewed as developmental modules (i.e. quasi-autonomous parts, QAPs) induced out of their natural context. They develop all their defining features in locations of the body where they usually do not occur, demonstrating that development of the QAPs is locally controlled (Wagner in Bock and Cardew 1999, Rutishauser and Moline 2005). Thus, *Ledermanniella letouzeyi* and other African podostemads have a natural capacity to regenerate flowers from the adult stem cortex (see Xu *et al.* 2006).

Root identity crisis: root–shoot indistinction

Typical roots of vascular plants are non-photosynthetic, endogenously branching organs with an apical meristem protected by a cap. Developmental geneticists have pointed out that there are vascular plants that do not always show a clear differentiation into root and shoot. They have also stressed the fact that roots and shoots may have important regulatory mechanisms (including CLAVATA signalling pathways) in common (Scheres *et al.* 1996, Byrne *et al.* 2003, Birnbaum and Benfey 2004, Friedman *et al.* 2004, Stahl and Simon 2005). Lacking a better term for 'relevant organogenetic properties', Barlow *et al.* (2001) have spoken of properties of 'rootiness' guiding an uncommitted primordium towards the 'root' developmental pathway whereas properties of 'shootiness' are needed for the primordial commitment towards 'shoot' (i.e. leafy stem) development.

Root–shoot indistinction is consistent with the phylogenetic hypothesis that in vascular plants the root evolved from an ancestral shoot (Raven and Edwards 2001, Schneider *et al.* 2002). Thus, it becomes understandable that various flowering plants such as the river-weeds may suffer from a root identity crisis (Figure 11.6A–G; Rutishauser and Moline 2005). The Podostemaceae are peculiar angiosperms confined to tropical waterfalls and river-rapids. Their 'roots' (or what are usually called 'roots') are fixed to submerged rocks mainly with adhesive hairs (Figure 11.6A). Most podostemaceous roots are

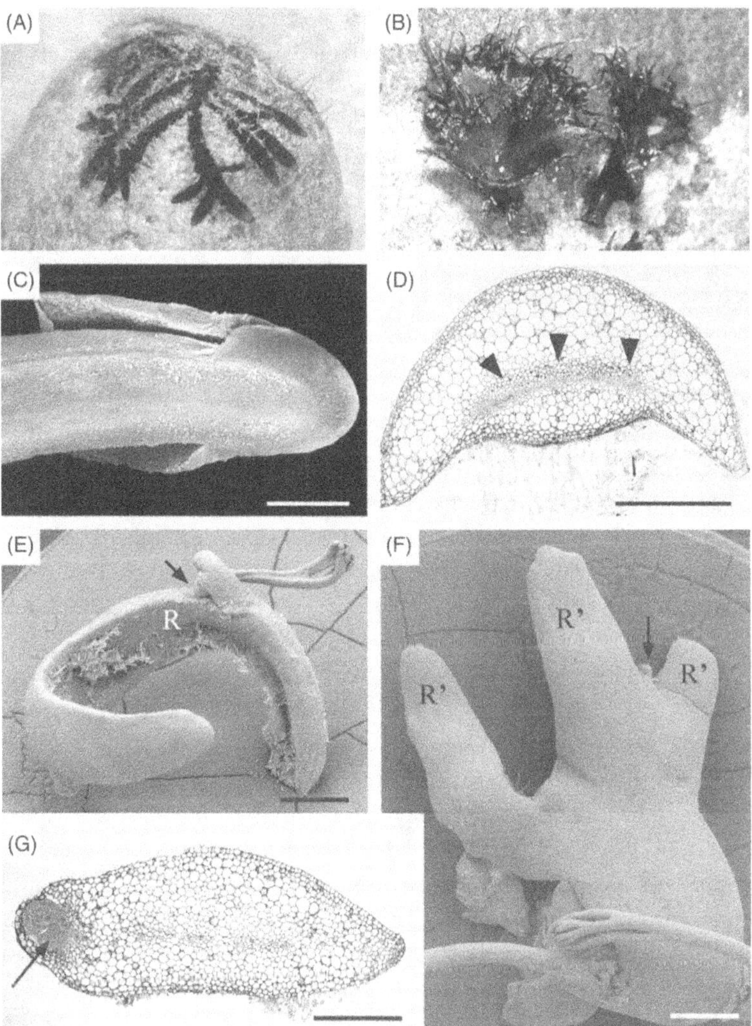

Figure 11.6 Podostemaceae (river-weeds) as flowering plants adapted to tropical waterfalls. Their roots are green, flattened and fixed to the rock. These roots are used to having an identity crisis. A–B, *Griffithella hookeriana* (S India) with root polymorphism: Roots are 1 cm broad ribbons and completely attached to the rock, or they are 2 cm high cups and fixed to the rock with a foot only. Needle-like leaves arise from endogenous buds along the root margins. C–D, *Thelethylax minutiflora* (Madagascar) with ribbon-like root. C, Root tip with asymmetrical cap, seen from below. D, Dorsiventral cross-section of ribbon-like root with concave lower side (fixed to the rock) and convex upper side. Arrow-heads point to lens-shaped vascular tissue replacing the central cylinder of typical roots.

(Cont.)

dorsiventrally flattened, forming ribbons or even broad crusts resembling foliose lichens. They are green and photosynthetic. They contain a lens-shaped vascular complex (Figure 11.6D) or a planar network of inconspicuous vascular bundles (Figure 11.6G). In various Podostemaceae the roots are provided with an asymmetrical cap (Figure 11.6C). The ribbon-like roots of other podostemads even lack such a cap and branch exogenously into daughter roots (Figures 11.6E, 11.6F). In the South Indian *Griffithella*, for example, the roots are highly polymorphic; there are either broad ribbons (Figure 11.6A) or cup-like structures with an anchoring foot (Figure 11.6B). Arber (1950: 134) came to the conclusion: 'The urge toward whole-shoot characters, which we have recognised in the leaf, may be detected, though less frequently, in the root. The root-thallus of the podostemads sometimes shows remarkably shoot-like features . . .'. This ambiguity of root organisation may explain why there are botanists who avoid the term 'root' for the flattened structures in Podostemaceae, using instead more neutral terms such as 'thallus' or even seemingly contradictory terms such as flattened 'stem' (Ota *et al.* 2001, Sehgal *et al.* 2002, Koi and Kato 2003).

Difficulties in distinguishing 'root identity' and 'shoot identity' (leafy stolons) are also known from Lentibulariaceae such as bladderworts (*Utricularia*) and butterworts (*Pinguicula*). Usually it is said that *Pinguicula* has roots and the sister genus *Utricularia* lacks them. However, *Utricularia* stolons may have arisen from what are called 'roots' in *Pinguicula* just by adding exogenous leaves to the root surface, as proposed by Rutishauser and Isler (2001). This would explain the high degree of similarity of the stolons (stems) of various bladderworts with the roots of some butterworts: lack of calyptra (root-cap), positive geotropic growth and 'awkward' phyllotaxis patterns found in bladderwort stolons, e.g. orthomonostichy with all leaves arranged along a single stem sector. Genes homologous to WUSCHEL in *Arabidopsis* may be involved in the ectopic induction of leaf development when *Pinguicula* 'roots' turn into *Utricularia* 'stolons' (see Gallois *et al.* 2004).

Caption for Figure 11.6 (Cont.)   Scale bars = 250 μm and 500 μm, respectively. E–G, *Stonesia ghoguei* (Cameroon) with narrow to broad ribbons, lacking root caps. Roots (R) branch exogenously giving rise to daughter roots (R'). Arrows point to shootlets arising from endogenous buds along the root margin. Vascular root tissue (as shown in cross-section) is divided up into a planar network. Scale bars = 1 mm (E, F) and 300 μm (G).

CONCLUDING REMARKS

An organ or structure is called heterotopic when it develops in an unusual position within the body plan. Concepts such as 'heterotopy', 'homeosis', 'homocracy' and 'ectopic expression of organ identity' coincide with the concept of 'quasi-autonomous parts' which can be induced out of their natural context (Sattler 1994, Shubin *et al.* 1997, Wagner in Bock and Cardew 1999, Baum and Donoghue 2002, Svensson 2004, Rutishauser and Moline 2005, Jaramillo and Kramer 2007). All these concepts describe the transformation of parts into structures normally found elsewhere according to the body plan. Developmental geneticists are invited to study some of the abovementioned plants in order to explain the decanalisation (relaxation) of their body plans as compared with typical vascular plants. Which key regulatory genes (including homeotic genes such as KNOX and MADS-box genes) are involved in the identity crises of these plants?

Vergara-Silva (2003: 260) gave a preliminary explanation for the frequently occurring identity crises of meristems and organs in vascular plants: 'Distinct groups of genes that in principle act in one categorical structure, are actually also expressed in another, and . . . the consequence that this overlapping pattern has on cell differentiation is an effective blurring of the phenotypic boundary between the structures themselves.'

ACKNOWLEDGEMENTS

This chapter is part of a research project supported by the Swiss National Science Foundation (Grant No. 3100 AO-105974/1).

REFERENCES

Arber, A. 1947. Analogy in the history of science. In M. F. Ashley Montagu (ed.) *Studies and Essays in the History of Science and Learning, Offered in Homage to G. Sarton.* New York: Henry Schuman, pp. 221–233.

Arber, A. 1950. *The Natural Philosophy of Plant Form.* Cambridge: Cambridge University Press.

Baluska, F., Volkmann, D. & Barlow, P. W. 2004. Cell bodies in a cage. *Nature* **428**, 371.

Barlow, P. W., Lück, H. B. & Lück, J. 2001. The natural philosophy of plant form: cellular autoreproduction as a component of a structural explanation of plant form. *Annals of Botany* **88**, 1141–1152.

Baum, D. A. & Donoghue, M. J. 2002. Transference of function, heterotopy and the evolution of plant development. In Q. C. B. Cronk, R. M. Bateman & J. A. Hawkins (eds.) *Developmental Genetics and Plant Evolution.* London: Taylor & Francis, pp. 52–69.

Bey, M., Stüber, K., Fellenberg, K. *et al.* 2004. Characterization of *Antirrhinum* petal development and identification of target genes of the class B MADS box gene *DEFICIENS*. *Plant Cell* **16**, 3197–3215.

Bharathan, G., Goliber, T. E., Moore, C. *et al.* 2002. Homologies in leaf form inferred from KNOX1 gene expression during development. *Science* **296**, 1858–1860.

Birnbaum, K. & Benfey, P. N. 2004. Network building: transcriptional circuits in the root. *Current Opinion in Plant Biology* **7**, 582–588.

Blochlinger, K., Jan, L. Y. & Jan, Y. N. 1991. Transformation of sensory organ identity by ectopic expression of *Cut* in *Drosophila*. *Genes & Development* **5**, 1124–1135.

Bock, G. R. & Cardew, G. (eds.) 1999. *Homology (Novartis Foundation Symposium 222)*. Chichester: John Wiley & Sons.

Brower, A. V. Z. 2000. Homology and the inference of systematic relationships: some historical and philosophical perspectives. In R. Scotland & R. T. Pennington (eds.) *Homology and Systematics*. London: Taylor & Francis, pp. 10–21.

Byrne, M. E., Kidner, C. A. & Martienssen, R. A. 2003. Plant stem cells: divergent pathways and common themes in shoots and roots. *Current Opinion in Genetics and Development* **13**, 551–557.

Charlton, W. A. 1991. Homoeosis and shoot construction in *Azara microphylla* Hook. (Flacourtiaceae). *Acta Botanica Neerlandica* **40**, 329–337.

Coen, E. 1999. *The Art of Genes: How Organisms Make Themselves*. Oxford: Oxford University Press.

Cronk, Q. C. B. 2001. Plant evolution and development in a post-genomic context. *Nature Reviews Genetics* **2**, 607–619.

Elledge, S. J. 1996. Cell cycle checkpoints: preventing an identity crisis. *Science* **274**, 1664–1672.

Endress, P. K. 2006. Angiosperm floral evolution: morphological developmental framework. *Advances in Botanical Research* **44**, 1–61.

Fisher, J. B. 2002. Indeterminate leaves of *Chisocheton* (Meliaceae): survey of structure and development. *Botanical Journal of the Linnean Society* **139**, 207–221.

Friedman, W. E., Moore, R. C. & Purugganan, M. D. 2004. The evolution of plant development. *American Journal of Botany* **91**, 1726–1741.

Fukuda, T., Yokoyama, J. & Tsukaya, H. 2003. Phylogenetic relationships among species in the genera *Chisocheton* and *Guarea* that have unique indeterminate leaves as inferred from sequences of chloroplast DNA. *International Journal of Plant Science* **164**, 13–24.

Gallois, J.-L., Nora, F. R., Mizukami, Y. & Sablowski, R. 2004. WUSCHEL induces shoot stem cell activity and developmental plasticity in the root meristem. *Genes & Development* **18**, 375–380.

Geuten, K., Becker, A., Kaufmann, K. *et al.* 2006. A petal identity crisis – MADS-domain proteins and the evolution of petaloidy in balsaminoid Ericales. *First Meeting of the European Society for Evolutionary Developmental Biology, Prague*. Abstracts, pp. 102–103.

Grob, V., Moline, P., Pfeifer, E., Novelo, A. R. & Rutishauser, R. 2006. Developmental morphology of branching flowers in *Nymphaea prolifera*. *Journal of Plant Research* **119**, 561–570.

Hawkins, J. A. 2002. Evolutionary developmental biology: impact on systematic theory and practice, and the contribution of systematics. In Q. C. B. Cronk, R. M. Bateman & J. A. Hawkins (eds.) *Developmental Genetics and Plant Evolution*. London: Taylor & Francis, pp. 32–51.

Hay, A. & Mabberley, D. J. 1994. On perception of plant morphology: some implications for phylogeny. In D. S. Ingram & A. Hudson (eds.) *Shape and Form in Plants and Fungi*. London: The Linnean Society of London, pp. 101–117.

Hofer, J. M. I., Gourlay, C. W. & Ellis, T. H. N. 2001. Genetic control of leaf morphology: a partial view. *Annals of Botany* **88**, 1129–1139.

Jackson, D. 1996. Plant morphogenesis: designing leaves. *Current Biology* **6**, 917–919.

Jaramillo, M. A. & Kramer, E. M. 2007. The role of developmental genetics in understanding homology and morphological evolution in plants. *International Journal of Plant Sciences* **168**, 61–72.

Kirchoff, B. K. 1991. Homeosis in the flowers of the Zingiberales. *American Journal of Botany* **64**, 833–837.

Koi, S. & Kato, M. 2003. Comparative developmental anatomy of the root in three species of *Cladopus* (Podostemaceae). *Annals of Botany* **91**, 927–933.

Lacroix, C., Jeune, B. & Purcell-Macdonald, S. 2003. Shoot and compound leaf comparisons in eudicots: dynamic morphology as an alternative approach. *Botanical Journal of the Linnean Society* **143**, 219–230.

Lugassi, N., Nakayama, N., Irish, V. F. & Zik, M. 2006. Identity crisis: a novel floral organ mutant. *XV Congress of the Federation of European Societies of Plant Biology, Lyon*. Abstracts, DEVO1–014.

Marx, G. A. 1987. A suite of mutants that modify pattern formation in pea leaves. *Plant Molecular Biology Reporter* **5**, 311–335.

Minelli, A. 1998. Molecules, developmental modules and phenotypes: a combinatorial approach to homology. *Molecular Phylogenetics and Evolution* **9**, 340–347.

Minelli, A. & Fusco, G. 2004. Evo-devo perspectives on segmentation: model organisms, and beyond. *Trends in Ecology and Evolution* **19**, 423–429.

Nagasawa, N., Miyoshi, M., Sano, Y. *et al.* 2003. *SUPERWOMAN1* and *DROOPING LEAF* leaf genes control floral organ identity in rice. *Development* **130**, 705–718.

Ota, M., Imaichi, R. & Kato, M. 2001. Developmental morphology of the thalloid *Hydrobryum japonicum* (Podostemaceae). *American Journal of Botany* **88**, 382–390.

Parcy, F., Nilsson, O., Busch, M. A., Lee, I. & Weigel, D. 1998. A genetic framework for floral patterning. *Nature* **395**, 561–566.

Prusinkiewicz, P. 2004. Self-similarity in plants: integrating mathematical and biological perspectives. In M. Novak (ed.) *Thinking in Patterns. Fractals and Related Phenomena in Nature*. Singapore: World Scientific, pp. 103–118.

Raven, J. A. & Edwards, D. 2001. Roots: evolutionary origins and biogeochemical significance. *Journal of Experimental Botany* **52**, 381–401.

Reiser, L., Sánchez-Baracaldo, P. & Kake, S. 2000. Knots in the family tree: evolutionary relationships and functions of knox homeobox genes. *Plant Molecular Biology* **42**, 151–166.

Rutishauser, R. 1995. Developmental patterns of leaves in Podostemaceae as compared to more typical flowering plants: saltational evolution and fuzzy morphology. *Canadian Journal of Botany* **73**, 1305–1317.

Rutishauser, R. 1999. Polymerous leaf whorls in vascular plants: developmental morphology and fuzziness of organ identity. *International Journal of Plant Science* **160**, S81–S103.

Rutishauser, R. & Isler, B. 2001. Fuzzy Arberian Morphology: *Utricularia*, developmental mosaics, partial shoot hypothesis of the leaf and other FAMous ideas of Agnes Arber (1879–1960) on vascular plant bauplans. *Annals of Botany* **88**, 1173–1202.

Rutishauser, R. & Moline, P. 2005. Evo-devo and the search for homology ("sameness") in biological systems. *Theory in Biosciences* **124**, 213–241.

Rutishauser, R. & Sattler, R. 1986. Architecture and development of the phyllode-stipules whorls in *Acacia longipedunculata*: controversial interpretations and continuum approach. *Canadian Journal of Botany* **64**, 1987–2019.

Sattler, R. 1994. Homology, homeosis, and process morphology in plants. In B. K. Hall (ed.) *The Hierarchical Basis of Comparative Biology*. New York: Academic Press, pp. 423–475.

Sattler, R. 1996. Classical morphology and continuum morphology: opposition and continuum. *Annals of Botany* **78**, 577–581.

Sattler, R. 2001. Some comments on the morphological, scientific, philosophical and spiritual significance of Agnes Arber's life and work. *Annals of Botany* **88**, 1215–1217.

Sattler, R. & Jeune, B. 1992. Multivariate analysis confirms the continuum view of plant form. *Annals of Botany* **69**, 249–262.

Sattler, R. & Rutishauser, R. 1990. Structural and dynamic descriptions of the development of *Utricularia foliosa* and *U. australis*. *Canadian Journal of Botany* **68**, 1989–2003.

Scheres, B. 2001. Plant cell identity. The role of position and lineage. *Plant Physiology* **125**, 112–114.

Scheres, B., McKhann, H. I. & van den Berg, C. 1996. Roots redefined: anatomical and genetic analysis of root development. *Plant Physiology* **111**, 959–964.

Schichnes, D., Schneeberger, R. & Freeling, M. 1997. Induction of leaves directly from leaves in the maize mutant *Lax midrib I-zero*. *Developmental Biology* **186**, 36–45.

Schneider, H., Pryer, K. M., Cranfill, R., Smith, A. R. & Wolf, P. G. 2002. Evolution of vascular plant body plans: a phylogenetic perspective. In Q. C. B. Cronk, R. M. Bateman, J. A. Hawkins (eds.) *Developmental Genetics and Plant Evolution*. London: Taylor & Francis, pp. 1–14.

Sehgal, A., Sethi, M. & Mohan Ram, H. Y. 2002. Origin, structure, and interpretation of the thallus in *Hydrobryopsis sessilis* (Podostemaceae). *International Journal of Plant Science* **163**, 891–905.

Shubin, N., Tabin, C. & Carroll, S. 1997. Fossils, genes and the evolution of animal limbs. *Nature* **388**, 639–648.

Sinha, N. R. 1999. Leaf development in angiosperms. *Annual Review of Plant Physiology and Plant Molecular Biology* **50**, 419–446.

Soltis, D. E., Albert, V. A., Kim, S. *et al.* 2005. Evolution of the flower. In R. J. Henry (ed.) *Plant Diversity and Evolution: Genotypic and Phenotypic Variation in Higher Plants*. Cambridge, MA: CAB International, pp. 165–200.

Stahl, Y. & Simon, R. 2005. Plant stem cell niches. *International Journal of Developmental Biology* **49**, 479–489.

Steeves, T. A., Hicks, G., Steeves, M. & Retallack, B. 1993. Leaf determination in the fern *Osmunda cinnamomea*: a reinvestigation. *Annals of Botany* **71**, 511–517.

Svensson, M. E. 2004. Homology and homocracy revisited: gene expression patterns and hypotheses of homology. *Development Genes & Evolution* **214**, 418–421.

Sylvester, A. W., Smith, L., & Freeling, M. 1996. Acquisition of identity in the developing leaf. *Annual Review of Cell and Developmental Biology* **12**, 257–304.

Tattersall, A. D., Turner, L., Knox, M. R. *et al.* 2005. The mutant *crispa* reveals multiple roles for *PHANTASTICA* in pea compound leaf development. *The Plant Cell* **17**, 1046–1060.

Theissen, G. 2000. Plant Breeding: *Flo*-like meristem identity genes: from basic science to crop plant design. *Progress in Botany* **61**, 167–183.

Theissen, G. 2005. Birth, life and death of developmental control genes: new challenges for the homology concept. *Theory in Biosciences* **124**, 199–212.

Tsukaya, H. 1995. Developmental genetics of leaf morphogenesis in dicotyledonous plants. *Journal of Plant Research* **108**, 407–416.

Tsukaya, H. 2002. Interpretation of mutants in leaf morphology: genetic evidence for a compensatory system in leaf morphogenesis that provides a new link between cell and organismal theories. *International Review of Cytology* **217**, 1–39.

Vergara-Silva, F. 2003. Plants and the conceptual articulation of evolutionary developmental biology. *Biology and Philosophy* **18**, 249–284.

Walker-Larsen, J. & Harder, L. D. 2000. The evolution of staminodes in angiosperms: patterns of stamen reduction, loss, and functional re-invention. *American Journal of Botany* **87**, 1367–1384.

Williams-Carrier, R. E., Lie, Y. S., Hake, S. & Lemaux, P. G. 1997. Ectopic expression of the maize *kn1* gene phenocopies the *Hooded* mutant of barley. *Development* **124**, 3737–3745.

Xu, J., Hofhuis, H., Heidstra, R. *et al.* 2006. A molecular framework for plant regeneration. *Science* **311**, 385–388.

Yagil, E. & Stebbins, G. L. 1969. The morphogenetic effects of the hooded gene in barley. II. Cytological and environmental factors affecting gene expression. *Genetics* **62**, 307–319.

Yaxley, J. L., Jablonski, W. & Reid, J. B. 2001. Leaf and flower development in pea (*Pisum sativum* L.): Mutants *cochleata* and *unifoliata*. *Annals of Botany* **88**, 225–234.

Yu, D., Kotilainen, M., Pöllänen, E. *et al.* 1999. Organ identity genes and modified patterns of flower development in *Gerbera hybrida*. *Plant Journal* **17**, 51–62.

Part III   Evolving diversity

Are there clades whose particular origin or evolutionary history are more adequately explained when considering the possibilities offered by changes at the level of developmental processes, instead of thinking in terms of the unceasing interplay between gene mutation and natural selection? Yes, this is exactly the field where evo-devo offers its best performances. These are stories rooted deep in time, where one must also consider the possibility that in due course even the 'rules of the game', such as the role of Hox genes, or more generally the genotype–phenotype relationships, have evolved along with their products. Or, they may be stories where adaptive explanations are unsatisfactory, and the course of evolution can appear to be driven more by the nature of variation that is produced at each generation than by adaptive necessities.

Jaume Baguñà, Pere Martinez, Jordi Paps and Marta Riutort (Chapter 12) address the problem of early bilaterian evolution. Their work is based on the most recent molecular phylogenies and on new data on Hox/ParaHox and microRNA sets that identify acoelomorphs as the earliest branching extant bilaterians. Evidence for axial homologies in gene expression between cnidarians and bilaterians and the evidence that cnidarians were at their origin bilaterally symmetric all point to an older last common ancestor for bilaterians. Thus, what under different phylogenetic hypotheses appear to be a number of phylogenetically coincident character changes (the complex Urbilateria hypothesis) turn into a series of nodes connected by stem ancestors along which new characters were progressively acquired.

Staying with the problem of bilaterian origins and early evolution, but at the level of molecular genetics, Jean Deutsch and Philippe Lopez (Chapter 13) present a novel hypothesis that argues that the expansion of the Hox complex at the base of the bilaterian clade was produced by a series of transposition events, and that the Hox genes themselves originated from transposons. The authors support their hypothesis by a discussion of the similarity between the homeodomain and the DNA-binding domain of bacterial integrases and eukaryotic transposases, and through the investigation of some rearrangements of the Hox complex in the drosophilid lineage. This results in a scenario for the evolution of the Hox complex from the basic complement of Hox genes in the common ancestor of cnidarians and bilaterians that accounts for several properties of the extant Hox genes.

Diversity in embryogenesis of nematode worms is much higher than what would be expected on the basis of the degree of variation among the juvenile phenotypes. The abundance of early developmental variations appears somehow paradoxical, as these do not have any obvious impact on the structure or performance of the resulting worms. Einhard Schierenberg and Jens Schulze (Chapter 14) ask why there are so many different developmental pathways to reach essentially the same goal. Did the special body plan of nematodes prevent a degree of morphological diversification like in arthropods or vertebrates? The authors explore possible explanations, among which is the interplay between the genetic program and external conditions (inside or outside the organism) that determines the chance for deviations from an original developmental pattern to arise and to succeed.

A demonstration that evo-devo does not reduce to comparative developmental genetics is provided by Nigel Hughes, Joachim Haug and Dieter Waloszek (Chapter 15) who offer a palaeo-evo-devo perspective on the evolution of basal euarthropod development based on the fossil record. The chapter reviews the morphological development of early arthropods as reflected by the ontogenetic series of several species of trilobites and a basal crustacean lineage known as the 'Orsten'-type fauna. Particular attention is paid to the segmentation schedule. Patterns of segment generation shared by these primitive groups may provide insights into the developmental mode of basal Euarthropoda, and thus into the evolution of arthropod ontogeny.

The origin and early evolution of land plants from aquatic 'algae' is discussed by Jane Langdale and Jill Harrison (Chapter 16). They review a selection of major steps in land plant evolution from an evo-devo perspective. These steps include the passage from haplontic to haplodiplontic life cycle, the emergence of apical growth and branching development, the evolution of vascular and root systems, the advances in energy storage and reproduction strategies. Developmental data, both at molecular-genetic and morphological level, concur to reconstruct a scenario for these key evolutionary transitions.

12

Unravelling body plan and axial evolution in the Bilateria with molecular phylogenetic markers

JAUME BAGUÑÀ, PERE MARTINEZ, JORDI PAPS AND MARTA RIUTORT

SETTING THE PROBLEM

The emergence of dramatic morphological differences (disparity) and the ensuing bewildering increase in the number of species (diversity) documented in the fossil record at key stages of animal and plant evolution have defied, and still defy, the explanatory powers of Darwin's theory of evolution by natural selection. Among the best examples that have captured the imagination of the layman and the interest of scores of scientists for 150 years are the origins of land plants from aquatic green plants, of flowering plants from seed plants, of chordates from non-chordates and of tetrapod vertebrates from non-tetrapods; and the conquest of the land by amphibians; the emergence of endotherms from ectotherm animals; the recurrent invention of flight (e.g. in arthropods, birds and mammals) from non-flying ancestors; and the origin of aquatic mammals from four-legged terrestrial ancestors.

Key morphological transitions pose a basic difficulty: reconstruction of ancestral traits of derived clades is problematic because of a lack of transitional forms in the fossil record and obscure homologies between 'ancestral' and derived groups. Lack of transitional forms, in other words gaps in the fossil record, brought into question one of the basic tenets of Darwin's theory, namely gradualism, as Darwin himself acknowledged. Since Darwin, however, and especially in the past 50 years, numerous examples that may reflect transitional stages between major groups of organisms have accumulated. Good examples

Evolving Pathways: Key Themes in Evolutionary Developmental Biology, ed. Alessandro Minelli and Giuseppe Fusco. Published by Cambridge University Press.

are the numerous fossils that connect whales, sirenians, seals and sea lions with different lineages of terrestrial mammals, the converse transitional series from swimming tetrapods to land tetrapods, the many fossils showing the transition from dinosaurs to birds illuminating the origin and early functions of feathers and flight, and those fossils illustrating the intermediate changes during the transition from aquatic green plants to land plants and from these to vascular plants.

Back in geological time, the last and potentially crippling example to the acceptance of the Darwinian theory is the advent of bilaterally symmetrical animals and its coincidence with the abrupt appearance of large-bodied skeletonised remains of most extant phyla. The event is usually referred to as the Cambrian 'explosion'. A great deal has been written about it, namely the recent reviews by Budd (2003), Valentine (2004), Conway Morris (2006) and Marshall (2006), to which readers are referred. In the writings of Gould (1989) the Cambrian 'explosion' has been considered the pivotal event in animal evolution for which special mechanisms have been sought, e.g. in terms of macro-evolutionary events. However, because the Cambrian 'explosion' mainly refers to the 'explosion' of bilaterally symmetrical body plans, we will argue that an understanding of the origin of bilateral organisms is even more important than the so-called Cambrian 'explosion', as well as a necessary step to explain it.

TRACKING DOWN THE EARLIEST EXTANT BILATERIANS: A SIMPLE OR A COMPLEX LAST COMMON ANCESTOR (LCA)?

By any standard, the appearance of bilateral organisms is the most thrilling success in animal evolution: 34 out of the 38 living phyla and over 99% of described living animal species are bilaterians, far more complex in structure and far more diverse in morphology and ecology than their radial forebears. A brief glimpse at any bilaterian organism, however simple, uncovers the main reasons for their evolutionary success: two oriented body axes and directed locomotion. The main or primary axis (antero-posterior, or A-P) distinguishes 'front' from 'back' of the body and is associated with the direction of locomotion, with the mouth, brain and sensory structures located at or near the anterior end, and the anus and other structures located at or near the posterior end. The second axis (dorso-ventral or D-V), orthogonal to the first, identifies the 'top' from the 'bottom' of the body, the latter usually related to locomotion, while the 'top' or dorsal bears sensory and defensive structures to avoid predation. Oriented locomotion was the key to the colonisation by the pre-Cambrian benthos and thereafter the plankton, and aided by the development of sensory structures and feeding organs at the anterior/ventral end that increased predatory and escape capabilities. Another key feature of bilaterians is the presence of a third embryonic layer, the mesoderm, between the ectoderm and endoderm. In combination with either the ecto- or endoderm, the mesoderm provides an extraordinary variety of new tissues and organs not seen in any radial organism. Finally, other features often considered to be present in the first bilaterians are a true brain, a through-gut, excretory system, body cavities (coelom), segments, and even appendages and simple hearts and eyes (Table 12.1).

Current views suggest that the bilaterians arose from ancestors that were radially symmetric instead of bilateral and, therefore, had a single body axis (the oral-aboral, or O-AB) and no mesoderm (hence diploblastic). In addition, they had a decentralised nerve net and a blind gut. These features are maintained by the extant members of the phylum

Table 12.1 *Character states of the main morphological and developmental components at the dawn of bilaterians.*

The simple Urbilateria scenario assumes a structurally simple organism. The alternative complex Urbilateria scenario considers that most morphological and developmental components of extant bilaterians were also functionally conserved in the bilaterian ancestor.

Developmental and morphological characters	Simple Urbilateria	Complex Urbilateria
1 – A-P axis	Present	Present
2 – D-V axis	Present	Present
3 – Mesoderm	Present	Present
4 – Nervous system	Present (slightly centralised)	Present (centralised; CNS)
5 – Hox cluster	Basic (3–4 genes)	Expanded (7–9 genes)
6 – Brain	Clumps of cells	Present (true brain)
7 – Gut	Blind gut	Through-gut
8 – Excretory system	Absent	Present
9 – microRNAs	? (few)	? (some)
10 – Body cavities (coelom)	Absent	Present
11 – Segmentation	Absent	Present
12 – Heart	Absent	Present
13 – Appendages	Absent	Present?
14 – Body size	Small	Large
15 – Life cycle	Direct	Indirect (+larvae)

Cnidaria (corals, sea anemones, hydras and jellyfish) and Ctenophora (comb jellies). However, whenever a hypothetical early bilaterian with the first, second or third set of apomorphic (derived) characters (Table 12.1) is compared with a radial organism bearing none of them, and from which it is assumed to originate, one is left wondering how this actually took place.

Since Haeckel's Gastraea, scores of theories have tried to answer this key evolutionary question (see Willmer 1990, for a historical review, and Holland 2003, for details on the evolution of the nervous system). In a first major set of hypotheses, ancestral bilaterian traits such as body axes and mesoderm appeared concurrently with advanced characters such as coelom and segments. Hence, non-segmented, non-coelomate cnidarians with blind guts, either under larval or adult appearance, were directly transformed to coelomate segmented bilaterians, bearing through-guts and complex nervous systems (Archicoelomate Theories) (for a recent critical update, see Holland 2003). A second major set of hypotheses (see Salvini-Plawen 1978 for a thorough review) featured a more gradual scenario from sexually reproducing, bottom-pelagic organisms (protoplanula or archiplanula), akin to present cnidarian planula larva, already exhibiting bilateral symmetry. From such organisms originated the cnidarian polyps, which settled onto the substratum, as well as the early bilaterians which resembled present day acoel and nemertodermatid flatworms (Planula-Acoeloid Theory). Accordingly, the first bilaterians were non-segmented, non-coelomate (acoelomate) organisms with a blind gut from which pseudo-coelomate and coelomate, segmented and non-segmented protostomes and deuterostomes evolved.

The phylogenetic consequences of these conflicting scenarios, in terms of character changes necessary between ancestors and descendants, are very different. Under the archicoelomate scenario, the number of coincident characters clumping at the Last Common Ancestor (LCA) node of the bilaterians is large. This makes it difficult to place them into any temporal order along the stem leading to the LCA (Figure 12.1A). Also, it implies either a large number of extinctions of intermediary taxa and, consequently, major gaps in our knowledge, or a wholesale correlated transformation from one life form (radial) to another (bilateral). Under this hypothesis, the LCA appears as a rather complex organism (dubbed complex Urbilateria; Kimmel 1996). In contrast, the planuloid-acoeloid scenario posits a reduced number of characters at the stem leading to the LCA (Figure 12.1B), and features fewer and simpler stem ancestors and a simple LCA. Under both scenarios,

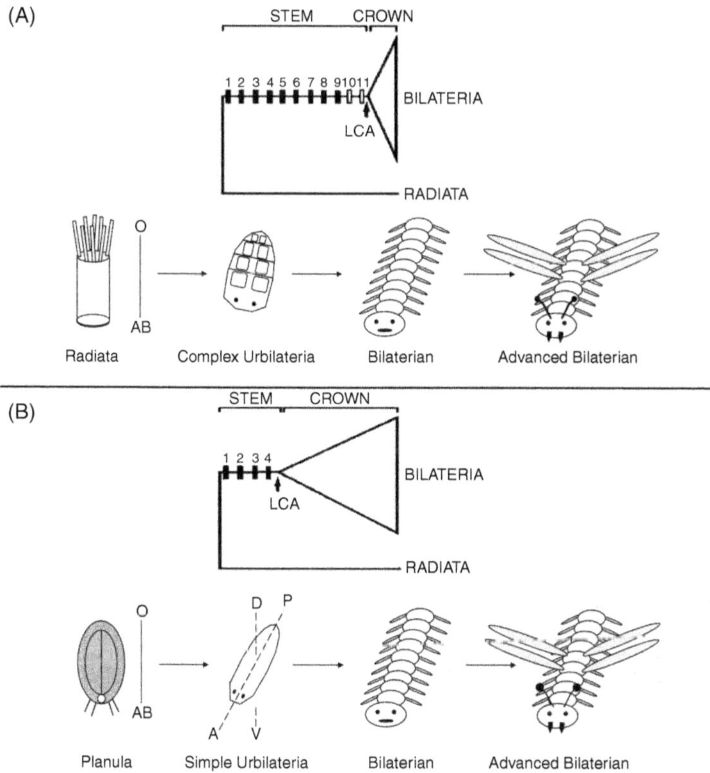

Figure 12.1 Conflicting phylogenies and scenarios on the nature and origin of the Last Common Ancestor (LCA) of the Bilateria, also featuring the extent of stem and crown groups. A, The complex Urbilateria scenario features a large, complex ancestor bearing most characters of present-day bilaterians (characters 1–9, and eventually characters 10–11 in Table 12.1). This ancestor originated from either an adult (polyp) or a larval radial cnidarian (archicoelomate theory, originally proposed by Sedgwick 1884). From this LCA evolved the more complex protostomes and deuterostomes. Note that all characters leading to the LCA are clumped at the stem. The large triangle indicates the diversification of crown bilaterians and its short height shows that its rate was fast (Cambrian 'explosion'?). B, The simple Urbilateria scenario features a small, simple LCA, similar to present-day acoelomorph flatworms, bearing a reduced set of characters (1–4 of Table 12.1) of extant bilaterians. This ancestor originated from radial planuloid ancestors similar to the planula larva of extant cnidarians (planuloid-acoeloid theory; for main references see Salvini-Plawen 1978 and Willmer 1990). From this ancestor evolved more complex bilaterians to be followed by the most advanced protostomes and deuterostomes. Note that the number of characters leading to the LCA are few, that time of diversification of crown bilaterians was longer and its rate slower than in the alternative scenario. A: anterior; AB: aboral; D: dorsal; O: oral; P: posterior; V: ventral.

however, phylogenetic advances may discover extinct (or hopefully extant) clades that break coincident character changes at the stem. The intercalation of these new clades will distribute inferred character changes across a series of branches instead of having them distributed solely at the LCA node (Donoghue 2005, Butterfield 2006).

In the 1990s, molecular phylogenies based on sequences of the ribosomal gene 18S and the Hox gene clusters bolstered the Archicoelomate scenario (and the complex Urbilateria). Both sets of data split the Bilateria into three superclades, the classical Deuterostomia and the protostomes divided into Ecdysozoa (Aguinaldo *et al.* 1997) and Lophotrochozoa (Halanych *et al.* 1995). The Ecdysozoa clustered several pseudocoelomate groups with arthropods, while the Lophotrochozoa joined most acoelomates (e.g. Platyhelminthes) to coelomate spiralians and lophophorates. Acoelomates and pseudocoelomates were displaced to more derived positions inside the tree and, therefore, had to originate by morphological simplification from complex coelomate segmented ancestors. Moreover, the amazing conservation of the genetic toolkit across the Bilateria, together with the apparently homologous expression of key developmental genes (e.g. segmentation and nervous system genes) in disparate bilaterian clades (annelids, insects, vertebrates; De Robertis and Sasai 1996, P. W. H. Holland 1998, L. Z. Holland 2000) were taken as evidence for the existence of similar developmental programs and their ensuing morphological characters in the Urbilateria ancestor. Finally, the lack of resolution of branching phyla within the three superclades gave support to the Cambrian 'explosion' as a real, sudden, cladogenetic event. In summary, hopes of finding extant 'intermediates' in the bilaterian stem lineage were considered doomed (Adoutte *et al.* 1999), the gradist interpretation of early bilaterian evolution dismissed, and the complex Urbilateria enthroned (Carroll *et al.* 2001).

THE ACOELOMORPHA, A LIKELY CANDIDATE FOR THE EARLIEST BRANCHING EXTANT BILATERIANS

Whereas the splitting of the Bilateria into the three superclades was corroborated by further data, other tenets of the new phylogeny proved unfounded. First, most new phylogenies were heavily pruned, leaving out several 'minor' phyla, namely 'basal' ecdysozoans and lophotrochozoans, to which most pseudocoelomates and acoelomates belong (Jenner 2000). Phylogenies of both superclades which include these 'minor' phyla (e.g. Gastrotricha, Gnathostomulid, Rotifera, Priapulid, Kinorhyncha, Rhabditophora, Chaetognatha) show them to branch at or near the base of the tree (Glenner *et al.* 2004, Peterson *et al.* 2005,

Mallatt and Giribet 2006). That makes untenable the proposal that most pseudocoelomate and acoelomate groups are secondarily derived from more complex ancestors. Second, similar expression patterns of key developmental genes (De Robertis and Sasai 1996, P. W. H. Holland 1998, L. Z. Holland 2000), taken as evidence of deep 'functional' homologies across the Bilateria, were found to be rather variable and it remained unclear whether they refer to cell-type specification or morphogenetic processes (Erwin and Davidson 2002, Nielsen and Martinez 2003). Moreover, they were not coded as characters and tested in a wide phylogenetic-cladistic analysis (Hübner 2006). Finally, molecular trees of the phylum Platyhelminthes showed it to be polyphyletic (Ruiz-Trillo *et al.* 1999, 2002). Indeed, the platyhelminth orders Acoela and Nemertodermatida branched at the base of the bilaterians while the rest of the phylum (Catenulida + Rhabditophora) fell at variable positions within the Lophotrochozoa (Ruiz-Trillo *et al.* 1999, Jondelius *et al.* 2002, Baguñà and Riutort 2004). Such a basal position was corroborated from sequences of other nuclear genes (Ruiz-Trillo *et al.* 2002, Telford *et al.* 2003) including Hox genes (Cook *et al.* 2004), mitochondrial genes (Ruiz-Trillo *et al.* 2004), and from the first microRNA (miRNA) gene tested in a large set of metazoans and found absent in diploblasts and acoels (Pasquinelli *et al.* 2003) (see below).

The proposal of Acoelomorpha (Acoela + Nemertodermatida) as the extant earliest branching bilaterians divides the Bilateria into two inclusive groups: a broad Bilateria including acoelomorphs, and a more derived Bilateria, named Eubilateria (Baguñà and Riutort 2004) or Nephrozoa (Jondelius *et al.* 2002), excluding this clade. The new phylogenetic proposal is fairly close to the planuloid-acoeloid scenario of Figure 12.1B. It puts back in time and reduces the number of character states leading to the LCA of bilaterians, and suggests that the LCA was small, acoelomate, unsegmented and a direct developer. However, it is very important to stress that Acoelomorpha, and acoels in particular, are by no means equivalent to the bilaterian LCA. They bear, among others, several autapomorphic characters (e.g. duet-spiral cleavage, an interconnecting ciliary rootlet system and bent cilia at terminal ends) which makes them a rather specialised group (Ax 1996).

Nuclear genes

The 18S and 28S ribosomal genes and the myosin heavy chain gene, together with 10 new nuclear genes from a large taxon sample

(63 species belonging to 19 phyla) have been used to further test the basal position of acoelomorphs. Combined 18S + 28S trees and concatenated datasets totalling 13 genes gave similar results (Figure 12.2 for the 13 gene dataset; J. Paps, J. Baguñà and M. Riutort, unpublished data). Acoels and nemertodermatids branch in sequence with high support at the base of the bilaterians. Further, the three superclades are well resolved and some interesting internal clusterings suggested (e.g.

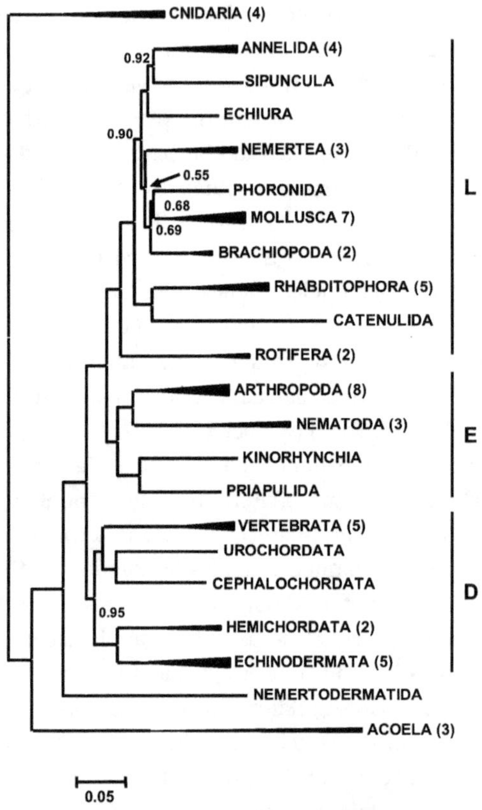

Figure 12.2 Phylogeny of bilaterians determined by Bayesian inference (MRBAYES using a GTR model and gamma distribution) from concatenated sequences of 13 genes (18 and 28S rDNA and 11 nuclear genes, 8446 nucleotides) from 63 species belonging to 19 metazoan phyla. All nodes show a maximum BPP (Bayesian Posterior Probability, obtained from 1000.000 replicates analysis) value of 1.00, except those at some specific nodes. In brackets, number of species per phylum, except those with single representatives. D: Deuterostomia; E: Ecdysozoa: L: Lophotrochozoa. Scale bar indicates the number of substitutions per position (from J. Paps, J. Baguñà and M. Riutort, unpublished data).

Priapulida, Kynorhyncha and Nematoda at the base of the Ecdysozoa, and Rotifera and Platyhelminthes (Catenulida + Rhabditophora) at the base of the Lophotrochozoa).

EST (Expressed Sequence Tags) collections

Complete genomes of several model systems (e.g. yeasts, *Drosophila*, *Caenorhabditis*, *Mus*, *Homo*) have been used to gather large numbers (>100) of homologous genes to examine the basic tenets of the new molecular phylogeny. Surprisingly, the first phylogenies failed to recover the superclade Ecdysozoa (Blair *et al.* 2002, Dopazo *et al.* 2004). However, while a large number of genes reduces the impact of stochastic errors of single-gene phylogenies, it does not deal with systematic errors. Such errors plagued early genome-derived phylogenies because sampling was poor (four to six species) and species had high/very high rates of nucleotide substitution (Jeffroy *et al.* 2006). To overcome these problems, a large number of both genes and species was used, and the new animal phylogeny and the clade Ecdysozoa were recovered again (Philippe *et al.* 2005). Rather than waiting for complete genomes of taxa from each phylum, the most convenient and less expensive approach is to sequence a small number of Expressed Sequence Tags (1000–5000 ESTs per species) from as many taxa as possible (Philippe and Telford 2006).

EST collections from 60 metazoan species belonging to 13 phyla, and an EST collection from the acoel *Convoluta pulchra*, have been used to test the basal position of acoels (H. Philippe, J. Baguñà, M. Riutort and P. Martinez, unpublished results). To avoid long-branch problems caused by fast-evolving clades (*Convoluta pulchra* among them), we introduced a site-heterogeneous mixture model (CAT; Lartillot *et al.* 2007) instead of standard, site-homogeneous models. Preliminary trees run under PhyloBayes (11 000 amino acid positions) resolve the bilaterians into the three big superclades, with sponges and cnidarians branching earlier, Platyhelminthes within the Lophotrochozoa, and acoels in an unstable position as a basal clade to bilaterians, protostomes or deuterostomes. Although the final position of acoels is unresolved (probably because *Convoluta pulchra* is a very fast-evolving species), it confirms clearly that acoels are not members of the Platyhelminthes.

Hox cluster genes

The Hox and ParaHox genes code for transcription factors that regulate A-P patterning in many bilaterian phyla. Most bilaterians have a Hox cluster comprising at least seven to eight distinct genes, or paralogy groups (PGs), and a ParaHox set bearing three genes usually not clustered. Therefore, finding a full set of Hox cluster genes in acoelomorphs would confirm they are not basal bilaterians; conversely, finding a reduced gene set, intermediate between those of cnidarians and bilaterians, would support their position as early branching bilaterians.

Hox and ParaHox genes have been isolated and analysed from five species of acoels and a single nemertodermatid (Cook *et al.* 2004, Jiménez-Guri *et al.* 2006; M. Q. Martindale, personal communication; P. Martinez and J. Baguñà, unpublished data). All acoels examined have a reduced complement of Hox genes: one anterior gene (PG1; an additional anterior gene exists in *Convoluta pulchra*; P. Martinez and J. Baguñà, unpublished data), one central gene (G4-5; Cook *et al.* 2004), and one posterior (PG9-10; a second posterior gene is present in *Paratomella rubra*; Cook *et al.* 2004), and one posterior ParaHox gene (*Cdx*). The nemertodermatid *Nemertoderma westbladi* bears two central Hox genes (PG4-5 and PG6-8) and one posterior (PG9-10), and two ParaHox: an *Xlox*-PG3 and a *Cdx* (Jiménez-Guri *et al.* 2006). In summary, assuming that anterior and posterior additional Hox genes are species-specific duplications, acoelomorphs do have one anterior, one or two central, and one posterior Hox genes, and one representative each of the *Xlox*-PG3 and *Cdx* ParaHox genes.

If a simple Hox gene cluster is substantiated in other acoelomorphs and found (or not) to be structurally collinear (E. Moreno, J. Baguñà and P. Martínez, work in progress) it might represent a simple Hox cluster intermediate between the simpler set of Hox/ParaHox genes in cnidarians and the expanded set (at least 7/8 PGs) of most bilaterians. Recent genome-wide analyses of two cnidarians (*Nematostella vectensis* and *Hydra magnipapillata*; Chourrout *et al.* 2006, Kamm *et al.* 2006) found anterior-like and extremely divergent 'posterior'-like Hox genes, no representatives of central genes, and a cluster of anterior and central/posterior ParaHox. This contradicts early claims of a ProtoHox cluster of four genes and a ParaHox cluster of three genes prior to cnidarian branching from which two Hox and one ParaHox were subsequently lost in the lineage leading to cnidarians (Brooke *et al.* 1998, Finnerty and Martindale 1999).

MicroRNA (miRNA) sets

MicroRNAs (miRNAs) are non-coding RNAs that control gene expression by decreasing the stability of translation of target mRNAs (reviewed by

Wienholds and Plasterk 2005). MicroRNAs and their mRNA targets are usually expressed in mutually exclusive domains; in other words, repression of mRNAs in cell types where the miRNA is expressed suggests that miRNAs stabilises and confers robustness to cell differentiation (Stark *et al.* 2005). From this, it follows that the diversity of miRNAs might be correlated with the number of cell types and, hence, with biological complexity, both features having steadily increased along animal evolution.

Recently, it has been reported that the number of different miRNAs roughly correlates with both the hierarchy of metazoan relationships and with the origination of metazoan morphological innovations through geological time (Sempere *et al.* 2006). The phylogenetic history (presence/absence) of 243 human and 70 fruit fly non-paralogous miRNAs was traced along a wide range of taxa from sponges to humans using Northern blots. Twenty-one miRNAs were found common to protostomes and deuterostomes (Figure 12.3) of which none is present in sponges and just two in cnidarians. Protostomes had 12 additional specific miRNAs and deuterostomes seven. Platyhelminthes, represented

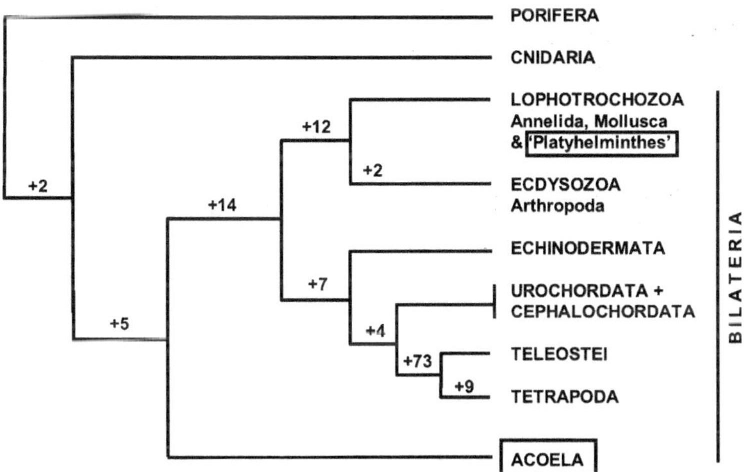

Figure 12.3 An abbreviated phylogenetic tree depicting some metazoan clades with, above the nodes, the number of new miRNAs appearing at each cladogenetic event. The number of different miRNAs in Acoela is low (7 miRNAs) whereas that of 'Platyhelminthes' (Catenulida + Rhabditophora) is similar (33 miRNAs) to those of other lophotrochozoans like annelids and molluscs. This supports previous work suggesting the polyphyly of Platyhelminthes and the basal position of Acoelomorpha (Ruiz-Trillo *et al.* 1999, 2002). Redrawn in a very modified form from Sempere *et al.* 2006.

by a marine polyclad, had almost all protostome miRNAs excluding the two ecdysozoan-specific miRNAs so far detected, confirming that they are lophotrochozoan protostomes.

If acoels are early-branching bilaterians, they should bear a reduced subset of the 21 miRNAs conserved across protostomes and deuterostomes. Consistently, only six miRNAs were found in the acoel *Childia* sp. (Sempere *et al.* 2006). Additional species of Platyhelminthes (including parasitic species) have most protostome-specific miRNAs as well as those shared by protostomes and deuterostomes (L. F. Sempere, P. Martinez, J. Baguñà and K. J. Peterson, unpublished data). Instead, a second acoel examined, *Symsagittifera roscoffensis*, has the same six miRNAs as *Childia* sp. Again, these data strongly support the idea that acoels are early-branching bilaterians and not members of the Platyhelminthes.

GENE EXPRESSION AND AXIAL HOMOLOGIES BETWEEN CNIDARIANS AND BILATERIANS

A major breakthrough in biology during the second half of the twentieth century has been the demonstration that, while animal phyla are morphologically very disparate, they are fundamentally similar genetically. While the genetic composition of extinct taxa (e.g. the LCA of bilaterians) cannot be directly determined, we can use the phylogenetic distribution of developmental genes in extant species to infer the 'genetic toolkit' of the bilaterian LCA. Within the framework of the new molecular phylogeny (Figure 12.1A), the bilaterian LCA is seen as endowed with scores of genes controlling, for example, body axiality, coelom formation and segmentation, photoreception, circulation and body appendages (Carroll *et al.* 2001). Such a constellation of genes had to be assembled at the dawn of the Bilateria from radial ancestors not bearing them.

The way we look at the origin of bilaterality changed recently when it was found that the morphologically simple and symmetrically 'radial' anthozoan cnidarians possess, besides genes involved in A-P polarity (*Hox/ParaHox, otx, ems, gsc*), gastrulation (*twist* [*twi*], *snail* [*sna*], *brachyury* [*Bra*], *forkhead* [*fkh*]), endodermal (GATA) and germ-cell (*nanos* [*nos*], *vasa* [*vas*]) specification, orthologues to bilaterian gene families previously thought to be absent in 'radial' organisms. Prominent among them are genes involved in mesoderm specification (*Nk2, mef2, MyoD*), D-V axial polarity (*Wnt-ß-catenin, dpp/bmp*; *Chordin/noggin* [*chd/nog*], *Gsh/ind, Msh, vnd*), nerve tissue and sensory-organ formation (*Notch/Delta* [*N/Dl*], *Achete/Scute* [*Ac/Sc*], *Netrin, Pax 3*) as well as in other cell signalling pathways (*hedgehog* [*hh*]), Receptor tyrosine kinases (*Egfr*, *Fgfr*) and *Jak/Stat* (for specific references, see Hayward *et al.* 2002, Finnerty *et al.* 2004, Martindale *et al.* 2004, Extavour *et al.* 2005, Martindale 2005, Matus *et al.* 2006, Rentzsch *et al.* 2006). The presence and expression in cnidarians of many of the genes involved in D-V patterning in bilaterians matched ideas (going back to Stephenson 1926, and held by Hyman 1951 and Salvini-Plawen 1978) of a second or directive axis in cnidarians (namely in anthozoans), perpendicular to the oral-aboral (O-AB) axis (Finnerty *et al.* 2004). Therefore, both cnidarians and bilaterians evolved from an ancestor already bilateral, putting the origin of the bilaterian LCA even further back in time.

Figure 12.4 summarises in a simplified form the A-P and D-V expression of selected developmental genes in cnidarians and bilaterians (for specific details see references above). Despite highly dynamic expressions, some A-P and D-V genes in cnidarians have patterns comparable to those of bilaterians. This seems so for gastrulation or 'posterior' genes such as *Wnt*, *bra*, *sna*, *twi*, *fkh*, for 'endodermal' or 'mesoendodermal' genes such as *GATA*, for 'mesodermal' genes like

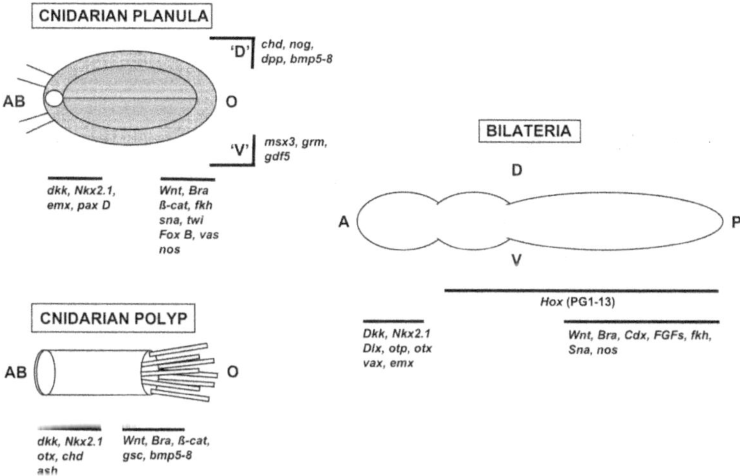

Figure 12.4 Comparative axial expression, in a simplified form, of key developmental genes between cnidarians (planula larva, top left, and polyp, bottom left) and bilaterians (right). In the planula larva genes expressed asymmetrically along the directive axis ('D-V' axis) are also depicted. A: anterior; AB: aboral; D: dorsal; O: oral; P: posterior; V: ventral. 'D' and 'V' imply the likely, but still undefined, DV character of the directive axis in cnidarians. For gene names and further details, see text.

NK2, *mef2* and *MyoD*, and for germ-cell genes such as *nos* and *vas*. However, the expression of key A-P genes such as *Hox/ParaHox*, *emx*, *otx*, *Nkx2.5*, and especially of key D-V genes such as *dpp/bmp* and *chd/ nog*, throws doubt on the existence of simple relationships between the A-P and D-V axes of bilaterians and the O-AB and directive axes of cnidarians, respectively (Chourrout *et al.* 2006, de Jong *et al.* 2006, Kamm *et al.* 2006). Patterns of expression of A-P genes differ dramatically between different species and those of D-V genes are complex and overlapping (de Jong *et al.* 2006). In particular, the bilaterian antagonist factors *dpp/sog* (or *bmp/Chd* in chordates) in *Nematostella* show asymmetric expression along the directive axis but, unexpectedly, also along the O-AB axis (Rentzsch *et al.* 2006).

A particularly vexing old problem, which may hold the key to axial homologies, is the correspondence between the O-AB axis of planula larva and polyp, and between these and the A-P axis of bilaterians. Planula larvae swim with the aboral or apical poles in front and the oral (bearing sometimes a transitory mouth) poles trailing. It is currently assumed that the aboral/oral (AB-O) axis in a planula corresponds to the A-P axis of bilaterians, and taking into account its directed locomotion, then AB = A and O = P. After settling with the anterior pole, the larva transforms into a polyp having the oral end up and the aboral end at the bottom. If axiality between planula and polyp is conserved, the oral (mouth) of the polyp would correspond to the P pole of bilaterians while the aboral (basal disk and foot) would correspond to the A pole. This interpretation is backed by traditional morphological arguments and by the striking similarities between the oral region in cnidarians and the organiser region of chordates and other gastrulation sites of bilaterians which corresponds to either the posterior or ventral pole of modern-day bilaterians (Arendt *et al.* 2001, Technau 2001). Alternatively, the oral pole of the polyp may correspond to the bilaterian anterior pole (Martindale 2005). This would entail, however, the inversion of the A-P axis between planula and polyp.

Gene expression in planula larvae (Figure 12.4) does not provide a definitive answer, but gives interesting clues. The best come from sets of genes in the oral region of both planula and polyp. *Wnt*, *ß-cat*, *Bra*, *sna*, *twi*, *fkh*, *vas* and *nos* are expressed in the posterior (oral) half of the planula larva and (some) in the hypostome area (oral pole) of the polyp. In bilaterians, such genes are expressed in posterior regions (including the posterior endoderm and germ cells) of the embryo and are involved in gastrulation and axial polarity. A second group of genes, *Dickkopf* (*Dkk*) and *Nkx2.1*, are expressed in the anterior (aboral) half of the planula and in the peduncle and basal disk (aboral pole) of the polyp. *Dkk* is particularly interesting because it antagonises the *Wnt* signalling pathway in both cnidarians and bilaterians. Activation of *Wnt* signalling in bilaterians enlarges posterior structures and inhibits anterior structures; in cnidarians, it results in extra heads and tentacles (Guder *et al.* 2006). Conversely, depletion of *Wnt* activity in bilaterians expands anterior structures, whereas in cnidarians it gives rise to extra feet and basal discs. In vertebrates, *Dkk1* is expressed in anterior regions and, when ectopically expressed, induces secondary heads (Glinka *et al.* 1997). In cnidarians, *Dkk* is expressed at the aboral end in the planula and polyp (Lee *et al.* 2006). *Wnt* is expressed at the oral end, and when *Dkk* is depleted, oral structures are expanded (Guder *et al.* 2006). If *Wnt* is considered a posterior marker in bilaterians and its antagonist *Dkk* an anterior marker, their expression in cnidarians and the results of over-expression/inhibition suggest that the aboral end of the planula (= foot of polyp) is homologous to the anterior region of bilaterians (Meinhardt 2002), whereas the oral end of a planula (= hypostome of polyp) is homologous to the posterior region of bilaterians. Under this scenario, the postulated inversion of axial polarity between planula and polyp is neither necessary nor tenable.

THE PLANULA-ACOELOID THEORY REVISITED WITH A CRITIQUE TO
AMPHISTOMIC SCENARIOS OF BILATERIAN EVOLUTION

New molecular phylogenies (Figure 12.2), new data on Hox/ParaHox and microRNA sets confirming the acoelomorphs as earliest extant branching bilaterians (Figure 12.3), the finding that all animal phyla (sponges included; Nichols *et al.* 2006) share a complex 'genetic toolkit', the evidence for axial homologies in gene expression between cnidarians and bilaterians (Figure 12.4), and the evidence that cnidarians are bilateral in origin, all converge to an older LCA for bilaterians (Figure 12.5), better named the CBA (Cnidarian-Bilaterian Ancestor). In turn this resembled more closely the ancestor envisaged in the Planula-Acoeloid Theory, with an axial and bilateral, bentho-pelagic sexual archiplanula with directed locomotion, anterior sensory pole, posterior mouth and a rudimentary gut. From this ancestor, both cnidarians and 'true' bilaterals emerged. This scenario has affinities with the early ideas of Metchnikoff, further elaborated by Hyman, Beklemishev, Ivanov and later on by Salvini Plawen (1978), to which readers are referred.

  Leaving aside sponges and placozoans, the primitive mode of feeding in metazoans appears to be grazing in and on the benthos,

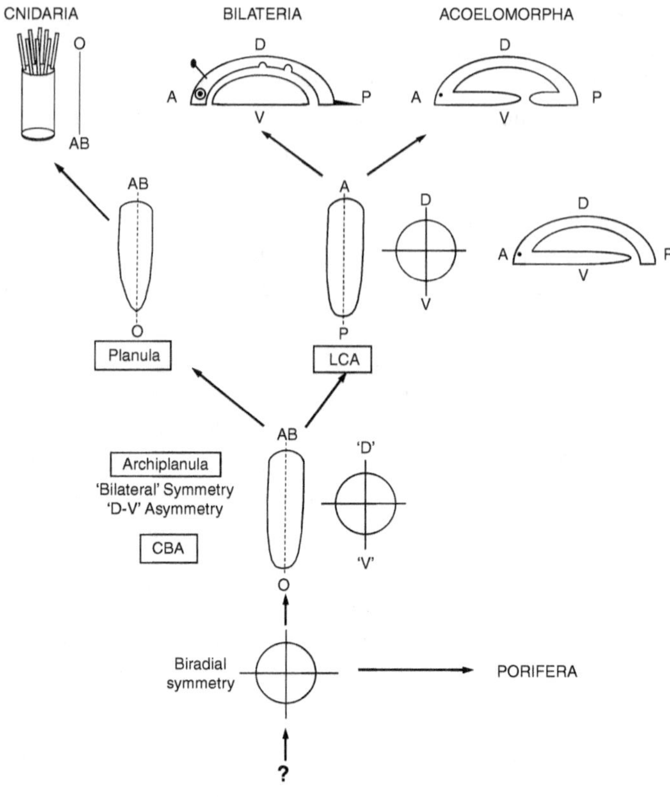

Figure 12.5 Phylogenetic hypothesis for lower Metazoa, from biradial ancestors (cross-section) up to the archiplanula or CBA (Cnidarian-Bilaterian Ancestor) with loose 'bilateral' symmetry and 'dorsoventral' ('D-V') asymmetry and from the latter to both extant cnidarians via planula forms and to the Last Common Ancestor (LCA) of bilaterians. The LCA showed defined antero-posterior (A-P) and dorso-ventral (D-V) axes and from them derived both the acoelomorphs (right) and the rest of the bilaterians (centre). Drawn from concepts, ideas and phylogenetic schemes of Salvini-Plawen (1978), Willmer (1990), Baguñà and Riutort (2004) and Martindale (2005). For further details, see text.

feeding upon organisms smaller than themselves such as bacteria, algae and other animals (Peterson *et al.* 2005). In other words, suspension feeding or active pelagic feeding, as in extant cnidarian polyps and ctenophores, was unlikely to be primitive. In both groups, it could only have occurred after the evolution of cnidoblasts (cnidarians) and colloblasts (ctenophores) which are no older than the Cambrian. This was concurrent with the appearance of appropriate food sources, namely mesozooplankton (Peterson 2005). Earliest cnidarians were probably small benthic grazers or burrowers with a main A-P axis (equivalent to AB-O), the oral end (mouth/anus) at the rear, and a cryptic D-V axis. Once planulas of stem cnidarians developed a rudimentary pair of tentacles with primitive cnidocysts, and settled with the anterior pole to the substrate, ancestral archipolyps emerged ready to penetrate into the vacant ecological niche of sessile predators (Salvini-Plawen 1978).

Another group of bentho-pelagic sexual archiplanulas gave rise to stem bilaterians (Figure 12.5). Given that A-P and 'D-V' axes were already in place, key apomorphies leading to the LCA were the reinforcement of the D-V axis, probably helped by the appearance or 'segregation' of mesoderm from endomesoderm, and concentration at the anterior end of clumps of nerve cells to form a first primitive brain. A further or concurrent important development included the shift of the blasto-pore/oral opening to different positions on the ventral side (one of the most basal acoel genera, *Diopisthoporus*, has a posteriorly positioned mouth/anus; Salvini-Plawen 1978). The evolution of bilaterians with through gut (mouth + anus), which comprise all bilaterians except the acoelomorphs, the Platyhelminthes and *Xenoturbella* spp., was another key item in bilaterian evolution. According to van den Biggelaar and Dictus (2005), this might have occurred from cnidarian-like organisms in three different ways: (1) the blastopore maintained its posterior position becoming the anus, and a mouth developed later (Deuterostomia); (2) the posterior dorsal side of the blastopore extended (probably by proliferation as in some extant molluscs) shifting the mouth anteriorly towards the ventral side while the anus formed later (Protostomia); and (3) the body axis extended only along the dorsal side associated with the transformation of the blastopore into a longitudinal slit whose margins later fused in the middle, giving a tube with an anterior mouth opening and a posterior anal opening. This mode of blastopore closure, called amphistomy, has been proposed several times as a way to derive at a stroke the typical bilaterian body-plan features from a radial Gastraea from cnidarian adults (enterocoel theory of Sedgwick 1884), from benthic bilaterogastreas (Jägersten 1955) or from trochophora-type primary ciliary larvae (Arendt *et al.* 2001). In the last case, the expression of *otx* and *Bra* was considered sufficient evidence to derive both mouth and anus from blastoporal regions. There is a general consensus, however, that primary larvae are not primitive but derived, not truly homologous, and prone to convergence (Sly *et al.* 2003). Moreover, *Bra* and *otx*, besides their clear roles in gastrulation and in specifying anterior body regions respectively, are also activated anew in any invagination movements (e.g. *Bra* in stomodeum formation) and in all sorts of ciliary bands (*Otx*); therefore their expression in larvae is probably due to convergence and needs to be reassessed. Further, in most embryos of molluscs, the blastopore does not contribute to the formation of the anus as required by the amphistomy concept (see van den Biggelaar and Dictus 2005 for references). Finally, according to the concept of amphistomy in its original formulation, head formation is expected at one side of the blastopore, and the opposite side should be posterior. Thus, the animal-vegetal axis of eggs and embryos which is parallel to the A-P axis now becomes parallel to the D-V axis, whereas the A-P axis is made orthogonal to it. The main consequence is that the orientation of the expression domains of axial patterning genes is not in register between ancestor and descendant. Altogether, whereas amphistomic mechanisms may fit the specific developmental features of some lophotrochozoans (e.g. annelids) it cannot be extrapolated as a general mechanism, as in the original enterocoel theory (Sedgwick 1884) and variations thereof (Jägersten 1955, Arendt *et al.* 2001), to explain bilaterian evolution.

CONCLUSIONS AND PROSPECTS

New molecular phylogenies, in particular the proposal that acoelomorph flatworms are the earliest extant bilaterians, and the realisation that radial cnidarians have the axial features of bilaterians, are currently helping to unravel the sequential evolution of what once appeared to be a number of phylogenetically coincident character changes. Thus, key changes in bilaterian evolution are spread along several steps, which allow character states to be polarised. This argues against the complex Urbilateria hypothesis and helps us to see the evolution of the Bilateria as a series of successive Last Common Ancestor (LCA) nodes connected by stem ancestors along which new characters were acquired (Valentine 2006).

Refinements in data acquisition, evolution models, fossil record, molecular phylogenies, gene expression data (in particular forthcoming data on the expression of developmental genes in embryos and adults of acoelomorphs) and functional evo-devo will in the next few years be instrumental in unravelling the sequential evolution of clades at the base of the Deuterostomia, the Ecdysozoa and the Lophotrochozoa.

ACKNOWLEDGEMENTS

We thank Alessandro Minelli and Giuseppe Fusco for kindly inviting one of us (J.B.) to participate and deliver a talk at the Evo-Devo Meeting in

Venice, from which this chapter is a summary. Constructive criticism, corrections and text editing by one reviewer are highly appreciated. These studies were supported by grants from the Generalitat de Catalunya and from CICYT (Ministerio de Ciencia y Tecnologia).

REFERENCES

Adoutte, A., Balavoine, G., Lartillot, N. & de Rosa, R. 1999. Animal evolution: the end of intermediate taxa? *Trends in Genetics* **15**, 104–108.

Aguinaldo, A. M. A., Turbeville, J. M., Linford, L. S. *et al.* 1997. Evidence for a clade of nematodes, arthropods and other moulting animals. *Nature* **387**, 489–493.

Arendt, D., Technau, U. & Wittbrodt, J. 2001. Evolution of the bilaterian larval foregut. *Nature* **409**, 81–85.

Ax, P. 1996. *Multicellular Animals. A New Approach to the Phylogenetic Order in Nature*, Vol. 1. Berlin: Springer.

Baguñà, J. & Riutort, M. 2004. The dawn of bilaterian animals: the case of the acoelomorph flatworms. *BioEssays* **26**, 1046–1057.

Blair, J. E., Ikeo, K., Gojobori, T. & Hedges, S. B. 2002. The evolutionary position of nematodes. *BMC Evolutionary Biology* **2**, 7.

Brooke, N. M., García-Fernández, J. & Holland, P. W. H. 1998. The ParaHox gene cluster is an evolutionary sister of the Hox cluster. *Nature* **392**, 920–922.

Budd, G. E. 2003. The Cambrian fossil record and the origin of phyla. *Integrative & Comparative Biology* **43**, 157–165.

Butterfield, N. J. 2006. Hooking some stem-group 'worms': fossil lophotrochozoans in the Burgess Shale. *BioEssays* **28**, 1161–1166.

Carroll, S. B., Grenier, H. J. K. & Weatherbee, S. D. 2001. *From DNA to Diversity*. Malden: Blackwell Science.

Chourrout, D., Delsuc, F., Chourrout, P. *et al.* 2006. Minimal ProtoHox cluster inferred from bilaterian and cnidarian Hox complements. *Nature* **442**, 684–687.

Conway Morris, S. 2006. Darwin's dilemma: the realities of the Cambrian 'explosion'. *Philosophical Transactions of the Royal Society of London B* **361**, 1069–1083.

Cook, C. E., Jiménez, E., Akam, M. & Saló, E. 2004. The *Hox* gene complement of acoel flatworms, a basal bilaterian clade. *Evolution & Development* **6**, 154–163.

de Jong, D. M., Hislop, N. R., Hayward, D. C. *et al.* 2006. Components of both major axial patterning systems of the Bilateria are differentially expressed along the primary axis of a 'radiate' animal, the anthozoan cnidarian *Acropora millepora*. *Developmental Biology* **298**, 632–643.

De Robertis, D. M. & Sasai, Y. 1996. A common plan for dorsoventral patterning in Bilateria. *Nature* **380**, 37–40.

Donoghue, M. J. 2005. Key innovations, convergence, and success: macroevolutionary lessons from plant phylogeny. *Paleobiology* **31**, 77–93.

Dopazo, H., Santoyo, J. & Dopazo, J. 2004. Phylogenomics and the number of characters required for obtaining an accurate phylogeny of eukaryote model species. *Bioinformatics* **20** (supplement 1), i116–i121.

Erwin, D. H. & Davidson, E. H. 2002. The last common bilaterian ancestor. *Development* **129**, 3021–3032.

Extavour, C. G., Pang, K., Matus, D. Q. & Martindale, M. Q. 2005. *vasa* and *nanos* expression patterns in a sea anemone and the evolution of bilaterian germ cell specification mechanisms. *Evolution & Development* **7**, 201–215.

Finnerty, J. R. & Martindale, M. Q. 1999. Ancient origins of axial patterning genes: Hox genes and ParaHox genes in the Cnidaria. *Evolution & Development* **1**, 16–23.

Finnerty, J. R., Pang, K., Burton, P., Paulson, D. & Martindale, M. Q. 2004. Origins of bilateral symmetry: Hox and *dpp* expression in a sea anemone. *Science* **304**, 1335–1337.

Glenner, H., Hansen, A. J., Sorensen, M. V. *et al.* 2004. Bayesian inference of the Metazoan phylogeny: a combined molecular and morphological approach. *Current Biology* **14**, 1644–1649.

Glinka, A., Wu, W., Onichtchouk, D., Blumenstock, C. & Niehrs, C. 1997. Head induction by simultaneous repression of Bmp and Wnt signalling in *Xenopus. Nature* **389**, 517–519.

Gould, S. J. 1989. *Wonderful Life: The Burgess Shale and the Nature of History.* New York: W.W. Norton.

Guder, C., Pinho, S., Nascak, T. G. *et al.* 2006. An ancient Wnt-Dickkopf antagonism in *Hydra. Development* **133**, 901–911.

Halanych, K. M., Bachelor, J., Aguinaldo, A. M. A. *et al.* 1995. 18S rDNA evidence that lophophorates are protostome animals. *Science* **267**, 1641–1643.

Hayward, D. C., Samuel, G., Pontynen, P. C., Catmull, J. & Saint, R. 2002. Localized expression of a dpp/BMP2/4 ortholog in a coral embryo. *Proceedings of the National Academy of Sciences of the USA* **99**, 8106–8111.

Holland, L. Z. 2000. Body-plan evolution in the Bilateria: early anteroposterior patterning and the deuterostome–protostome dichotomy. *Current Opinion in Genetics & Development* **10**, 434–442.

Holland, N. D. 2003. Early central nervous system evolution: an era of skin brains? *Nature Reviews Neuroscience* **4**, 1–11.

Holland, P. W. H. 1998. Major transitions in animal evolution: a developmental genetic perspective. *American Zoologist* **38**, 829–842.

Hübner, C. 2006. Hox genes, homology and axis formation: The application of morphological concepts to evolutionary developmental biology. *Theory in Biosciences* **124**, 371–396.

Hyman, L. H. 1951. *The Invertebrates*, Vol. 2. *Platyhelminthes and Rhynchocoela.* New York: McGraw-Hill.

Jägersten, G. 1955. On the early phylogeny of the Metazoa. The bilaterogastrea theory. *Zoologiska Bidrag från Uppsala* **30**, 321–354.

Jeffroy, O., Brinkmann, H., Delsuc, F. & Philippe, H. 2006. Phylogenomics: the beginning of incongruence? *Trends in Genetics* **22**, 225–231.

Jenner, R. A. 2000. Evolution of animal body plans: the role of metazoan phylogeny at the interface between pattern and process. *Evolution & Development* **2**, 208–221.

Jiménez-Guri, E., Paps, J., Garcia-Fernàndez, J. & Saló, E. 2006. *Hox* and *ParaHox* genes in Nemertodermatida, a basal bilaterian clade. *International Journal of Developmental Biology* **50**, 675–679.

Jondelius, U., Ruiz-Trillo, I., Baguñà, J. & Riutort, M. 2002. The Nemertodermatida are basal bilaterians and not members of the Platyhelminthes. *Zoologica Scripta* **31**, 201–215.

Kamm, K., Schierwater, B., Jakob, W., Dellaporta, S. L. & Miller, D. J. 2006. Axial patterning and diversification in the Cnidaria predate the Hox system. *Current Biology* **16**, 1–7.

Kimmel, C. B. 1996. Was *Urbilateria* segmented? *Trends in Genetics* **12**, 329–331.

Lartillot, N., Brinkmann, H. & Philippe, H. 2007. Suppression of long-branch attraction artefacts in the animal phylogeny using a site-heterogeneous model. *BMC Evolutionary Biology* **7**, supplement 1, S4.

Lee, P. N., Pang, D., Matus, D. Q. & Martindale, M. Q. 2006. A WNT of things to come: evolution of Wnt signalling and polarity in cnidarians. *Seminars in Cell & Developmental Biology* **17**, 157–167.

Mallatt, J. & Giribet, G. 2007. Further use of nearly complete 28S and 18S rRNA genes to classify Ecdysozoa: 37 more arthropods and a kinorhynch. *Molecular Phylogenetics and Evolution* **40**, 772–794.

Marshall, C. R. 2006. Explaining the Cambrian 'explosion' of animals. *Annual Review of Earth & Planetary Sciences* **34**, 355–384.

Martindale, M. Q. 2005. The evolution of metazoan axial properties. *Nature Reviews Genetics* **6**, 917–927.

Martindale, M. Q., Pang, K. & Finnerty, J. R. 2004. Investigating the origins of triploblasty: 'mesodermal' gene expression in a diploblastic animal, the sea anemone *Nematostella vectensis* (phylum, Cnidaria; class, Anthozoa). *Development* **131**, 2463–2474.

Matus, D. Q., Thomsen, G. H. & Martindale, M. Q. 2006. Dorso/ventral genes are asymmetrically expressed and involved in germ-layer demarcation during cnidarian gastrulation. *Current Biology* **16**, 499–505.

Meinhardt, H. 2002. The radial-symmetric hydra and the evolution of the bilateral body plan: an old body became a young brain. *BioEssays* **24**, 185–191.

Nichols, S. A., Dirks, W., Pearse, J. S. & King, N. 2006. Early evolution of animal cell signalling and adhesion genes. *Proceedings of the National Academy of Sciences USA* **103**, 12451–12456.

Nielsen, C. & Martinez, P. 2003. Patterns of gene expression: homology or homocracy? *Development, Genes & Evolution* **213**, 149–154.

Pasquinelli, A. E., McCoy, A., Jiménez, E. *et al.* 2003. Expression of the 22 nucleotide *let-7* heterochronic RNA throughout the Metazoa: a role in life history evolution? *Evolution & Development* **5**, 372–378.

Peterson, K. J. 2005. Macroevolutionary interplay between planktic larvae and benthic predators. *Geology* **33**, 929–932.

Peterson, K. J., McPeek, M. A. & Evans, D. A. D. 2005. Tempo and mode of early animal evolution: inferences from rocks, Hox, and molecular clocks. *Paleobiology* **31** (supplement 2), 36–55.

Philippe, H., Delsuc, F., Brinkmann, H. & Lartillot, N. 2005. Phylogenomics. *Annual Review of Ecology, Evolution and Systematics* **36**, 541–562.

Philippe, H. & Telford, M. J. 2006. Large-scale sequencing and the new animal phylogeny. *Trends in Ecology & Evolution* **21**, 614–620.

Rentzsch, F., Anton, R., Saina, M. *et al.* 2006. Asymmetric expression of the BMP antagonists *chordin* and *gremlin* in the sea anemone *Nematostella vectensis*: Implications for the evolution of axial patterning. *Developmental Biology* **296**, 375–387.

Ruiz-Trillo, I., Paps, J., Loukota, M. *et al.* 2002. A phylogenetic analysis of myosin heavy chain type II sequences corroborates that Acoela and Nemertodermatida are basal bilaterians. *Proceedings of the National Academy of Sciences of the USA* **99**, 11246–11251.

Ruiz-Trillo, I., Riutort, M., Fourcade, H. M., Baguñà, J. & Boore, J. L. 2004. Mitochondrial genome data support the basal position of Acoelomorpha and the polyphyly of the Platyhelminthes. *Molecular Phylogenetics & Evolution* **33**, 321–332.

Ruiz-Trillo, I., Riutort, M., Littlewood, D. T. J., Herniou, E. A. & Baguñà, J. 1999. Acoel flatworms: earliest extant bilaterian metazoans, not members of Platyhelminthes. *Science* **283**, 1919–1923.

Salvini-Plawen, L. V. 1978. On the origin and evolution of the lower Metazoa. *Zeitschrift für Zoologische Systematik und Evolutionsforschung* **16**, 40–88.

Sedgwick, W. 1884. On the origin of metameric segmentation and some other morphological questions. *Quarterly Journal of the Microscopical Society* **24**, 43–82.

Sempere, L., Cole, C. N., McPeek, M. A. & Peterson, K. J. 2006. The phylogenetic distribution of Metazoan microRNAs: insights into evolutionary complexity and constraint. *Journal of Experimental Zoology B (Molecular and Developmental Evolution)* **306**, 575–588.

Sly, B. J., Snoke, M. S. & Raff, R. R. 2003. Who came first – larvae or adults? Origins of bilaterian metazoan larvae. *International Journal of Developmental Biology* **47**, 623–632.

Stark, A., Brennecke, J., Bushati, N., Russell, R. B. & Cohen, S. M. 2005. Animal microRNAs confer robustness to gene expression and have a significant impact on 3'UTR evolution. *Cell* **123**, 1–14.

Stephenson, T. A. 1926. *British Sea Anemones.* London: The Ray Society.

Technau, U. 2001. Brachyury, the blastopore and the evolution of the mesoderm. *BioEssays* **23**, 788–794.

Telford, M. J., Lockyer, A. E., Cartwright-Finch, C. & Littlewood, D. T. J. 2003. Combined large and small subunit ribosomal RNA phylogenies support a basal position of the acoelomorph flatworms. *Proceedings of the Royal Society of London B* **270**, 1077–1083.

Valentine, J. W. 2004. *On the Origin of Phyla.* Chicago: University of Chicago Press.

Valentine, J. W. 2006. Ancestors and urbilateria. *Evolution & Development* **8**, 391–393.

van den Biggelaar, J. A. M. & Dictus, W. J. A. G.. 2005. Gastrulation in the molluscan embryo. In C. Stern (ed.) *Gastrulation*, Cold Spring Harbor Laboratory Press. pp. 63–77.

Wienholds, E. & Plasterk, R. H. A. 2005. MicroRNA function in animal development. *FEBS Letters* **579**, 5911–5922.

Willmer, P. 1990. *Invertebrate Relationships: Patterns in Animal Evolution.* Cambridge University Press: Cambridge.

13

Are transposition events at the origin of the bilaterian Hox complexes?

JEAN S. DEUTSCH AND PHILIPPE LOPEZ

The genome sequences of two non-bilaterian animals, the cnidarians *Nematostella vectensis* and *Hydra magnipapillata*, have been recently completed. These new data lead to the fascinating result that the complement of Hox genes in the cnidarian ancestor is considerably lower than that in the bilaterians, although the complexity of their genome is otherwise similar (Technau *et al.* 2005). Thus, there is a correlation between the radiation of the Bilateria and the expansion of the Hox complex.

In the first part of this chapter, we shall present and discuss these data. In the second part, we shall present a novel hypothesis accounting for this phenomenon. In short, we surmise that the expansion of the Hox complex at the base of the Bilateria was due to a series of transposition events. Indeed, we hypothesise that the Hox genes themselves originate from transposons. The main support for this hypothesis is provided by the similarity between the homeodomain and the DNA-binding domain of bacterial integrases and eukaryotic transposases. We also examine some very precise rearrangements of the Hox complex in the Drosophilidae lineage. In the third part, we propose a scenario for the evolution of the Hox complex from the basic complement of Hox genes in the common ancestor of cnidarian and bilaterian animals. This scenario, based on our transposition hypothesis, accounts for several properties of the extant Hox genes.

TO SET THE SCENE: THE HOX EXPLOSION

The homeobox is a conserved motif found in a huge variety of eukaryotic genes, encoding a DNA-binding domain. Although homeobox-containing genes are known from various branches of the eukaryote tree, such as fungi and plants, ANTP-class genes, to which Hox and paraHox genes belong, form a monophyletic group only known from animals (Bharathan *et al.* 1997, Holland and Takahashi 2005).

The Metazoa (i.e. Animalia) comprise non-bilaterian and bilaterally symmetric animals. Four extant non-bilaterian phyla are known: Porifera (sponges), Ctenophora (comb jellies), Cnidaria (corals, sea anemones, jellyfish and their kin) and Placozoa. There are no nerve cells in sponges, so that the presence of well-characterised nerve cells is a clear synapomorphy unifying Ctenophora and Cnidaria with the Bilateria, in a clade called 'Eumetazoa'. Placozoa is a problematic phylum, comprising a single species, *Trichoplax adhaerens*, with a very simple morphology. Schierwater (2005) strongly advocates for its origin from the base of the Metazoa, before the emergence of the Porifera. However, it shares certain synapomorphies, such as belt desmosomes and neuropeptides, with Eumetazoa (Schuchert 1993). We thus think that it is a derived, secondarily simplified, eumetazoan.

The so-called 'new phylogeny' dispatched the bilaterian phyla into three 'super-phyla', Deuterostomia, Lophotrochozoa and Ecdysozoa (Aguinaldo *et al.* 1997). Two phyla may not fit in this classification: the Chaetognatha and the Acoelomorpha. Chaetognatha have been long considered as '*incertae sedis*' (Ball and Miller 2006). They are now either regarded as a sister group to all Protostomes (Marletaz *et al.* 2006) or included within the Lophotrochozoa (Matus *et al.* 2006). Acoelomorpha, a phylum recently created following the exclusion of the Acoela and the Nemertodermatida from the phylum Platyhelminthes, is assumed to stem from the very base of the Bilateria, before the split into the three super-phyla (Baguñá and Riutort 2004). For both these problematic taxa, Chaetognatha and Acoelomorpha, more data are needed, on a larger panel of genes and taxa, until a firm phylogenetic conclusion can be drawn.

Almost all bilaterian phyla suddenly appear in the fossil record at the base of the Cambrian within a short length of time, *c.* 540 to 550 million years (Myr) before present. This was called the 'Cambrian explosion'. In parallel, there was a sudden increase in the number of Hox genes.

Indeed, there is no Hox gene in sponges (Manuel and Le Parco 2000). As for ctenophores, a single small fragment of a putative Hox gene isolated from *Beroe ovata* has been withdrawn from the data banks as a contaminant at the request of the authors. In contrast, Hox and Hox-related genes have been studied from several cnidarian species (Gauchat *et al.* 2000, Chourrout *et al.* 2006, Kamm *et al.* 2006, Ryan *et al.* 2006). Summarising, we can draw the following conclusions: (1) the number of Hox genes in the repertoire of the ancestral cnidarian is low, not more than two or three; (2) these primitive Hox genes are more similar to the Hox1–2, Hox3 and maybe the posterior Hox9–14 of so-called 'paralogy groups' (PG) of bilaterian Hox genes; (3) lineage-specific duplications have increased this primitive number in several cnidarian taxa; (4) the cnidarian Hox complex, if it ever existed primitively, has been disrupted and reorganised during the evolution of the Cnidaria.

Since the work of de Rosa *et al.* (1999), who examined the repertoire of Hox genes in a diversity of bilaterian taxa, a number of data have been added, all supporting the main conclusion of this pioneer work: on a qualitative and quantitative basis (type and number of Hox genes), the repertoire of Hox genes supports the classification of the Bilateria into the three 'super-phyla' first proposed by Aguinaldo *et al.* (1997). These data now allow confident conclusions to be drawn about what the complement of Hox genes was in the ancestor of several phyla and, in the best cases, what their genomic organisation was.

In the Arthropoda, the Hox genes' basic complement comprises 10 genes, the two sister genes *lab* and *pb*, orthologous to the paralogy groups (PG) PG1 and PG2, respectively; a single *zen* gene orthologous to PG3; six genes belonging to the 'median' group PG4–8, namely *Dfd*, *Scr*, *ftz*, *Antp*, *Ubx* and *abdA*, and a single 'posterior' gene, *AbdB*, corresponding to PG9–14 in deuterostomes. From all available genomic data, we can infer that these ten Hox genes were primitively grouped in a single cluster, despite some breaks and rearrangements that occurred during the evolution of long germ-band insects.

Among the Ecdysozoa, the Onychophora, a phylum closely related to the Arthropoda, and the Priapulida fit the arthropod scheme (de Rosa *et al.* 1999), with the possible exception of a duplication of the posterior gene. In contrast, in various nematodes the Hox complement appears quite reduced and derived: some genes are missing, others are derived, mosaic or duplicated, and the Hox complex is profoundly rearranged and disrupted (Aboobaker and Blaxter 2003). This seems specific to the Nematoda, since a species belonging to the Nematomorpha (probably the closest relatives to the nematodes) has a full complement of arthropod-like Hox genes. We can thus infer that the Hox complement present in the ecdysozoan ancestor comprised ten different genes, or at least nine if the duplication leading to the sister genes *Ubx*/*abdA* were specific to the Arthropoda. They were most probably arranged in a single complex.

As for the Lophotrochozoa, data are less complete than those for the Ecdysozoa, both in terms of the number of taxa studied and the structure of the genes and complexes. Lophotrochozoan species possess clear orthologues of the PG1/*lab*, PG2/*pb*, and PG3/*zen*. As 'median' genes, they share PG4/*Dfd* and PG5/*Scr*. Telford (2000) hypothesised an orthology relationship between the lophotrochozoan median gene *Lox5*, the arthropod *ftz* and PG6. They also possess clear orthologues of *Antp* (possibly a member of PG7). *Lox2* and *Lox4* are two sister genes, arising from a different duplication from the one that generated *Ubx* and *abdA* in the arthropod lineage (Wong *et al.* 1995). This amounts to six 'median' Hox genes. In addition, lophotrochozoans have two specific 'posterior' genes, *Post1* and *Post2* (de Rosa *et al.* 1999). The two latter genes, together with the couple *Lox2*/*Lox4*, constitute characteristic signatures of the lophotrochozoan lineage. Platyhelminths show a disturbed panel of Hox genes with derived and duplicated genes. In the parasite platyhelminth *Schistosoma mansoni*, the Hox complex is disintegrated and dispersed in the genome (Pierce *et al.* 2005). We can derive a figure of 11 Hox genes as the complement of Hox genes in the primitive lophotrochozoan. Whether they were clustered in a complex is still an open issue.

The Deuterostomia includes two branches: the Ambulacraria, uniting the Echinodermata and the Hemichordata, and the Chordata, comprising the Cephalochordata, the Urochordata and the Vertebrata. The vertebrates have undergone several whole genome duplications during their evolution, leading to up to four paralogous Hox clusters in the Tetrapoda and (primitively) up to eight in the Teleostei. From sequence comparisons between the four Hox clusters in mammals, a primitive complex of 13 Hox genes was derived (McGinnis and Krumlauf 1992). The discovery of a 14th Hox gene in the cephalochordate *Branchiostoma floridae* (Ferrier *et al.* 2000), in the coelacanth *Latimeria chalumnae* and in the shark *Heterodontus francisci* (Powers and Amemiya 2004) added one more posterior gene to the ancestral chordate Hox complex. The grouping in a single cluster of the Hox genes in the amphioxus and the tight clustering of the Hox genes in vertebrates led to the hypothesis of a single complex of 14 Hox genes in the chordate ancestor. In the Urochordata, losses and rearrangements yielded a disorganised and derived Hox cluster, variable among taxa.

Summing up data from a number of species belonging to diverse classes among the hemichordates and the echinoderms, a Hox complex orthologous to that of the chordates can be derived in the common ancestor of the Ambulacraria, possibly with a smaller number of posterior genes. The Hox cluster of the sea urchin *Strongylocentrotus purpuratus* is profoundly perturbed, maybe in line with the huge modification of the echinoderm body plan (Cameron *et al.* 2006). In total, a single cluster of 14 genes comprising at least six posterior genes can reliably be postulated to have been present in the deuterostome ancestor (Monteiro and Ferrier 2006).

In both Chaetognatha and Acoelomorpha, a reduced number of Hox genes have been reported, with some of them showing no clear orthology with known Hox genes from other bilaterian taxa (Papillon *et al.* 2003, Cook *et al.* 2004). This has been attributed to the 'primitive' nature of these Hox genes and phyla. However, 'mosaic' homeodomain sequences could result from divergent evolution after loss of some Hox genes as well. This kind of evolution of remaining Hox genes is exemplified in the case of echinoderms, such as *Hox4* and *Hox5* from starfish compared with Hox genes from echinoid species that have lost one median gene (Long *et al.* 2003).

The most parsimonious figure for the number of Hox genes present in the common ancestor of the three bilaterian branches is nine genes: two anterior genes, orthologous to PG1 and PG2, one anterior-median (PG3), five median (PG4 to PG8) and one posterior gene, to which we refer in the following as PG9*, being the ancestor of PG9 to PG14. Hence, the number of Hox genes suddenly jumped from two to three genes as present in the common ancestor of the Cnidaria and the Bilateria (Ferrier and Holland 2001, Garcia-Fernàndez 2005, Chourrout *et al.* 2006) to nine in the common ancestor of extant bilaterians, with a further increase to 10 in the Ecdysozoa (duplication of PG8 to *Ubx* and *abdA*), to 11 in the Lophotrochozoa (duplication of PG8 to *Lox2* and *Lox4* and of PG9* to *Post1* and *Post2*) and to 14 in the deuterostome ancestor (duplications of PG9* in PG9 to PG14). This sudden increase is what we call 'the Hox explosion' that paralleled the radiation of the Bilateria.

This observation needs an explanation. Here we hypothesise that the 'Hox explosion' is due to a burst of transposition events and that the Hox genes themselves are primitive transposons that have been 'domesticated' during further evolution of the metazoan genome.

THE HOMEO DOMAIN PROTEIN AS A TRANSPOSASE

We shall now present the first piece of evidence supporting the transposition mechanism that, as we suggest, has operated at the origin of the bilaterian Hox complex. We review the current literature presenting evidence that the *RAG1* gene, involved in the recombination events leading to the diversity of the vertebrates' immune response, is derived from a transposon. We show that metazoan homeodomains are very similar to the DNA-binding domain of the RAG1 protein, similarity being the greatest for Hox homeodomains. Hence we surmise that the Hox genes are also issued from a transfer of DNA by a transposon of a similar kind.

Schatz *et al.* (1989) discovered the RAG1 protein as a main player for V(D)J recombination of immunoglobin and receptor genes. RAG1 interacts with its partner RAG2, encoded by a neighbouring gene, and with HMG proteins in a multimeric complex. In this complex, both the DNA-binding domains and the critical DDE acidic residues active in recombination are located within the RAG1 moiety (De and Rodgers 2004).

Thompson (1995) first suggested that the *RAG* locus has evolved from a transposase. Spanopolou *et al.* (1996) discussed the parallel between V(D)J and bacterial recombination. Bernstein *et al.* (1996) underlined the structural similarity between RAG1 and bacterial integrases. This 'transposon hypothesis' on the origin of the V(D)J system has gained support from evidence provided by Hiom *et al.* (1998) and Reddy *et al.* (2006) that the RAG proteins are able to generate transpositions in vitro and in vivo. Last but not least, Kapitonov and Jurka (2005) revealed sequence similarity between the RAG1 protein and the transposase and between the V(D)J recombination signals and the target of a new DNA transposon, called '*Transib*'.

We aligned the DNA nonamer-binding domain of RAG1, highly conserved throughout gnathostome evolution, to a sample of homeodomains from metazoan proteins representative of the diversity of this group of transcription factors. A part of this alignment is shown in Figure 13.1, with homeodomains of Hox genes from *Drosophila melanogaster* (fly) as an ecdysozoan representative, *Nereis virens* (polychaete worm) for lophotrochozoans and *Mus musculus* (mouse) for deuterostomes. So-called 'posterior genes', i.e. *AbdB*, *Post1*, *Post2* and *Hox9* to *Hox13*, have been discarded because of rapid evolutionary rate (Chourrout *et al.* 2006). For the same reason, *Drosophila* genes corresponding to *Hox3* (i.e. *zen*, *zen2*, *bicoid*) as well as *ftz* have not been taken into account.

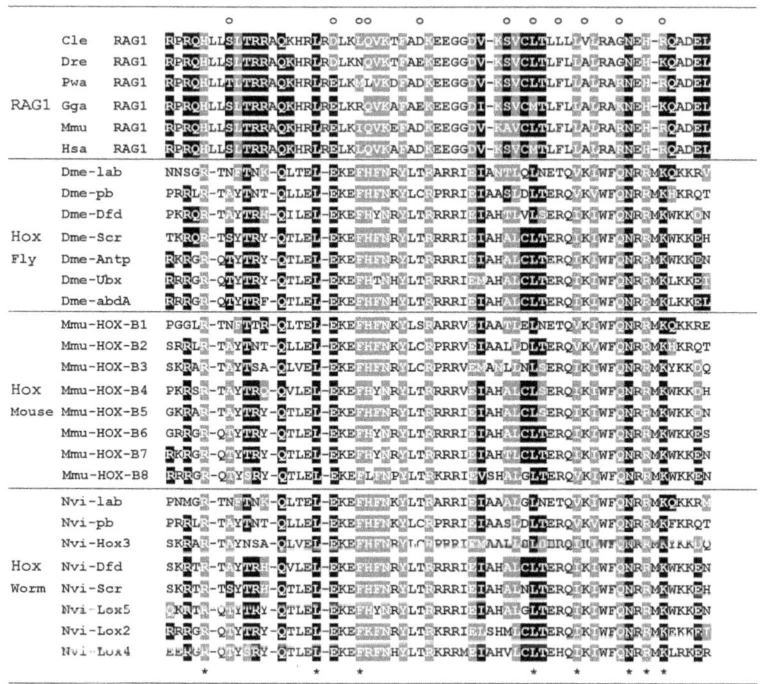

Figure 13.1 Comparison between the RAG1 DNA-binding domain and Hox homeodomains. An alignment is shown between the RAG1 DNA-binding domain and Hox homeodomains. Upper panel: RAG1 proteins of vertebrate species, chosen to represent the diversity of gnathostomes. Cle: *Carcharhinus leucas*, shark [U62645]; Dre: *Danio rerio*, teleost fish [NM_131389]; Pwa: *Pleurodeles waltl*, newt [AJ010258]; Gga: *Gallus gallus*, chicken [NM_001031188]; Mmu: *Mus musculus*, mouse [AY413840]; Hsa: *Homo sapiens*, man [AY413838]. Only 11 sites out of 61 show changes, most often for similar amino acids (open circles above the alignment). Other panels: comparison between Hox homeodomains and RAG1 proteins. Species representing the three bilaterian superphyla: Dme: *Drosophila melanogaster*, fly, Mmu: mouse and Nvi: *Nereis virens*, polychaete worm. Only identical and similar amino acids between Hox homeodomains and RAG1 have been scored, not those between Hox homeodomains themselves. White letters, black shadowed: identical amino acids. White letters, grey shadowed: similar amino acids. Two amino acids are here taken as similar when they show some similarity in structure and function (e.g. basic) *and* when the corresponding codons differ by a single base. Stars below the alignment correspond to residues conserved in all Hox genes and other homeobox-containing genes analysed for similarity to RAG1 (see Table 13.1).

Table 13.1 *Scores of identical and similar aminoacids between RAG1 and homeodomains. Same alignment and same definition of similarity for amino acids as in Figure 13.1.*

			Identical	Similar	Total
Hox	fly	*lab*	9	18	27
		pb	11	15	26
		Dfd	11	17	28
		Scr	14	15	29
		Antp	15	14	29
		Ubx	14	15	29
		abdA	16	14	30
	mouse	*Hox-b1*	10	15	25
		Hox-b2	10	15	25
		Hox-b3	8	18	26
		Hox-b4	11	17	28
		Hox-b5	11	17	28
		Hox-b6	13	14	27
		Hox-b7	14	14	28
		Hox-b8	12	15	27
paraHox	fly	*caudal*	7	15	22
		ind	9	16	25
	mouse	*Cdx1*	8	14	22
		Cdx2	7	16	23
		Cdx4	6	17	23
		Pdx1	11	14	25
		Gsx1	8	16	24
		Gsx2	8	16	24
Emx/ems	fly	*ems*	8	18	26
		E5	9	17	26
	mouse	*Emx1*	9	16	25
		Emx2	10	15	25
Evx/eve	fly	*eve*	9	15	24
	mouse	*Evx1*	9	15	24
		Evx2	9	15	24
NK	fly	*slou*	11	12	23
		ladybird-e	8	17	25
		ladybird-l	6	17	23
		msh	11	14	25
		tinman	7	20	27
		bagpipe	10	17	27
	mouse	*Sax1*	11	12	23
		Lbx1	9	18	27
		Lbx2	8	17	25

Table 13.1 (cont.)

			Identical	Similar	Total
		Msx1	11	14	25
		Msx2	11	14	25
		Msx3	11	14	25
		NK2.5	8	18	26
		NKx3.2	10	17	27
PAX	fly	*gsb*	11	13	24
		gsb-n	10	13	23
	mouse	*Pax3*	10	13	23
POU	fly	*vvl*	9	13	22
	mouse	*POU-III*	9	11	20
TLX	fly	*C15*	9	15	24
	mouse	*TLX1*	10	16	26
		TLX2	11	15	26
		TLX3	11	16	27
LIM	fly	*apterous*	7	18	25
	mouse	*Lhx2*	7	18	25
		Lhx1	7	12	19
		Lhx3	8	10	18
TALE	fly	*exd*	6	17	23
	mouse	*Pbx1*	7	15	22
		Pbx2	7	16	23
		Pbx3	7	16	23
	fly	*hth*	6	13	19
	mouse	*Meis1*	8	12	20

Using a rather restrictive definition of amino acid similarity (see legend of Figure 13.1), we scored aminoacids identical and similar to RAG1 for every protein aligned (Table 13.1). Surprisingly, a relatively high score of about 20 amino acids similar and identical to RAG1 was obtained for all homeodomains, despite their diversity. In fact, as many as 8 out of 60 residues (stars in Figure 13.1) are conserved in all proteins examined here and these are identical or similar to those of the RAG1 DNA-binding domain. The other residues similar to RAG1 vary according to the homeodomain family. Among all homeodomains scanned, the median Hox proteins present the best score, up to 30 (Figure 13.1 and Table 13.1). In particular, they are unique in presenting a (V/L)CL(T/S) motif quite similar to the homologous VCLT motif of RAG1. This motif is located at the loop between helix2 and helix3 of the homeodomain.

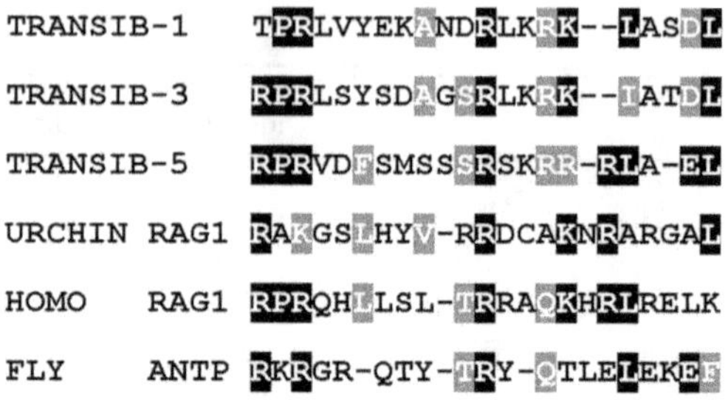

```
TRANSIB-1    TPRLVYEKANDRLKRK--LASDL
TRANSIB-3    RPRLSYSDAGSRLKRK--IATDL
TRANSIB-5    RPRVDFSMSSSRSKRR-RLA-EL
URCHIN RAG1  RAKGSLHYV-RRDCAKNRARGAL
HOMO   RAG1  RPRQHLLSL-TRRAQKHRLRELK
FLY    ANTP  RKRGR-QTY-TRY-QTLELEKEE
```

Figure 13.2 Similarity between the NH$_2$ arm of the Transib transposase, RAG1 and Hox proteins. Same grey tones as in Figure 13.1.

This location and the presence of the highly reactive cysteine residue are indicative of a putative protein–protein interaction motif.

As expected from the similarity with RAG1, metazoan homeodomains are also similar with *Transib* transposases (Figure 13.2). The similarity-domain-1 defined by Kapitonov and Jurka (2005) corresponds to the NH$_2$ part of the RAG1 DNA-binding domain and to the NH$_2$ arm of metazoan homeodomains, which is known to make contact with DNA.

Not all transposases belonging to DNA type II transposons present a helix–loop–helix DNA-binding domain. Not all metazoan transcription factors possess a homeodomain. Thus we find that the similarities observed here are more likely to result from common descent, that is, represent true homologies, rather than convergences. This means that RAG1 and metazoan homeobox genes are issued from transposons of the same family, not necessarily from the same and single lateral transfer event. Their common ancestor, as ancestor of two different transposons, might be far more distant than the common ancestor of all metazoans.

We know that certain homeobox genes are present outside the metazoans, in particular in fungi and plants. When phylogenetic analyses include fungi and plant homeobox genes, animal homeobox genes appear polyphyletic (Bharathan *et al.* 1997). Some clades are metazoan-specific, such as the 'ANTP super-class' including Hox, paraHox, NK, *engrailed*, *BarH* and related genes, others pre-date the divergences between plants, animals and fungi, such as a clade comprising *Knotted* from plants, *Cup* genes from fungi and *exd*/*Pbx* from animals. A very ancient homeobox gene might have been present in an ancestor eukaryote, giving rise to the present-day diversity of homeobox genes in the three lineages by multiple duplications followed by diversification. This 'classical' scenario requires that a number of the different gene families thus generated have been lost in each lineage (Bharathan *et al.* 1997).

Alternatively, if homeobox genes are issued from transposons, a scenario involving multiple transfers of related but not identical transposons at different times reconciles the gene tree with the species tree more parsimoniously. In particular, although a number of homeobox genes belonging to different families, including the ANTP-class, have been isolated from sponges, no Hox or paraHox genes have been detected (Richelle-Maurer *et al.* 2006). We thus suggest that an 'ANTP-super' transposon has invaded the common ancestor of all metazoans, and that in a second event, after the divergence between Porifera and Eumetazoa, the same or a related transposon has invaded the eumetazoan ancestor, generating the 'Hox-extended' family, comprising the Hox, paraHox, *Mox* and *Evx/eve* genes. After transfer, these transposons would have been 'domesticated' during evolution (Volff 2006). The role of transposable elements as a source of genetic evolutionary novelties is now better acknowledged (Biémont and Vieira 2006). Domestication must have involved a reduction of transposase activity, but it does not need to be rapid. Maintenance of some transposase activity may account for the high numbers of duplications and rearrangements of the Hox-extended family observed in the Cnidaria (Chourrout *et al.* 2006).

POSSIBLE TRANSPOSITION EVENTS IN THE HOX COMPLEX IN THE DROSOPHILIDAE LINEAGE

We can expect that during the course of transposon domestication, the transposase/recombinase activity of metazoan homeodomains has been progressively reduced or lost. Nevertheless, we wondered whether some transposition events could have occurred. We focused on Hox genes because, as they are not present in sponges, they might be the more recent and less derived homeobox genes. However, we surmised that transposition of Hox genes would have been lethal in organisms that develop progressively by posterior addition (Hughes and Jacobs 2005), because temporal collinearity would require the integrity of the Hox complex (Duboule 1994, Deutsch and Le Guyader 1998, Monteiro and Ferrier 2006).

Progressive development has been lost in certain bilaterian taxa, among which long-germ band insects. In these animals the Hox complex has a disturbed structure, as shown in the silk moth

Bombyx mori (Lepidoptera) (Yasukochi *et al.* 2004) and in the Drosophilidae (Negre *et al.* 2003). In *Drosophila pseudoobscura*, the *Deformed* Hox gene is in the same orientation in the ANT complex as the other Hox genes, whereas it is inverted in *D. melanogaster* (Randazzo *et al.* 1993). Despite this rearrangement, expression and function of *Dfd* and those of the neighbouring genes do not differ between the two *Drosophila* species. We thus inferred that the inversion must involve a segment larger than the mere Dfd transcript, including all relevant *cis*-regulatory sequences. Figure 13.3 shows the inversion by plotting the sequences of the *Dfd* region from these two species against each other.

We have drawn the structure of the Hox cluster on a phylogenetic tree of 12 *Drosophila* species whose complete genome sequence is currently available (Figure 13.4). It reveals that the *Dfd* gene has been inverted twice independently: once during the evolution leading to the *melanogaster* subgroup, once during the evolution to the *willistoni* group.

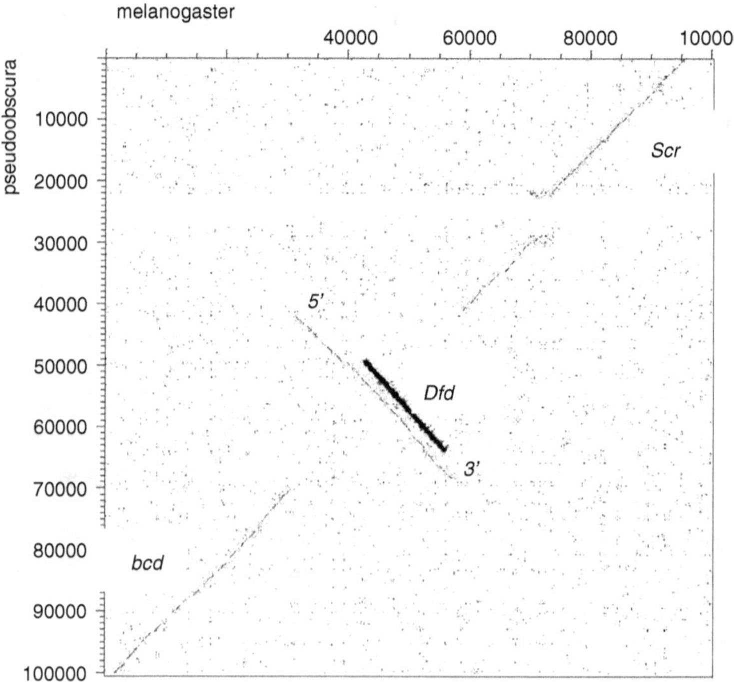

Figure 13.3 Dot-plot of the sequences of the *Dfd* region of *Drosophila melanogaster* (horizontal) vs. *D. pseudoobscura* (vertical). Solid bar: Dfd transcribed segment. *Scr*: *Sex combs reduced* gene; *bcd*: *bicoid* gene.

Figure 13.4 Conserved sequences close to the breakpoints of the
Dfd inversion. A, Dfd-up (5′ to the transcript); B, Dfd-dn (3′ to the
transcript). ANC: derived ancestral sequence. Small letters: bases
differing from the ancestral sequence. *sim*: *Drosophila simulans*; *sec*: *D.
sechellia*; *mel*: *D. melanogaster*; *yak*: *D. yakuba*; *ere*: *D. erecta*; *ana*: *D. ananassae*;
pse: *D. pseudoobscura*; *per*: *D.persimilis*; *wil*: *D. willistoni*; *moj*: *D. mojavensis*;
vir: *D. virilis*; *gri*: *D. grimshavi*. Below the Dfd-up sequences: ° (left
arm) and * (right arm) show the palindromic sequence.
Underlined on the Dfd-dn ancestral sequence: the ATTA motif, a putative
Hox target.

On each side of the inverted region, as close to the breakpoints
as we could determine, we found two segments of about 40 bp, here
called Dfd-up and Dfd-dn, highly conserved over 40 to 60 Myr of evolution in all 12 species (Figure 13.5). An imperfect palindrome is
located within the Dfd-up sequence (noted in Figure 13.5A). This
kind of secondary structure is indicative of a putative target of a
DNA-binding molecule. At the centre of the Dfd-dn segment, we
found a perfectly conserved ATTA motif (underlined in Figure 13.5B),
known to be the core of the Hox target (Ekker *et al.* 1994). Yet the Dfd-
up and Dfd-dn segments are distinct from all previously known conserved sequences located in this region, *cis*-regulatory enhancers
(Bergson and McGinnis 1990, Zeng *et al.* 1994, Lou *et al.* 1995), Polycomb Responsive Elements (Ringrose *et al.* 2003) and conserved repetitive sequences (Quesneville *et al.* 2005). They may have been preserved
from random drift because they are targets of the certain proteins,
possibly including Hox proteins. The presence of these motifs may
have favoured the advent of two independent inversions of the *Dfd*
locus during the evolution of the Drosophilidae.

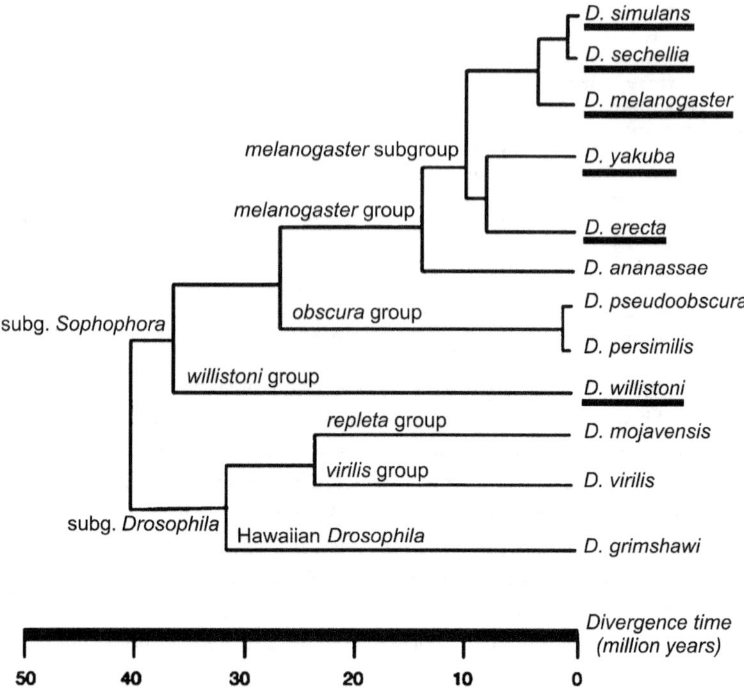

Figure 13.5 Phylogenetic tree of the Drosophilidae and *Dfd* inversion events. This phylogeny has been established by analysis of the complete genome sequences of these 12 *Drosophila* species (from FlyBase@flybase. bio.indiana.edu). The primitive 5′ to 3′ orientation of the *Dfd* gene is similar to that of the other Hox genes of the complex. Species where *Dfd* is inverted are underlined.

IMPLICATIONS OF THIS HYPOTHESIS: A SCENARIO FOR THE ORIGIN AND EARLY EVOLUTION OF BILATERIAN HOX COMPLEXES

The number of Hox genes in the common ancestor of Cnidaria and Bilateria was low, not more than two or three. It is not easy to derive the precise relations between these ancestral Hox genes and present bilaterian PGs. There are several reasons for that: (1) phylogenetic analyses on Hox-like genes depend on a low number of informative sites; (2) there have been *c.* 1000 Myr of evolution between this ancestor and extant species; (3) a more rapid rate of evolution is observed for the 'posterior' Hox genes (Chourrout *et al.* 2006); and (4) independent duplications occurred in various cnidarian lineages. Consensus exists on the presence of an ancestral *Hox1*-like gene, but disagreements are found about whether a second ancestral Hox gene was related to *Hox2* (Ryan *et al.*

2006) or *Hox3* (Chourrout *et al.* 2006), and whether the 'posterior type'
was present. Here we shall take as a start the more diversified hypothesis, that is, a primitive complex of three Hox genes, *Hox1/2*, *Hox3* and
*Hox9**, the latter being the ancestor of all rapidly evolving 'posterior'
Hox genes, PG9 to PG14. We can then draw a suite of steps in our scenario (Figure 13.6). Note that these steps did not necessarily happen in the

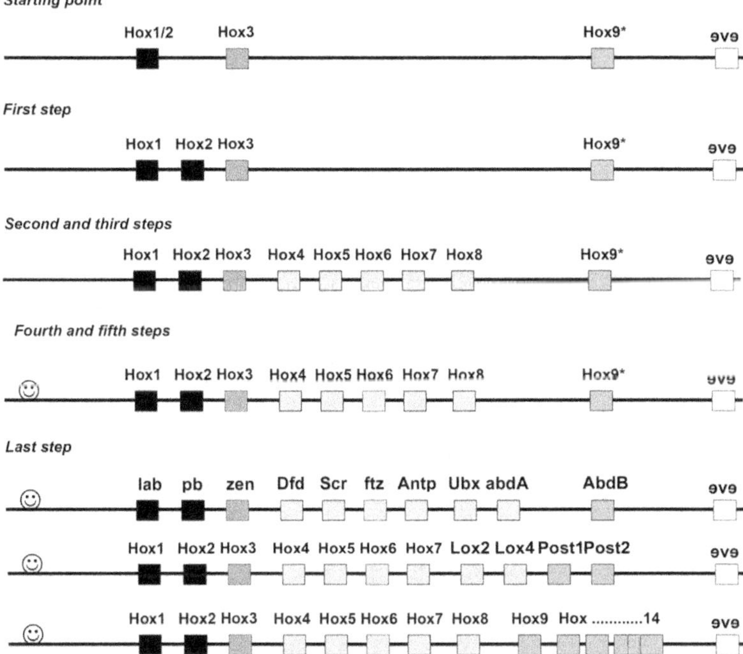

Figure 13.6 A scenario for the birth and early evolution of bilaterian Hox
complexes. *Starting point*: one among several hypotheses of what might
have been the Hox complex in the common ancestor of Cnidaria and
Bilateria (note that the *eve* gene has a reversed orientation relative to the
Hox genes). *First step*: Duplication of *Hox1/2* into *Hox1* and *Hox2*. *Second step*:
Burst of transposition events leading to the formation of 5 'median' Hox
genes. *Third step*: Repression of Hox expression due to repeat-induced gene
silencing. The Pc-G proteins could already play a role in this
heterochromatin-like repression. *Fourth step*: A *cis*-acting sequence (☺)
participates in relieving heterochromatin repression, thus creating temporal collinearity. *Fifth step*: Hox differentiation leading to posterior
prevalence and to the takeover of a quantitative mechanism by a qualitative mechanism. *Final step*: A few new transpositions/duplications,
specific to each superphylum, Ecdysozoa, Lophotrochozoa and Deuterostomia generate the present-day Hox complexes. See text.

order in which they are presented. Here they are drawn in a linear sequence for sake of clarity. These events occurred during the evolution from the common ancestor of the Cnidaria and Bilateria to the common ancestor of extant Bilateria (Valentine 2006).

The first step is the duplication of *Hox1/2* into *Hox1* and *Hox2*.

The second step is the burst of transposition events leading to the generation of *Hox4* to *Hox8*, the so-called 'median' Hox. This Hox explosion would generate a complex of nine genes, i.e. *Hox1* to *Hox9**.

Why should the transposed elements be close to one another? There is a trend for DNA type II transposons to transpose close to previously located transposons, as shown for the *Drosophila P*-element (Golic 1994). What might have initiated a sudden burst of transposition events? Knoll and Carroll (1999) stressed the possible importance of the increase in atmospheric oxygen in the generation of large animals at the early Cambrian. The metabolic stress thus provoked might have triggered the explosion of transpositions.

The then newborn Hox complex was poorly differentiated, with at most four types: anterior (*Hox1* and *Hox2*), anterior-median (*Hox3*), median (a series of identical *Hox4* to *Hox8*) and *Hox9**. This lack of differentiation between genes/transposons would result in partial repression of such a complex. Indeed, it is known that series of transposons repeated in tandem are prone to repression by heterochromatin formation. This phenomenon, called repeat-induced gene silencing (RIGS), has been observed in mammals and in other vertebrates (Henikoff 1998). A related phenomenon, dependent on Polycomb-group genes (Pc-G), is the so-called 'co-suppression' (Pal-Bhadra *et al.* 1997). The *Polycomb* gene and functionally related Pc-G genes are well known for their activity as repressors of the Hox genes, leading to poly-homeotic phenotypes that permitted their discovery (Lewis 1978). The mere formation of a poorly differentiated complex of multiple Hox genes/transposons is thus at the origin of the Pc-G mediated repression.

The third step is molecular evolution leading to a switch of Hox proteins from transposases to transcription factors (domestication). The primitive Hox coding frame and protein has likely encoded both functions, as seen in other DNA transposons, such as the *P*-element (Kaufman and Rio 1991). Selective pressure for preventing damage caused by transpositions would have favoured the repressor side of the protein's function.

The fourth step is the innovation of a genetic mechanism able to relieve Pc-G repression. Such an 'open for transcription' mechanism was first postulated by Peifer *et al.* (1987). Most bilaterians use the primitive mode of growing by posterior addition. In those organisms, spatial collinearity is the mere consequence of temporal collinearity. Chromatin opening of the Hox complex is progressive through time, from the 3′ end (*Hox1*) to the 5′ end ('posterior' genes). The opening device must rely on *trans*-acting factors and *cis*-acting sequences. Chromatin remodelling factors, such as the GAGA factor and other products of the trithorax-group (Trx-G) genes, are able to counteract the repressing activity of Pc-G proteins. Some of them are present throughout the eukaryotes and could have been recruited for this new function in bilaterians. Zákány *et al.* (2004) and Tarchini *et al.* (2006) in Duboule's laboratory recently provided evidence in the mouse for a remote *cis*-sequence involved in the progressive opening of the HoxD complex from the 3′ end.

Once the third step (turning transposase into a transcription factor) and fourth step (progressive opening of the chromatin) were completed, the primitive, yet poorly differentiated, Hox complex was ready to perform its present-day function, i.e. determining differentiation of the bilaterian body along the A-P axis. Indeed, in this primitive bilaterian, growing by terminal addition, the anterior head is set up at once, then parts of the 'trunk' (in a broad sense) are progressively added. Progressive Hox complex opening (temporal collinearity) would produce more Hox proteins as the trunk is developing. The various parts of the trunk would thus be *genetically* differentiated as a result of a Hox-dose effect. Such a Hox-dose effect can still be seen between Hox paralogues in the mouse: replacement of the original *Hoxa3* and *Hoxd3* genes by two *Hoxa3* or two *Hoxd3* results in equally viable mice (Greer *et al.* 2000).

The fifth step is differentiation of the Hox genes and proteins. In the primitive Hox complex, the 'median' genes PG4 to PG8 were identical. Although all present-day Hox proteins possess a similar DNA-binding homeodomain, they do not show identical transcriptional activity. This differential activity depends in part on interaction with co-factors, such as the so-called 'PBC' proteins, exd/PBX and hth/Meis1 (van Dijk and Murre 1994). Hox proteins interact with PBC proteins through their hexapeptide motif (Neutcboom *et al.* 1995). The difference in transcriptional activity between Hox paralogous proteins is engraved in their primary sequences, in part in the hexapeptide and in the N-terminal arm of the homeodomain. In addition, the various paralogues present a striking differential property, called 'phenotypic suppression' or 'posterior prevalence': when two Hox proteins are present in the same cell, the 'posterior' one prevails (Duboule and Morata 1994). Posterior prevalence brings evidence that random drift during about 600 Myr of evolution cannot account for Hox differentiation. Positive selection has driven Hox molecular evolution in order to replace a quantitative mechanism for trunk differentiation along the A-P axis based on Hox-dose effect with a qualitative mechanism based on differentiated Hox proteins. The latter would be more efficient, in terms of developmental accuracy and fidelity. Posterior prevalence has been the device required for suppressing quantitative effects.

The mechanisms underlying posterior prevalence are conserved between *Drosophila* and mouse (Duboule and Morata 1994). We can thus infer that Hox sequence differentiation, including important features for posterior prevalence, was already completed in the common ancestor of extant bilaterians.

The last step occurred during the length of time that lasted between the common ancestor of all the Bilateria and the ancestors of each of the three superphyla, i.e. Ecdysozoa, Lophotrochozoa, Deuterostomia.

From a complex comprising nine differentiated Hox genes, a small number of new duplication/transposition events occurred. In the ecdysozoan lineage, the *Hox8* gene duplicated into *Ubx* and *abdA*. Akam *et al.* (1988), providing the first phylogenetic analysis of Hox genes, showed that these two genes are younger than the other Hox genes of *Drosophila*. In the lophotrochozoan lineage, Wong *et al.* (1995) demonstrated that an independent duplication of *Hox8* generated the *Lox2* and *Lox4* genes. De Rosa *et al.* (1999) showed that two posterior genes, *Post1* and *Post2*, were specific to this super-phylum. They clearly belong to the 'posterior' Hox genes issued from the ancestral *Hox9**. In the deuterostome lineage, an explosion of duplications/transpositions affected the likely unique ancestral gene *Hox9**, leading to a basic number of six (*Hox9* to *Hox14*) 'posterior' Hox genes.

After this last burst of duplications/transpositions occurring in parallel within the three super-phyla, the evolution of the Hox complexes seems to have undergone a stasis. Indeed, from that point up to present, the basic complement of Hox genes in a complex has been stabilised. With rare exceptions, there was no further addition of Hox genes. Evolution of the Hox complexes evolved almost exclusively by duplications of entire complexes, probably owing to whole genome duplications, and losses of Hox genes within a pre-existing complex. Under our hypothesis, repression of the transposition mechanism, directly by the activity of a repressor or indirectly by loss of the transposase function or both, provides an explanation to the stasis of the Hox complexes after the split of the Bilateria in the three 'super-phyla'.

CONCLUSION

We present a novel hypothesis, i.e. Hox genes issuing from a transposon, to account for the origin and formation of the bilaterian Hox complex. We think that this is not a 'just so story'. Indeed, it is based on facts that are available in the scientific literature. In addition, it provides explanations for some properties of the Hox complex and some aspects of its evolution (and stasis) that remain obscure. We also believe that some features of our hypothesis (e. g. transposase activity) could be experimentally tested. And last, as more genomic data will be available in the near future, they may vindicate or invalidate the hypothesis.

ACKNOWLEDGEMENTS

Because of space limitations, we cite review papers rather than reports of the original works. We apologise to the authors and warmly advocate the readers to go and read the original papers referenced in the reviews. We gratefully acknowledge Dominique Anxolabéhère, Romain Derelle, Alexandre Hassanin, Michael Manuel, Danielle Nouaud, Hadi Quesneville and Stéphane Ronsseray for discussions. We thank Peter Holland very much for insightful comments on a previous version of this manuscript, which led us to undertake new sequence analyses. We are grateful to Alessandro Minelli and Giuseppe Fusco for inviting one of us (J.D.) to the evo-devo meeting held in Venice in May 2006 and for giving us the opportunity to publish our hypothesis in the present volume.

REFERENCES

Aboobaker, A. A. & Blaxter, M. L. 2003. Hox gene loss during dynamic evolution of the nematode cluster. *Current Biology* **13**, 37–40.

Aguinaldo, A. M., Turbeville, J. M., Linford, L. S. *et al.* 1997. Evidence for a clade of nematodes, arthropods and other moulting animals. *Nature* **387**, 489–493.

Akam, M., Dawson, I. & Tear, G. 1988. Homeotic genes and the control of segment diversity. *Development* **104** (supplement), 123–133.

Baguñà, J. & Riutort, M. 2004. The dawn of bilaterian animals: the case of acoelomorph flatworms. *BioEssays* **26**, 1046–1057.

Ball, E. E. & Miller, D. J. 2006. Phylogeny: the continuing classificatory conundrum of chaetognaths. *Current Biology* **16**, R593–R596.

Bergson, C. & McGinnis, W. 1990. An autoregulatory enhancer element of the *Drosophila* homeotic gene *Deformed*. *EMBO Journal* **9**, 4287–4297.

Bernstein, R. M., Schluter, S. F., Bernstein, H. & Marchalonis, J. J. 1996. Primordial emergence of the recombination activating gene 1 (*RAG1*): sequence of the complete shark gene indicates homology to microbial integrases. *Proceedings of the National Academy of Sciences of the USA* **93**, 9454–9459.

Bharathan, G., Janssen, B. J., Kellogg, E. A. & Sinha, N. 1997. Did homeodomain proteins duplicate before the origin of angiosperms, fungi, and metazoa? *Proceedings of the National Academy of Sciences of the USA* **94**, 13749–13753.

Biémont, C. & Vieira, C. 2006. Junk DNA as an evolutionary force. *Nature* **443**, 521–524.

Cameron, R. A., Rowen, L., Nesbitt, R. *et al.* 2006. Unusual gene order and organization of the sea urchin Hox cluster. *Journal of Experimental Zoology B (Molecular and Developmental Evolution)* **306**, 45–58.

Chourrout, D., Delsuc, F., Chourrout, P. *et al.* 2006. Minimal ProtoHox cluster inferred from bilaterian and cnidarian Hox complements. *Nature* **442**, 684–687.

Cook, C. E., Jiménez, E., Akam, M. & Saló, E. 2004. The Hox gene complement of acoel flatworms, a basal bilaterian clade. *Evolution & Development* **6**, 154–163.

De, P. & Rodgers, K. K. 2004. Putting the pieces together: identification and characterization of structural domains in the V(D)J recombination protein RAG1. *Immunological Reviews* **200**, 70–82.

de Rosa, R., Grenier, J., Andreeva, T. *et al.* 1999. Hox genes in brachiopods and priapulids and protostome evolution. *Nature* **399**, 772–776.

Deutsch, J. & Le Guyader, H. 1998. The neuronal zootype: a hypothesis. *Comptes Rendus de l'Académie des Sciences (Paris), série III* **321**, 713–719.

Duboule, D. 1994. Temporal colinearity and the phylotypic progression: a basis for the stability of a vertebrate bauplan and the evolution of morphologies through heterochrony. *Development 1994 supplement*, 135–142.

Duboule, D. & Morata, G. 1994. Colinearity and functional hierarchy among genes of the homeotic complexes. *Trends in Genetics* **10**, 358–364.

Ekker, S. C., Jackson, D. G., von Kessler, D. P. *et al.* 1994. The degree of variation in DNA sequence recognition among four *Drosophila* homeotic proteins. *EMBO Journal* **13**, 3551–3560.

Ferrier, D. E. & Holland, P. W. 2001. Ancient origin of the Hox gene cluster. *Nature Reviews Genetics* **2**, 33–38.

Ferrier, D. E., Minguillon, C., Holland, P. W. & Garcia-Fernàndez, J. 2000. The amphioxus Hox cluster: deuterostome posterior flexibility and *Hox14*. *Evolution & Development* **2**, 284–293.

Garcia-Fernàndez, J. 2005. The genesis and evolution of homeobox gene clusters. *Nature Reviews Genetics* **6**, 881–892.

Gauchat, D., Mazet, F., Berney, C. *et al.* 2000. Evolution of Antp-class genes and differential expression of *Hydra* Hox/paraHox genes in anterior patterning. *Proceedings of the National Academy of Sciences of the USA* **97**, 4493–4498.

Golic, K. G. 1994. Local transposition of *P* elements in *Drosophila melanogaster* and recombination between duplicated elements using a site-specific recombinase. *Genetics* **137**, 551–563.

Greer, J. M., Puetz, J., Thomas, K. R. & Capecchi, M. R. 2000. Maintenance of functional equivalence during paralogous Hox gene evolution. *Nature* **403**, 661–665.

Henikoff, S. 1998. Conspiracy of silence among repeated transgenes. *BioEssays* **20**, 532–535.

Hiom, K., Melek, M. & Gellert, M. 1998. DNA transposition by the RAG1 and RAG2 proteins: a possible source of oncogenic translocations. *Cell* **94**, 463–470.

Holland, P. W. & Takahashi, T. 2005. The evolution of homeobox genes: implications for the study of brain development. *Brain Research Bulletin* **66**, 484–490.

Hughes, N. C. & Jacobs, D. K. 2005. The end of everything: metazoan terminal addition. *Evolution & Development* **7**, 497.

Kamm, K., Schierwater, B., Jakob, W., Dellaporta, S. L. & Miller, D. J. 2006. Axial patterning and diversification in the Cnidaria predate the Hox system. *Current Biology* **16**, 920–926.

Kapitonov, V. V. & Jurka, J. 2005. RAG1 core and V(D)J recombination signal sequences were derived from Transib transposons. *PLoS Biology* 3, e181.

Kaufman, P. D. & Rio, D. C. 1991. *Drosophila P*-element transposase is a transcriptional repressor in vitro. *Proceedings of the National Academy of Sciences of the USA* **88**, 2613–2617.

Knoll, A. H. & Carroll, S. B. 1999. Early animal evolution: emerging views from comparative biology and geology. *Science* **284**, 2129–2137.

Lewis, E. B. 1978. A gene complex controlling segmentation in *Drosophila*. *Nature* **276**, 565–570.

Long, S., Martinez, P., Chen, W. C., Thorndyke, M. & Byrne, M. 2003. Evolution of echinoderms may not have required modification of the ancestral deuterostome HOX gene cluster: first report of PG4 and PG5 Hox orthologues in echinoderms. *Development Genes & Evolution* **213**, 573–576.

Lou, L., Bergson, C. & McGinnis, W. 1995. *Deformed* expression in the *Drosophila* central nervous system is controlled by an autoactivated intronic enhancer. *Nucleic Acids Research* **23**, 3481–3487.

Manuel, M. & Le Parco, Y. 2000. Homeobox gene diversification in the calcareous sponge, *Sycon raphanus*. *Molecular Phylogenetics and Evolution* **17**, 97–107.

Marletaz, F., Martin, E., Perez, Y. *et al.* 2006. Chaetognath phylogenomics: a protostome with deuterostome-like development. *Current Biology* **16**, R577–R778.

Matus, D. Q., Copley, R. R., Dunn, C. W. *et al.* 2006. Broad taxon and gene sampling indicate that chaetognaths are protostomes. *Current Biology* **16**, R575–R576

McGinnis, W. & Krumlauf, R. 1992. Homeobox genes and axial patterning. *Cell* **68**, 283–302.

Monteiro, A. S. & Ferrier, D. E. 2006. Hox genes are not always colinear. *International Journal of Biological Sciences* **2**, 95–103.

Negre, B., Ranz, J. M., Casals, F., Caceres, M. & Ruiz, A. 2003. A new split of the Hox gene complex in *Drosophila*: relocation and evolution of the gene *labial*. *Molecular Biology and Evolution* **20**, 2042–2054.

Neuteboom, S. T., Peltenburg, L. T., van Dijk, M. A. & Murre, C. 1995. The hexapeptide LFPWMR in Hoxb-8 is required for cooperative DNA binding with Pbx1 and Pbx2 proteins. *Proceedings of the National Academy of Sciences of the USA* **92**, 9166–9170.

Pal-Bhadra, M., Bhadra, U. & Birchler, J. A. 1997. Cosuppression in *Drosophila*: gene silencing of *Alcohol dehydrogenase* by *white-Adh* transgenes is Polycomb dependent. *Cell* **90**, 479–490.

Papillon, D., Perez, Y., Fasano, L., Le Parco, Y. & Caubit, X. 2003. Hox gene survey in the chaetognath *Spadella cephaloptera*: evolutionary implications. *Development Genes & Evolution* **213**, 142–148.

Peifer, M., Karch, F. & Bender, W. 1987. The Bithorax Complex: control of segmental identity. *Genes & Development* **1**, 891–898.

Pierce, R. J., Wu, W., Hirai, H. *et al.* 2005. Evidence for a dispersed Hox gene cluster in the platyhelminth parasite *Schistosoma mansoni*. *Molecular Biology and Evolution* **22**, 2491–2503.

Powers, T. P. & Amemiya, C. T. 2004. Evidence for a Hox14 paralog group in vertebrates. *Current Biology* **14**, R183–R184.

Quesneville, H., Bergman, C. M., Andrieu, O. *et al.* 2005. Combined evidence annotation of transposable elements in genome sequences. *PLoS Computational Biology* **1**, 166–175.

Randazzo, F. M., Seeger, M. A., Huss, C. A. *et al.* 1993. Structural changes in the Antennapedia-complex of *Drosophila pseudoobscura*. *Genetics* **134**, 319–330.

Reddy, Y. V., Perkins, E. J. & Ramsden, D. A. 2006. Genomic instability due to V(D)J recombination-associated transposition. *Genes & Development* **20**, 1575–1582.

Richelle-Maurer, E., Boury-Esnault, N., Itskovich, V. B. *et al.* 2006. Conservation and phylogeny of a novel family of non-Hox genes of the Antp class in Demospongiae (Porifera). *Journal of Molecular Evolution* **63**, 222–230.

Ringrose, L., Rehmsmeier, M., Dura, J. M. & Paro, R. 2003. Genome-wide prediction of Polycomb/Trithorax response elements in *Drosophila melanogaster*. *Developmental Cell* **5**, 759–571.

Ruiz-Trillo, I., Riutort, M., Littlewood, D. T., Herniou, E. A. & Baguñà, J. 1999. Acoel flatworms: earliest extant bilaterian Metazoans, not members of Platyhelminthes. *Science* **283**, 1919–1923.

Ryan, J. F., Burton, P. M., Mazza, M. E. *et al.* 2006. The cnidarian-bilaterian ancestor possessed at least 56 homeoboxes. Evidence from the starlet sea anemone, *Nematostella vectensis*. *Genome Biology* **7**, R64.

Schatz, D. G., Oettinger, M. A. & Baltimore, D. 1989. The V(D)J recombination activating gene, RAG-1. *Cell* **59**, 1035–1048.

Schierwater, B. 2005. My favorite animal, *Trichoplax adhaerens*. *BioEssays* **27**, 1294–1302.

Schuchert, P. 1993. *Trichoplax adhaerens* (Phylum Placozoa) has cells that react with antibodies against the neuropeptide RFamide. *Acta Zoologica* **74**, 115–117.

Spanopoulou, E., Zaitseva, F., Wang, F. H. *et al.* 1996. The homeodomain region of Rag-1 reveals the parallel mechanisms of bacterial and V(D)J recombination. *Cell* **87**, 263–276.

Tarchini, B. & Duboule, D. 2006. Control of HoxD genes' collinearity during early limb development. *Developmental Cell* **10**, 93–103.

Technau, U., Rudd, S., Maxwell, P. *et al.* 2005. Maintenance of ancestral complexity and non-metazoan genes in two basal cnidarians. *Trends in Genetics* **21**, 633–639.

Telford, M. J. 2000. Evidence for the derivation of the *Drosophila fushi tarazu* gene from a Hox gene orthologous to lophotrochozoan *Lox5*. *Current Biology* **10**, 349–352.

Thompson, C. B. 1995. New insights into V(D)J recombination and its role in the evolution of the immune system. *Immunity* **3**, 531–539.

Valentine, J. W. 2006. Ancestors and Urbilateria. *Evolution & Development* **8**, 391–393.

van Dijk, M. A. & Murre, C. 1994. Extradenticle raises the DNA binding specificity of homeotic selector gene products. *Cell* **78**, 617–624.

Volff, J.- N. 2006. Turning junk into gold: domestication of transposable elements and the creation of new genes in eukaryotes. *BioEssays* **28**, 913–922.

Wong, V. Y., Aisemberg, G. O., Gan, W. B. & Macagno, E. R. 1995. The leech homeobox gene *Lox4* may determine segmental differentiation of identified neurons. *Journal of Neuroscience* **15**, 5551–5559.

Yasukochi, Y., Ashakumary, L. A., Wu, C. *et al.* 2004. Organization of the Hox gene cluster of the silkworm, *Bombyx mori*: a split of the Hox cluster in a non-*Drosophila* insect. *Development Genes & Evolution* **214**, 606–614.

Zákány, J., Kmita, M. & Duboule, D. 2004. A dual role for Hox genes in limb anterior-posterior asymmetry. *Science* **304**, 1669–1672.

Zeng, C., Pinsonneault, J., Gellon, G., McGinnis, N. & McGinnis, W. 1994. Deformed protein binding sites and cofactor binding sites are required for the function of a small segment-specific regulatory element in *Drosophila* embryos. *EMBO Journal* **13**, 2362–2377.

14

Many roads lead to Rome: different ways to construct a nematode

EINHARD SCHIERENBERG AND JENS SCHULZE

It has been well established that considerable differences exist in the developmental pattern among animal taxa, for instance with respect to how blastomeres perform their early cleavages, how they acquire different fates or how symmetry is formed (Gilbert and Raunio 1997). Even among relatively closely related species, for instance within sea urchins or tunicates, impressive differences can be found in the pattern of development (Jeffery *et al.* 1999, Raff 1999).

Nematodes appear to be excellent candidates for a comparative study of early embryogenesis (Schierenberg 2005a). The phylum Nematoda is very old, its origin dating back to the Cambrian (Douzery *et al.* 2004), and has many different species (estimates range from tens of thousands to several millions); eggs can develop outside the mother from the first cleavage onward, they are transparent (although to a variable degree), the freshly hatched juveniles appear to have essentially invariant species-specific cell numbers of around 600 cells (for those species tested so far), many strains can be cultured in the laboratory on simple agar plates, and, last but not least, one of them, *Caenorhabditis elegans*, has become one of the best-studied model systems.

In this chapter, selected aspects of the early embryogenesis of five representatives from different branches of the phylogenetic tree are compared with *C. elegans* and the impact of the observed differences for evolutionary considerations are discussed. Following a brief reference to phylogeny, basic features of early embryogenesis of *C. elegans* will be summarised to aid in appreciating the data from other nematodes reported subsequently.

NEMATODE PHYLOGENY

Based mainly on molecular sequence data, a modern nematode phylogeny was suggested by Blaxter *et al.* (1998), extended and modified by De Ley and Blaxter (2002), with five clades in three subclasses. Recently, from a larger set of species, 339 nearly full-length small-subunit rDNA sequences were analysed and revealed a backbone of 12 consecutive dichotomies that subdivide the phylum Nematoda into 12 clades (Holterman *et al.* 2006; Figure 14.1). The clade numbers used below refer to this work.

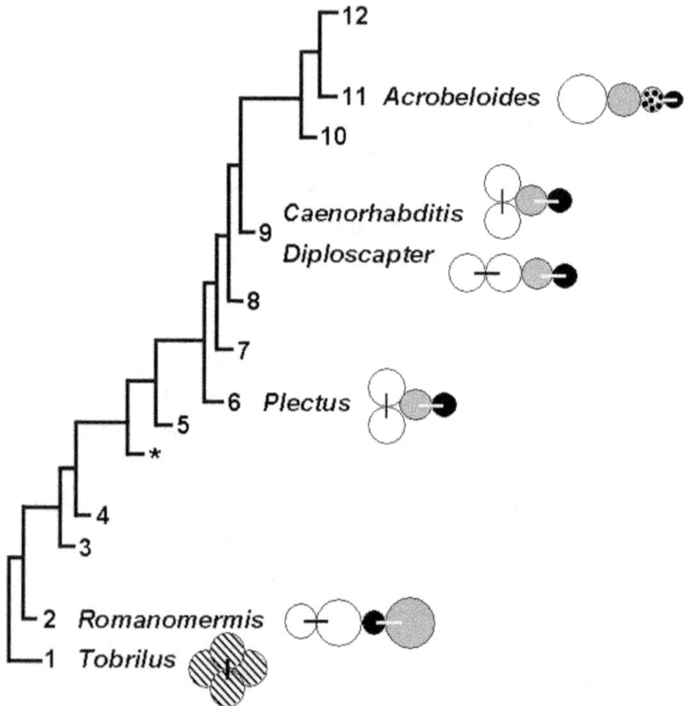

Figure 14.1 Simplified phylogenetic tree of nematodes. The tree is subdivided into 12 clades (1–12) and one unresolved branch (*) based primarily on DNA sequence data (Holterman *et al.* 2006). Branch lengths reflect substitution rates. Affiliations of the six representatives discussed here to individual clades and blastomere arrangements in four-cell stages are shown. The latter illustrate primary cell positions resulting from the orientation of cleavage spindles. Because of constraints of the egg envelope, rearrangements lead to a diamond-shaped pattern in *Caenorhabditis*, *Plectus* and *Romanomermis*. White, AB (S_1 in *Romanomermis*; for nomenclature, see legend to Figure 14.6) or AB daughters; grey, EMS (S_2); dotted, C (S_3); black, germ line (P_2; in *Acrobeloides*, P_3); striped, apparently equal cells of unknown fate. Connecting lines, sister cells.

CAENORHABDITIS ELEGANS EMBRYOGENESIS:
THE REFERENCE SYSTEM

Caenorhabditis elegans is a small (about 1 mm long) hermaphroditic soil nematode, which can be easily cultured in the laboratory on agar plates. Development from first cleavage to hatching is very rapid (12 h at 25°c) and eggs (size *c.* 55 × 35 µm) are remarkably transparent. The fact that rare males occur that can be mated to the hermaphrodites (male sperm is used preferentially) makes *C. elegans* a particularly amenable system for developmental geneticists (Brenner 1974; see also www.wormbook.org). A number of scientific milestones have been reached with *C. elegans*. It was the first metazoan whose genome was completely sequenced (The *C. elegans* Genome Consortium 1998); the complete wiring diagram of the nervous system has been described (White *et al.* 1986); ground-breaking methods like gene silencing with RNAi (Fire *et al.* 1998) and visualisation of gene expression in vivo with the GFP technique (Chalfie *et al.* 1994) were originally established in this system; and, finally, cell lineages of all 558 cells present at hatching have been documented (Sulston *et al.* 1983).

Figure 14.2A depicts the generation of five somatic founder cells via a series of unequal cleavages in the germ line and fates of their decendants. Upon fertilisation, immediately after fusion of the two pronuclei, the zygote divides into two unequal cells, a larger, anterior somatic cell AB and a smaller, posterior germline cell P_1. The AB cell divides with a transverse spindle orientation into ABa and ABp (Figure 14.2A). Both AB blastomeres are initially equipotent but nevertheless execute different developmental programs owing to inductive signals that they (and at least some of their descendants) receive from neighbouring cells (see below). The P_1 cell cleaves with a longitudinal spindle orientation unequally into a somatic cell EMS and a new germline cell P_2 (Figure 14.2B). Further unequal divisions of P_2 and its daughter P_3 generate the somatic founder cells C and D, respectively. Soon after the division of P_0, leading to the 24-cell stage, the two daughters of the gut precursor E initiate gastrulation by moving into the interior of the embryo. This important process will be considered in more detail at the end of this chapter.

From this brief synopsis the central role of the germ line with its stem-cell-like character from the first division of the embryo onward becomes obvious, leading to the stepwise generation of somatic founder cells. Germline cells contain specific cytoplasmic granules ('P granules') which can be visualized with antibodies (Strome and Wood 1983).

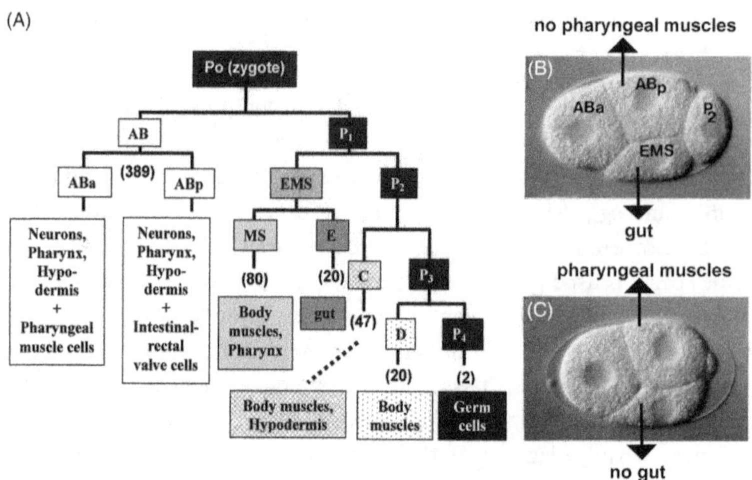

Figure 14.2 Cell lineages and inductive interactions in the early *C. elegans* embryo. A, Early cell lineage showing generation of five somatic founder cells (AB, MS, E, C, D) and the primordial germ cell P_4. Predominant or exclusive fates are given below individual lineage branches. Numbers in parentheses indicate cell numbers at hatching. B, Four-cell embryo with selected cell fates derived from ABp and EMS. C, After elimination of P_2 the developmental program of both neighbouring cells is altered because of missing inductions.

Other important features of early *C. elegans* embryogenesis require experimental interference (e.g. visualisation of gene expression with green fluorescent protein [GFP] constructs, mutant analysis, laser micro-manipulation or blastomere recombination) to become obvious. These include inductive events between individual blastomeres, just two of which will be mentioned here (for a more detailed description, see reviews by Basham and Rose 2001, Edgar 2001). In the four-cell stage (Figure 14.2B), the germline cell P_2 induces both of its neighbouring cells ABp and EMS via receptor–ligand interactions to execute specific developmental programs. While in the former case (ABp) homologues of Delta/Notch are involved, in the second case (EMS) genes of the Wnt/Frizzled signal cascade are active (Kimble and Simpson 1997; Rocheleau *et al.* 1997). Thus, in both cases mechanisms that are well conserved in the animal kingdom play a central role in embryonic cell specification. If the signalling source P_2 is eliminated (Figure 14.2C), ABp and EMS generate descendants with an altered fate and embryos arrest without reaching a vermiform stage (Priess and Thomson 1987, Schierenberg 1987).

OTHER NEMATODES SELECTED FOR EMBRYONIC STUDIES:
A BRIEF DESCRIPTION

Diploscapter coronatus is a close relative of *C. elegans*: both are members of clade 9 (Figure 14.1). However, *D. coronatus* is only about half the size of *C. elegans* and reproduces parthenogenetically. It lays its eggs prior to first cleavage. Eggs are only slightly smaller than those of *C. elegans* but embryogenesis takes about five times as long at room temperature.

Plectus sp. (strain ES 601; clade 6) and *Acrobeloides nanus* (clade 11) are similar to *D. coronatus* with respect to the features listed above. However, all three species can be easily distinguished on the basis of behaviour, body shape and a variety of anatomical features. While *Diploscapter* and *Acrobeloides* are cultured like *C. elegans*, all *Plectus* species we have studied require low-salt conditions and thus seem to occupy specific ecological niches.

Romanomermis culicivorax (clade 2) is a gonochoristic (male/female) parasitoid in mosquitos which leaves its host in the pre-reproductive phase and can then be kept in distilled water without food, where the animals copulate while forming prominent and permanent aggregates. Females grow to more than 2 cm in length and can lay more than 2000 one-cell stage eggs with a diameter of 80–90 µm. Embryogenesis takes about 10 times as long as in *C. elegans*.

Tobrilus diversipapillatus (clade 1) was found on the shores of lakes and small river banks. Although specimens can be kept in the laboratory for weeks, we are not yet able to culture them. Adults are about twice as long as *C. elegans* but eggs only about 30% longer than those of *C. elegans*. Compared with other representatives of this clade embryos are rather transparent and develop fast, i.e. only about two times slower than rate of *C. elegans*.

MODE OF REPRODUCTION AND ESTABLISHMENT OF
THE PRIMARY EMBRYONIC AXIS

Most higher organisms follow a gonochoristic mode of reproduction, which is thought to give at least long-term advantages because of the continuous recombination of alleles, resulting for instance in the loss of lethal mutations (Maynard-Smith 1978) and a better resistance to parasites (Hamilton *et al.* 1990). However, the advantages of sex are counterbalanced by at least short-term advantages of parthenogenetic species where each individual can reproduce and where the costs of mate search, courtship, intraspecific competition etc. can be saved. It is generally agreed that the gonochoristic mode is original and other variants like hermaphroditism and parthenogenesis are derived forms.

Parthenogenesis is frequently observed in certain free-living nematode taxa. Several such species are being cultured and studied in our laboratory (Skiba and Schierenberg 1992, Lahl *et al.* 2003, 2006), including the *Diploscapter*, *Acrobeloides* and *Plectus* species introduced above. This offers the opportunity to analyse in detail developmental peculiarities that accompany the parthenogenetic type of reproduction.

During oogenesis in the internally self-fertilising hermaphrodite *C. elegans*, oocytes arrest during meiosis and need to be induced by a sperm-derived signal to resume their meiotic program (Miller *et al.* 2001, Hajnal and Berset 2002) in order to become haploid and be ready for fertilisation. Egg cells lose their centrioles, and meiotic divisions take place without them (Albertson and Thomson 1993). The sperm then delivers the centriole necessary to generate embryonic cleavage spindles. In *C. elegans* it is also the sperm that induces formation of the primary embryonic axis: the area of its entrance into the egg defines the posterior pole (Goldstein and Hird 1996, Cowan and Hyman 2004).

These findings make clear that development of parthenogenetic nematodes must require certain modifications during oogenesis and/or early embryogenesis. These include: (1) establishment of egg polarity without fertilisation, i.e. either by random chance processes or via polarising cues acting in the mother; (2) preservation or restoration of diploidy without paternal contribution, either through absent or incomplete meiosis or via compensating postmeiotic processes; and (3) formation of cleavage spindles despite the absence of a sperm-derived centriole requiring either survival of the original centriole, a *de novo* synthesis in the egg cell, or formation of centrosomes without centrioles.

Here, we want to point out some peculiarities concerning aspects (1) and (2). By experimentally inhibiting egg-laying we determined the orientation of early-stage embryos within the uterus relative to the vulva (Figure 14.3; Lahl *et al.* 2006). In *C. elegans* oocytes are fertilised at the pole that enters the spermatheca first and thus embryos cleaving in the uterus point with their posterior pole toward the vulva. In *A. nanus* we found that embryos also showed a preferred orientation in the gonadal tube, but with opposite orientation to *C. elegans*. Thus, it appears that in *A. nanus* some external cue other than the one from sperm induces the direction of egg polarity. In eggs of *D. coronatus* we found that half of them point with their anterior pole and half with their posterior pole toward the vulva. Here, the fixation of anterior–posterior polarity seems to be independent of an external signal and determined randomly by chance.

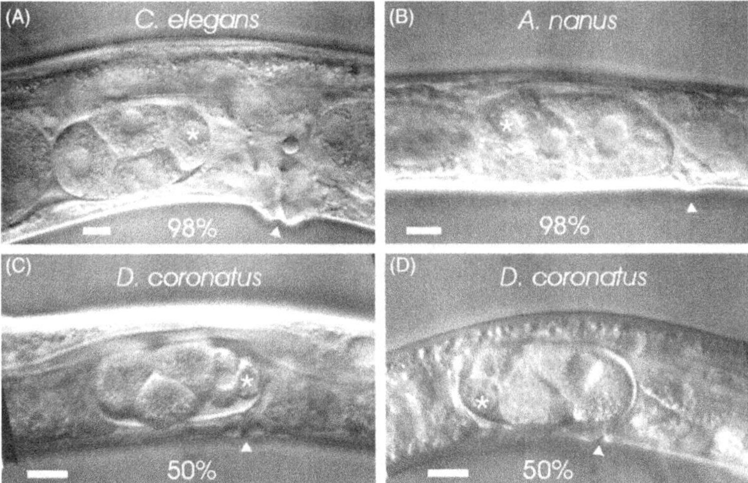

Figure 14.3 Establishment of axis polarity in parthenogenetic nematodes. Variations in the establishment of embryonic polarity. A, *C. elegans*, 98% of all embryos point with their posterior pole toward the vulva; B, *A. nanus*, 98% of all embryos point with their anterior pole toward the vulva; C and D, *D. coronatus*, equal proportions of embryos point with their anterior or posterior pole toward the vulva; arrowheads, position of the vulva; asterisk, germline cells P_2 (A, B) or P_3 (C, D). Scale bars, 10 μm. From Lahl *et al.* (2006).

Parthenogenetic species not only differ from *C. elegans* with respect to how diploidy is established but also differ among themselves. In *A. nanus* only one polar body (PB) is formed because the products of the second meiotic division fuse again. In *D. coronatus* two PBs are found, but these result from a cleavage of the first PB while the second meiotic divison is suppressed. In *Plectus* sp. two PBs are generated as well, but here in conjunction with two regular meiotic divisions. Circumstantial evidence suggests that *Plectus* restores its diploid status via an additional DNA replication round (Lahl *et al.* 2006).

Our preliminary studies on three *Acrobeloides* species with different modes of reproduction indicate that embryonic variances beyond meiosis and fertilisation are not correlated to parthenogenesis.

VARIATIONS IN EARLY LINEAGE AND PATTERN FORMATION

It is remarkable that the species considered here form a variety of different spatial patterns already from the four-cell stage onwards. However, even before or during gastrulation they all merge into a similar scheme (Schierenberg 2001, 2005a).

Diploscapter: a close relative of *C. elegans* with different early cell patterning

Instead of a diamond-shaped blastomere arrangement in the four-celled embryo, some nematode species show a linear grouping along the anterior-posterior axis (Malakhov 1994; Dolinski *et al.* 2001). Such an arrangement is also found in *D. coronatus* where not only P_1 but also the AB cell forms a longitudinally oriented cleavage spindle (Figure 14.1). This means much more than just a minor variation of a common pattern, as consequently P_2 never contacts ABp and contacts P_3 in only 50% of all embryos. Thus, an induction as found in *C. elegans* requiring cell–cell contacts (Priess and Thomson 1987) cannot take place here. Physical removal of P_1 through a laser-induced hole in the eggshell reveals that the unusual spindle orientation in AB is cell-autonomous. Cell lineage studies show that despite the absence of induction, like in *C. elegans* ABp descendants execute different fates from those of ABa descendants (V. Lahl, J. Schulze and E. Schierenberg, manuscript in preparation). Later, cells rearrange and reach a *C. elegans*-like pattern. Cell ablation experiments show that it is the EMS cell that takes the leading function in this process.

In conclusion, even close relatives of *C. elegans* may show considerable deviations during early development. In the case of *Diploscapter* it has been speculated that the differences may reflect a simplification of the developmental program (reduction of cell–cell interactions) at the cost of speed (necessary cell rearrangements). In addition, the linear array of blastomeres accompanied by an elongated eggshell may allow even a small species with a little vulva to produce relatively large eggs with an increased amount of nutritive or other maternal gene products (V. Lahl, J. Schulze and E. Schierenberg, manuscript in preparation).

Acrobeloides: an example for early embryonic plasticity

Developmental studies in *A. nanus* led to some unexpected findings (Wiegner and Schierenberg 1998, 1999). Overall embryogenesis proceeds about five times slower than in *C. elegans*, whereby initial cell cycles are particularly long. Inhibiting transcription shows that early cleavage requires zygotic gene activity while the *C. elegans* embryo reaches more than 100 cells under these conditions because of a generous maternal supply. Like in *C. elegans*, five somatic founder cells and a primordial germ cell are generated during early embryogenesis. However, the sequence of cleavages is different in that divisions in the germ line occur prematurely relative to mitoses in somatic cells (Figure 14.1). Thus, the primordial germ cell P_4 is already present in the six-cell stage while in *C. elegans* this occurs much later, at the 24-cell stage. In contrast to *C. elegans* no indication of germline-induced induction was found in *A. nanus*. For instance, any blastomere in the neighbourhood of the gut precursor cell can be removed and the remainder of the embryo will nevertheless form differentiated gut cells (Figure 14.4). However, the story goes further. Even when the gut precursor itself is eliminated the embryo compensates for this loss and partial embryos can even develop into hatching juveniles. This demonstrates that *A. nanus* carries a regulative potential absent in *C. elegans*.

Based on these data a model has been suggested according to which early blastomeres in *A. nanus* are multipotent and compete for a primary fate (Figure 14.5).

Plectus sp.: differences in symmetry formation and gastrulation

Although the four-cell stage of *Plectus* looks similar to that of *C. elegans* (Figure 14.1), soon afterwards peculiarities arise that appear to be typical for the whole family Plectidae.

In contrast to all other taxa mentioned here, gastrulation starts as early as the eight-cell stage (Figure 14.8A–C). This can be interpreted as a heterochronic shift giving the gut founder cell the premature ability to ingress while in other nematodes only its daughters or even granddaughters can do so. The migration of the EMS cell in *Diploscapter* (see above) can be understood along the same lines: that is, even another cell generation earlier the gut precursor cell becomes competent to migrate and/or neighbouring cells exhibit necessary cell surface molecules to do so. A second characteristic feature of *Plectus* is its early prominent bilateral symmetry which is formed within individual lineages via cell divisons with strict left–right spindle orientations (Lahl *et al.* 2003; Figure 14.8b, c).

Romanomermis: visible cytoplasmic segregation and different fate assignments

Embryonic cell lineage analyses have been performed in a variety of species from clades 6–12, while for the remaining clades only a few lineage studies exist (Malakhov 1994, Voronov 1999). Reasons are that only some of the latter are being cultured in the laboratory, development is slow and embryos are insufficiently transparent. One exception is *Romanomermis culicivorax* (Figure 14.1), whose development proceeds reasonably rapidly and

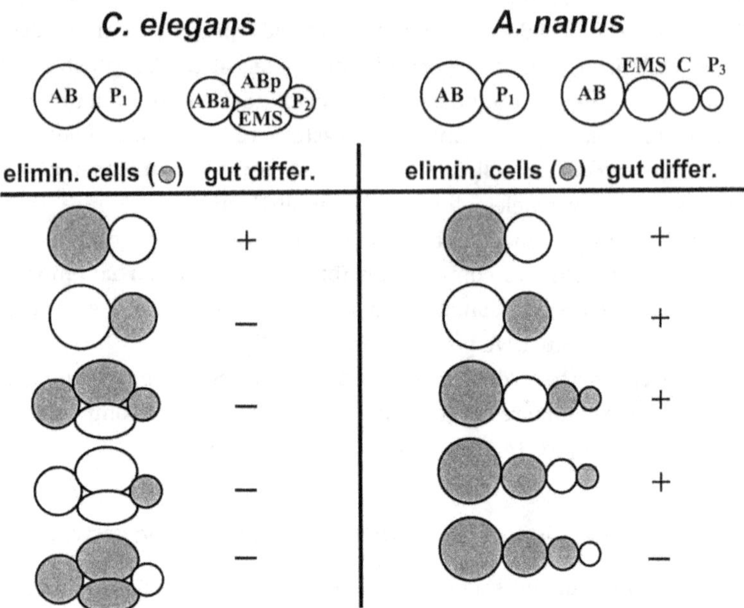

Figure 14.4 Differences in regulative behaviour between early *A. nanus* and *C. elegans* embryos. Top, intact two- and four-cell arrangements. Eliminated cells are marked in grey. '+', development of differentiated gut cells; '–', absence of differentiated gut cells at the terminal phenotype.

in which a moderate density of yolk granules allows detailed cell lineage studies (J. Schulze and E. Schierenberg, unpublished data).

In several respects *Romanomermis* differs from the species introduced above. The embryo contains – so far uniquely among nematodes – coloured cytoplasm segregated to the somatic founder cell S_2 (Figure 14.6), reminiscent of coloured myoplasm in some ascidian embryos (Jeffery 2001). However, here this blastomere appears to give rise to the complete hypodermis which eventually overgrows the remainder of the embryo. This process with the repeated duplication of cell groups (Figure 14.7) seems fundamentally different from the way in which hypodermis is generated in *C. elegans* (see concluding remarks). As another major difference to representatives of clades 6–12, we find that in *Romanomermis*, another early blastomere generates the complete alimentary tract, i.e. pharynx and gut. However, this is obviously true for other members of clades 1 and 2 as well (Malakhov 1994). In summary, our observations indicate that cell lineages and fate assignment in *Romanomermis* follow a less complex scheme than in *C. elegans*.

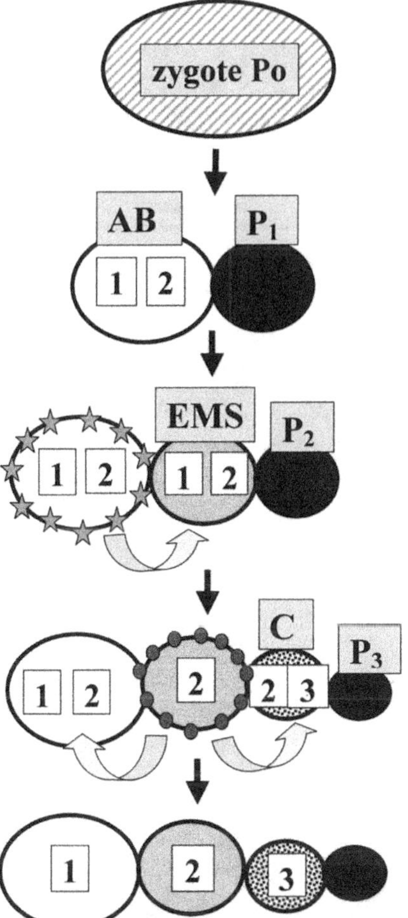

Figure 14.5 Model for cell specification in *A. nanus*. Early blastomeres can execute two alternative developmental programs (1 + 2 or 2 + 3; 1 = AB, 2 = EMS, 3 = C). Competing for a primary fate, inhibiting interactions (curved arrows) transmitted by specific cell surface molecules (stars and circles) between neighbouring cells lead to the restriction of developmental potential in a hierarchical manner. At least between AB and EMS, reciprocal interactions take place (after Schierenberg 2005a, modified).

Tobrilus: a nematode with unusual gastrulation

Gastrulation, the most dramatic process of reorganisation in the embryo, results in the formation of distinct germ layers. The classical type of gastrulation and probably the archaic one (Technau and Scholz 2003) starts with the formation of a hollow sphere (coeloblastula)

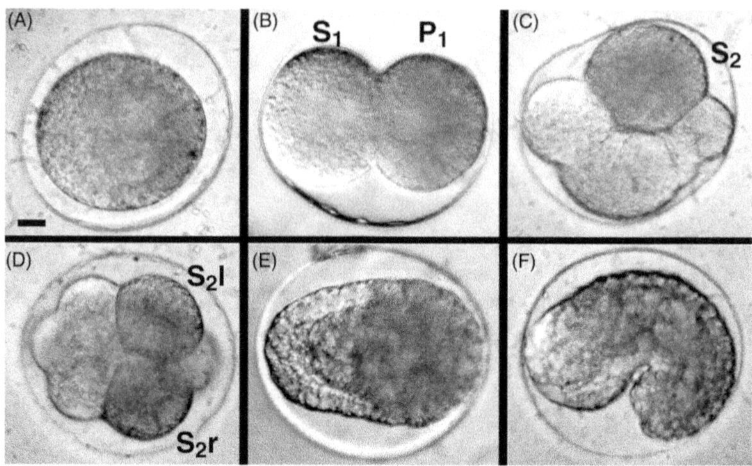

Figure 14.6 Segregation of coloured cytoplasm in *Romanomermis*. Translocation of brownish cytoplasm to the posterior pole prior to first cleavage (A) and consequent segregation into P_1 and later into S_2 (B, C). With the next division both daughter cells (S_2l and S_2r) receive the coloured components. During further development S_2 descendants expand from posterior to anterior (E, F). Note that nomenclature differs from *C. elegans* because of differences in cell position and fate. Formally, S_1 corresponds to AB and S_2 to EMS in *C. elegans*. A–C, F, Lateral view; D, E, dorsal view. Scale bar, 10 μm.

and subsequent invagination of endo- and mesodermal precursors. However, major variations exist even within the same phylum (Gilbert and Raunio 1997).

The nematodes studied in the past show a unique pattern of gastrulation not found elsewhere in the animal kingdom. Some key features of *C. elegans* gastrulation (Bucher and Seydoux 1994, Nance and Priess 2002) are briefly summarised here (Figure 14.8A–C). Soon after the primordial germ cell P_4 has been generated in the 24-cell stage, the two daughters of the gut precursor cell E, lying at a posterior-ventral position, start to ingress. Instead of a typical blastocoel, only a few small extracellular spaces are present at any time (von Ehrenstein and Schierenberg 1980, Nance and Priess 2002). Cells forming the mesoderm (i.e. body muscles and part of the pharynx) are derived from four different lineages. They immigrate in a piecemeal fashion at different times and places (von Ehrenstein and Schierenberg 1980, Sulston *et al.* 1983).

In contrast to all nematodes described so far, in *Tobrilus* a large blastocoel surrounded by a single layer of blastomeres forms

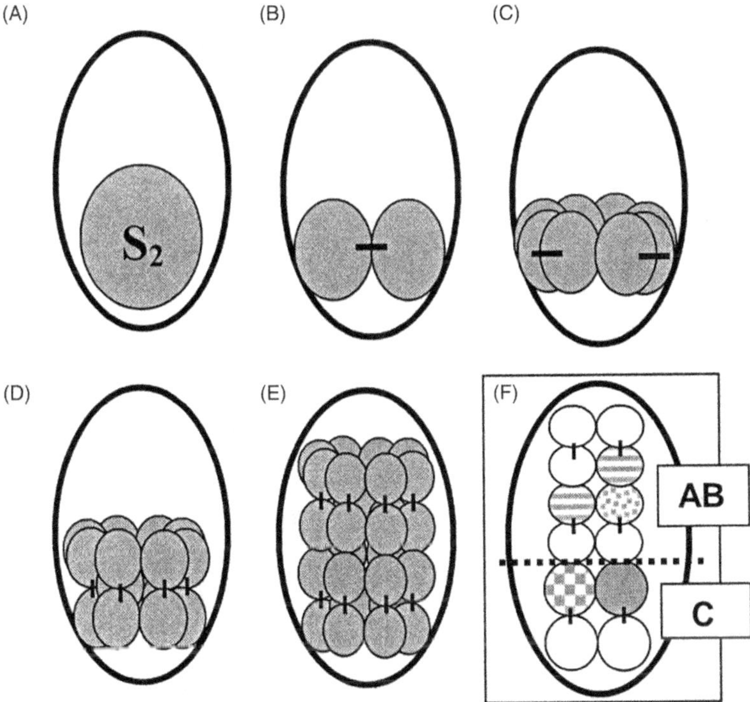

Figure 14.7 Hypodermis formation in *R. culicivorax* and *C. elegans*. Dorsal view. A–E, *Romanomermis*; owing to repeated divisions with transverse spindle orientation (B, C), descendants of S_2 (carrying brownish cytoplasm, see Figure 14.6), form a ring-like structure (C). As a result of consecutive divisions with longitudinal spindle orientations, repetitive units form that extend from posterior to anterior (D, E). F, *C. elegans*. Hypodermis is derived from two different lineages, AB (eight anterior cells) and C (four posterior cells). Colour code indicates to what extent the surviving descendants of the blastomere shown differentiate into hypodermis (Sulston *et al.* 1983). Grey, 100%; checkered, 70–80%; striped, 50–60%; dotted, 30–40%; white, 0–10%. Note that in reality 16 AB cells are already present when four C cells are formed.

(Figure 14.8a′), similar to blastula stages in other invertebrates (such as sea urchins; Figure 14.8a″) and also vertebrates. Around the 64-cell stage a small number of cells start to invaginate into the blastocoel (similar to sea urchin; Figure 14.8b′, b″). Movement and division of the internalised blastomeres result in their continuous extension and a corresponding decrease in blastocoel size (Figure 14.8c′, c″). It seems that these invaginated cells form not only intestine but also pharynx. From the 128-cell stage onward, a third layer of blastomeres invaginates and extends

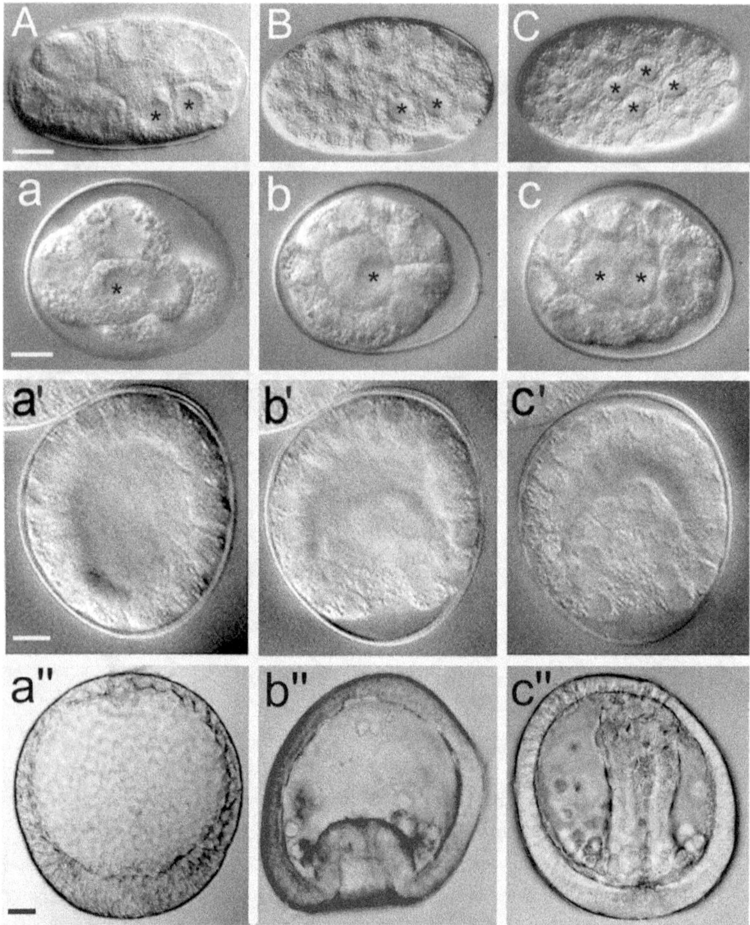

Figure 14.8 Gastrulation in nematodes and sea urchin. A–C, *C. elegans*, immigration of two gut precursors and subsequent division into four cells; blastocoel essentially absent. a–c, *Plectus* sp., immigration of one gut precursor and subsequent division into two cells, dorsal view. Note strict bilateral symmetry (b, c). a′–c′, *Tobrilus diversipapillatus*, large blastocoel, invagination of multiple cells. After formation of a large blastocoel (a′) invagination of endoderm (b′, c′) takes place. a″–c″, *Psammechinus miliaris* (sea urchin). Asterisks, gut precursors. Orientation: lateral view except *Plectus* (dorsal view). Scale bars, 10 μm.

between the compact mass of central cells and the surrounding ectoderm (Schierenberg 2005b).

The observations reported here demonstrate that a change took place within the phylum Nematoda in how three different germ layers are generated. It appears likely that gastrulation as seen in *Tobrilus* represents the original (plesiomorphic) state and that the standard nematode pattern is a derived condition.

CONCLUDING REMARKS

The data summarised in this chapter document that embryogenesis in nematodes is more variable than the final product, the hatching juvenile, would predict (for detailed lineage studies in addition to *C. elegans*, see Houthoofd *et al.* 2003, 2006). It has been suggested that the unexpectedly large genetic differences even between closely related nematode species (Fitch and Thomas 1997) is due to a 2–3 times higher nucleotide substitution rate compared with most other Metazoa (Aguinaldo *et al.* 1997). In addition, clades 8–12 (formerly indicated as Secernentea; including *Caenorhabditis, Diploscapter* and *Acrobeloides*) seem to have evolved considerably faster than clades 1–7 (formerly indicated as Adenophorea; including *Tobrilus, Romanomermis* and *Plectus*), possibly owing to higher metabolic rates and shortened generation times (Holterman *et al.* 2006). The special body plan of nematodes apparently prevented a corresponding degree of morphological diversification as found in other phyla like arthropods or vertebrates.

The wealth of early developmental variations appears paradoxical in a way, as these do not have any obvious impact on structure or performance of the resulting worms. Why then are there different ways to reach essentially the same goal? Two explanations can be offered. It could either be a result of neutral evolution, in which variations are due to system-inherent plasticity without any adaptive value, or the different ways may reflect alternative developmental strategies to increase fitness, for instance by making production faster or cheaper (Schierenberg 2001). Furthermore, it remains to be determined how dramatic the changes in the underlying genetic control must be to achieve apparently massive modifications on the cellular level. It appears rather difficult to imagine in terms of lineage transformations how the two variants of hypodermis formation as found in *Caenorhabditis* and *Romanomermis* (Figure 14.8) arose from a common pattern during evolution. However, if cell specification involves a position-based mechanism (e.g. 'all peripheral cells with no contact to the elementary tract shall form hypodermis') both species may only differ in the timing of when such a decision is made.

In addition to the different timing of gastrulation specified above, a number of other early embryonic peculiarities can be interpreted as heterochronic shifts (V. Lahl, J. Schulze and E. Schierenberg, manuscript in preparation). As heterochrony is often considered the single most important process of evolutionary change (Raff 1996) it would be interesting to pinpoint which of the numerous developmental variances among nematodes *cannot* be explained with such a mechanism.

The model of 'cell focusing' suggested by Schnabel *et al.* (2006) to illuminate the movement of blastomeres to specific embryonic regions in *C. elegans* according to their identity, and studies on the role of the germline as a polarising centre (Bischoff and Schnabel 2006), may also be helpful in imagining how species-specific modifications may have arisen during evolution.

It is not immediately obvious why early embryogenesis should be more variable than later phases. One argument has been that development is modular and integration of the emerging modules increases over time, putting fewer constraints on early development (Raff 1996). This seems reasonable for organisms where cells are specified relatively late, like vertebrates and possibly very slow-developing nematodes as found in clades 1 and 2 (Voronov and Panchin 1998). However, for the fast *C. elegans*-type of development, where essential decisions going along with specific cell–cell interactions take place in a very early phase, it must be questioned whether this argument is valid. Another reason for extended early variability could be the different role of maternal gene products during that period. As model systems have usually been selected because of their rapid development (Bolker 1995) maternal gene expression may be disproportionately high there. The huge differences between *C. elegans* and *A. nanus* with respect to maternal contribution during the early cleavage phase (Wiegner and Schierenberg 1998) support such a view.

In order to correlate ontogeny and phylogeny, embryonic variations may be useful heuristically as independent phylogenetic markers in addition to morphology and molecules. By looking at processes such as axis specification (Goldstein *et al.* 1998), cleavage pattern, arrangement of blastomeres (Dolinski *et al.* 2001, Houthoofd *et al.* 2003), germline behaviour and gastrulation (Schierenberg and Lahl 2004, Schierenberg 2005b), attempts have been made to trace the evolution of embryonic diversity in nematodes.

According to the Ecdysozoa hypothesis, nematodes are a neighbouring taxon to arthropods (Aguinaldo *et al.* 1997). Although we do not know what a last common ancestor of nematodes and arthropods might have looked like, it appears not unlikely that it was already segmented (or at least possessed some repetitive body elements) and that this feature was secondarily lost in conjunction with the reduction in cell numbers. It may therefore be attractive to look for potential remnants (or precursors) of segmentation in representatives positioned close to the basis of the nematode branch. Hypodermis formation in *Romanomermis* via generation of repetitive ring structures (Figure 14.7) is as close as we can come so far to something that is reminiscent of segmentation (J. Schulze and E. Schierenberg, unpublished results). The search for genes involved in segmentation (like *engrailed*) and their expression pattern in archaic nematodes may be helpful in determining whether such similarities are more than analogies and in general for the ongoing dispute about the phylogenetic position of nematodes.

Our studies have shown that different roads lead to Rome, i.e. to a juvenile ready to compete in the struggle for life. By extending comparative studies to a larger number of species and by identifying relevant genes, we should learn more about the intrinsic prerequisites for the implementation of embryonic novelty. In addition, we may better understand to what extent the interplay between the genetic program and external conditions (inside or outside the organism) determines the chance for deviations from an original developmental pattern to arise and to succeed. Finally, the question can be addressed of whether the establishment of modified embryonic cell behaviour as described here follows similar rules of variation and selection as assumed for so many morphological and physiological traits.

REFERENCES

Aguinaldo, A. M., Turbeville, J. M., Linford, L. S., *et al.* 1997. Evidence for a clade of nematodes, arthropods and other moulting animals. *Nature* **387**, 489–493.

Albertson, D. G. & Thomson, J. N. 1993. Segregation of holocentric chromosomes at meiosis in the nematode *Caenorhabditis elegans*. *Chromosome Research* **1**, 15–26.

Basham, S. E. & Rose, L. S. 2001. *Caenorhabditis elegans* embryo: establishment of asymmetry. In *Encyclopedia of Life Sciences*. Chichester:Wiley. www.els.net.

Bischoff, M. & Schnabel, R. 2006. A posterior centre establishes and maintains polarity of the *Caenorhabditis elegans* embryo by a Wnt-dependent relay mechanism. *PLOS Biology* **4**, e396.

Blaxter, M. L., De Ley, P., Garey, J. R., *et al.* 1998. A molecular evolutionary framework for the phylum Nematoda. *Nature* **392**, 71–75.

Bolker, J. A. 1995. Model systems in developmental biology. *BioEssays* **17**, 451–455.

Brenner, S. 1974. The genetics of *Caenorhabditis elegans*. *Genetics* **77**, 71–94.

Bucher, E. A. & Seydoux, G. 1994. Gastrulation in the nematode *Caenorhabditis elegans*. *Seminars in Developmental Biology* **5**, 121–130.

Chalfie, M., Tu, Y., Euskirchen, G., Ward, W. W. & Prasher, D. C. 1994. Fluorescent protein as a marker for gene expression. *Science* **263**, 802–805.

Cowan, C. R. & Hyman, A. A. 2004. Centrosomes direct cell polarity independently of microtubule assembly in *C. elegans* embryos. *Nature* **431**, 92–96.

De Ley, P. & Blaxter, M. L. 2002. Systematic position and phylogeny. In D. L. Lee (ed.) *The Biology of Nematodes*. London: Taylor & Francis, pp. 1–30.

Dolinski, C., Baldwin, J. G. & Thomas, W. K. 2001. Comparative survey of early embryogenesis of Secernentea (Nematoda), with phylogenetic implications. *Canadian Journal of Zoology* **79**, 82–94.

Douzery, E. J., Snell, E. A., Bapteste, E., Delsuc, F. & Philippe, H. 2004. The timing of eukaryotic evolution: does a relaxed molecular clock reconcile proteins and fossils? *Proceedings of the National Academy of Sciences of the USA* **101**, 15386–15391.

Edgar, L. G. 2001. *Caenorhabditis elegans* embryogenesis: genetic analysis of cell specification. In *Encyclopedia of Life Sciences*. Chichester: Wiley, www.els.net.

Fire, A., Montgomery, M. K., Kostas, S. A., Driver, S. E. & Mello, C. C. 1998. Potent and specific genetic interference by double-stranded RNA in *Caenorhabditis elegans*. *Nature* **391**, 806–811.

Fitch, D. H. & Thomas, W. K. 1997. Evolution. In D. L. Riddle, T. Blumenthal, B. J. Meyer & J. R. Priess (eds.) *C. elegans II*, New York: Cold Spring Harbor Laboratory Press, pp. 815–850.

Gilbert, S. F. & Raunio, A. M. 1997. *Embryology. Constructing the Organism.* Sunderland, MD: Sinauer Associates.

Goldstein, B., Frisse, L. M. & Thomas, W. K. 1998. Embryonic axis specification in nematodes: evolution of the first step in development. *Current Biology* **8**, 157–160.

Goldstein, B. & Hird, S. N. 1996. Specification of the anteroposterior axis in *Caenorhabditis elegans*. *Development* **122**, 1467–1474.

Hajnal, A. & Berset, T. 2002. The *C. elegans* MAPK phosphatase LIP-1 is required for the G(2)/M meiotic arrest of developing oocytes. *EMBO Journal* **21**, 4317–4326.

Hamilton, W. D., Axelrod, R. & Tanese, R. 1990. Sexual reproduction as an adaptation to resist parasites. *Proceedings of the National Academy of Sciences of the USA* **87**, 3566–3573.

Holterman, M., van der Wurff, A., van den Elsen, S. *et al.* 2006. Phylum-wide analysis of SSU rDNA reveals deep phylogenetic relationships among nematodes and accelerated evolution toward crown clades. *Molecular Biology and Evolution* **23**, 1792–1800.

Houthoofd, W., Jacobsen, K., Mertens, C. *et al.* 2003. Embryonic cell lineage of the marine nematode *Pellioditis marina*. *Developmental Biology* **258**, 57–69.

Houthoofd, W., Willems, M., Vangestel, S. *et al.* 2006. Different roads to form the same gut in nematodes. *Evolution & Development* **8**, 362–369.

Jeffery, W. R. 2001. Determinants of cell and positional fate in ascidian embryos. *International Review of Cytology* **203**, 3–62.

Jeffery, W. R., Swalla, B. J., Ewing, N. & Kusakabe, T. 1999. Evolution of the ascidian anural larva: evidence from embryos and molecules. *Molecular Biology and Evolution* **16**, 646–654.

Kimble, J. & Simpson, P. 1997. The LIN-12/Notch signaling pathway and its regulation. *Annual Review of Cell and Developmental Biology* **13**, 333–361.

Lahl, V., Halama, C. & Schierenberg, E. 2003. Comparative and experimental embryogenesis of Plectidae (Nematoda). *Development, Genes & Evolution* **213**, 18–27.

Lahl, V., Sadler, B. & Schierenberg, E. 2006. Egg development in parthenogenetic nematodes: variations in meiosis and axis formation. *International Journal of Developmental Biology* **50**, 393–398.

Malakhov, V. V. 1994. *Nematodes. Structure, Development, Classification and Phylogeny.* Washington: Smithsonian Institution Press.

Maynard-Smith, J. 1978. *The Evolution of Sex*. Cambridge: Cambridge University Press.

Miller, M. A., Nguyen, V. Q., Lee, M. H. *et al.* 2001. A sperm cytoskeletal protein that signals oocyte meiotic maturation and ovulation. *Science* **291**, 2144–2147.

Nance, J. & Priess, J. R. 2002. Cell polarity and gastrulation in *C. elegans*. *Development* **129**, 387–397.

Priess, J. R. & Thomson, J. N. 1987. Cellular interactions in early *C. elegans* embryos. *Cell* **48**, 241–250.

Raff, R. A. 1996. *The Shape of Life*. Chicago: University of Chicago Press.

Raff, R. A. 1999. Larval homologies and radical evolutionary changes in early development. In G. R. Bock & G. Cardew (eds.) *Homology (Novartis Foundation Symposia* 222 Chichester: Wiley, pp. 110–121.

Rocheleau, C. E., Downs, W. D., Lin, R. *et al.* 1997. Wnt signaling and an APC-related gene specify endoderm in early *C. elegans* embryos. *Cell* **90**, 707–716.

Schierenberg, E. 1987. Reversal of cellular polarity and early cell–cell interaction in the embryo of *C. elegans*. *Developmental Biology* **122**, 452–463.

Schierenberg, E. 2001. Three sons of fortune: early embryogenesis, evolution and ecology of nematodes. *BioEssays* **23**, 841–847.

Schierenberg, E. 2005a. Embryological variation during nematode development. In *WormBook*, www.wormbook.org.

Schierenberg, E. 2005b. Unusual cleavage and gastrulation in a freshwater nematode: developmental and phylogenetic implications. *Development, Genes & Evolution* **215**, 103–108.

Schierenberg, E. & Lahl, V. 2004. Embryology and phylogeny of nematodes. In R. C. Cook & D. J. Hunt (eds.) *Nematology Monographs & Perspectives*, Vol. 2. Leiden: Brill, pp. 667–679.

Schnabel, R., Bischoff, M., Hintze, A. *et al.* 2006. Global cell sorting in the *C. elegans* embryo defines a new mechanism for pattern formation. *Developmental Biology* **294**, 418–431.

Skiba, F. & Schierenberg, E. 1992. Cell lineages, developmental timing and spatial formation in embryos of free living soil nematodes. *Developmental Biology* **151**, 597–610.

Strome, S. & Wood, W. B. 1983. Generation of asymmetry and segregation of germ-line granules in early *Caenorhabditis elegans* embryos. *Cell* **35**, 15–25.

Sulston, J. E., Schierenberg, E., White, J. G. & Thomson, J. N. 1983. The embryonic cell lineage of the nematode *Caenorhabditis elegans*. *Developmental Biology* **100**, 64–119.

Technau, U. & Scholz, C. B. 2003. Origin and evolution of endoderm and mesoderm. *International Journal of Developmental Biology* **47**, 531–539.

The *C. elegans* Sequencing Consortium 1998. Genome sequence of the nematode *C. elegans*: a platform for investigating biology. *Science* **282**, 2012–2018.

von Ehrenstein, G. & Schierenberg, E. 1980. Cell lineages and development of *C. elegans* and other nematodes. In B. M. Zuckerman (ed.) *Nematodes as Biological Models*, Vol. 1: *Behavioral and Developmental Models*. New York: Academic Press, pp. 1–72.

Voronov, D. A. 1999. The embryonic development of *Pontonema vulgare* (Enoplida: Oncholaimidae) with discussion of nematode phylogeny. *Russian Journal of Nematology* **7**, 105–114

Voronov, D. A. & Panchin, Y. V. 1998. Cell lineage in marine nematode *Enoplus brevis*. *Development* **125**, 143–150.

White, J. G., Southgate, E., Thomson, J. N. & Brenner, S. 1986. The structure of the nervous system of *Caenorhabditis elegans*. *Philosophical Transactions of the Royal Society of London B* **314**, 1–340.

Wiegner, O. & Schierenberg, E. 1998. Specification of gut cell fate differs significantly between the nematodes *Acrobeloides nanus* and *Caenorhabditis elegans*. *Developmental Biology* **204**, 3–14.

Wiegner, O. & Schierenberg, E. 1999. Regulative development in a nematode embryo: a hierarchy of cell fate transformations. *Developmental Biology* **215**, 1–12.

15

Basal euarthropod development: a fossil-based perspective

NIGEL C. HUGHES, JOACHIM T. HAUG AND DIETER WALOSZEK

The morphological gap between Euarthropoda (the crown group that contains all extant arthropods) and living arthropod-like animals such as onychophorans, tardigrades and pentastomids is bridged by a number of fossils known primarily from rocks some 520 to 490 million years old (e.g. *Fuxianhuia*, *Chengjiangocaris*, *Shankouia*, see Figure 15.1; cf. Waloszek *et al.* 2005). These centimetre-scale fossil animals illuminate critical steps in early arthropod evolution (particularly head and limb development) but provide a limited amount of developmental information because of a lack of early ontogenetic stages. Small individuals that might represent pre-mature stages are scarce or absent, and the degree of allometry among the available individuals is generally modest. A limited number of early arthropod taxa do show more substantial ontogenetic information (Waloszek and Maas 2005). This chapter reviews the morphological development of early arthropods from two perspectives. The first is that provided by ontogenetic series based on the well-preserved biomineralised exoskeletons of trilobites, the best represented arthropod taxon in Palaeozoic rocks, but one whose development is seldom considered in broader comparative context. The second is that provided by 'Orsten'-type preserved faunas, in which the entire cuticle of numerous post-embryonic specimens of various species, mainly representatives of the crustacean evolutionary lineage (stem derivatives and Labrophora with phosphatocopines and members of the Eucrustacea; see Maas and Waloszek 2005) was replaced with spectacular fidelity by calcium phosphate in the absence of any compaction (Müller 1985). Such preservation has permitted detailed

Evolving Pathways: Key Themes in Evolutionary Developmental Biology, ed. Alessandro Minelli and Giuseppe Fusco. Published by Cambridge University Press.
© Cambridge University Press 2008.

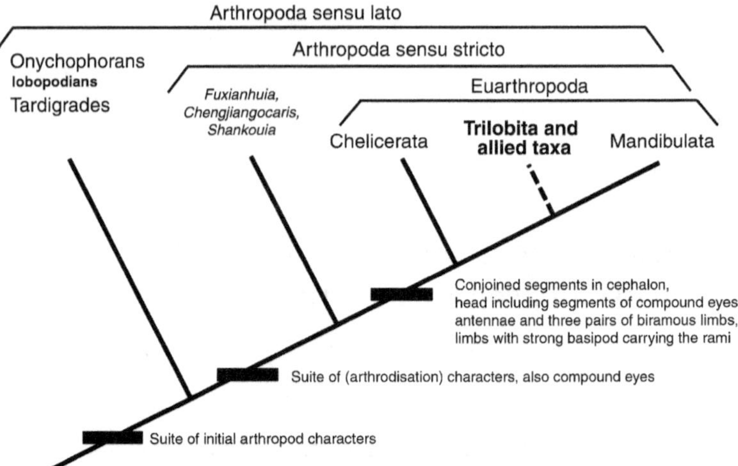

Figure 15.1 Phylogenetic relationships among major groups of living and fossil arthropod taxa (modified from Waloszek *et al.* 2005). We use the informal term 'allied taxa' to refer to a group of euarthropods with slim, long and feeler-like antennae and subsequent serial appendage pairs being biramous with lamellae-bearing exopods in cephalon and trunk, with a cephalic shield that covered basically three pairs of biramous limbs, and a trunk composed of largely homonomous exoskeletal segments commonly divided into an anterior region of freely articulating segments and a pygidium-like structure posteriorly.

reconstruction of portions of the ontogeny of several late Cambrian euarthropods normally absent from the fossil record. Patterns of segment generation common to these different sources may provide insight into the developmental mode of basal Euarthropoda, and thus into the evolution of arthropod ontogeny.

TRILOBITE STRUCTURE AND DEVELOPMENT

This discussion of trilobite segmentation focuses on the development of the segmentation of the biomineralised exoskeleton. Although appendages are known in some 20 species (see Hughes 2003a, Table 1, for a review) and there was a direct correlation between appendage pairs and dorsal exoskeletal segments in the non-terminal regions of the anterior–posterior (A-P) axis, ontogenetic information about trilobite development is almost entirely restricted to the development of the biomineralised exoskeleton. This was divided into two principal regions along the A-P axis, the cephalon and the trunk (Figure 15.2). Within

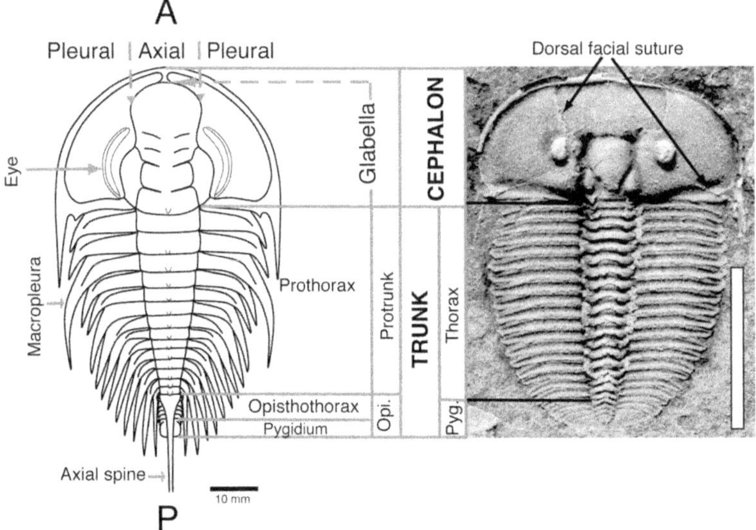

Figure 15.2 Basic dorsal morphology of two trilobites. A is anterior, P is posterior, Pyg. is the pygidium. Figure on the left, based on a generalised olenelloid trilobite, had a boundary between two distinct batches of segments located within the thorax, dividing the protrunk from the opisthotrunk (Opi.). *Aulacopleura konincki*, on the right, displayed the homonomous trunk condition in which all segments shared a similar morphology.

the cephalon exoskeletal segmentation was most evident in the axial region. The number of cephalic segments remained constant during the ontogenies of individual species, although in some cases adjacent segments differed markedly in shape. The number of segments within the cephalon (at least four) also appears to have been approximately constant throughout the Trilobita. It is not known whether these segments were specified simultaneously or sequentially. In striking contrast, the number of segments in the trunk region evidently varied both ontogenetically within species and phylogenetically among them. Trunk segmentation was expressed in both axial and pleural (lateral) regions, and displayed several different attributes whose variation was partially independent of one another. Such attributes include: (1) the number of trunk segments in the mature, segment invariant (epimorphic) phase of postembryonic development; (2) the number and development of functional articulations within the trunk region; (3) the form of trunk segments.

THE NUMBER AND GENERATION OF TRUNK SEGMENTS

The site at which new trunk segments were generated was a subterminal generative zone. This was located by studies of segments that were individualised from their first appearance, commonly by a unique axial or pleural spine (e.g. Stubblefield 1926, Chatterton 1994), which first appeared adjacent to the posterior end of the trunk, just as in the great majority of those arthropods in which segment expression is sequential (including those classified as having 'short germ-band' embryonic development). The development of trilobite trunk segmentation is also comparable to that of many other (eu)arthropods in that new trunk segments appeared sequentially through a series of postembryonic instars, a pattern that is termed anamorphic development.

In all species of trilobite in which ontogeny is well known, anamorphic instars were succeeded by an ontogenetic phase during which moulting and size increase continued, but which was invariant in the number of segments expressed in the biomineralised exoskeleton. This biphasic accretive-invariant segmentation pattern seen in trilobites, some myriapods and some crustaceans is termed 'hemianamorphic' development (Enghoff *et al.* 1993). In trilobites, following the general usage of specialists on myriapod biology, the second, 'segment invariant' phase is known as the 'epimorphic' phase (Hughes 2003b, Hughes *et al.* 2006). Hemianamorphic development evidently characterised the vast majority of the over 15 000 trilobite species known. It is not clear whether those rare trilobites with over 100 trunk segments (Paterson and Edgecombe 2006) achieved an epimorphic phase, and it is possible that some trilobites continued to add segments throughout life. The range in mature trunk segment number among trilobites varied widely, from forms bearing fewer than 10 trunk segments to those with over 100. Such a range of variation was evident even amongst early Cambrian species.

The rate of production of segments during the anamorphic phase was variable both ontogenetically within species and phylogenetically among them, and phases in which successive instars maintained a stable number of segments intercalated within other phases of anamorphic segment production have been reported (McNamara *et al.* 2003). Nevertheless, the number of segments expressed between successive instars was generally one or two segments, not the much larger numbers seen between instars in some derived myriapods (Fusco 2005) and crustaceans (Walossek 1993). Such a pattern of regular and gradual segment generation greatly aids in the reconstruction of trilobite ontogenies which are based on sclerites derived from multiple individuals and gives the impression that trilobite growth was 'track-like' (Hughes and Chapman 1995), i.e. channelled along a trajectory of progressive, modest change. Nevertheless, more striking ontogenetic changes have been recognised between particular instars and have been labelled as 'metamorphosis' in trilobites (Evitt 1961, but see Hughes *et al.* 2006). There are also cases in which juveniles cannot be linked with any co-occurrent mature forms, perhaps suggesting morphological changes too extreme to permit such association in the absence of the sequential instars of any individual (Chatterton and Speyer 1997).

ARTICULATIONS BETWEEN TRUNK SEGMENTS

In later ontogenetic stages, the trilobite trunk was divided into the thorax, an anterior set of segments each with a functioning joint along its entire anterior and posterior margins, and the pygidium, a posterior set of conjoined segments that did not functionally articulate with one another. The similarity of form in thoracic and pygidial segments partially reflected their common site of origin at the subterminal growth zone. An important aspect of trilobite development is that at first appearance all trunk segments were conjoined to segments immediately anterior to them. During early post-embryonic development all segments, whether part of the cephalon or trunk, were dorsally conjoined. The appearance of a functional joint between the rear of the cephalon and the anterior of the trunk marked the transition from the protaspid to the meraspid ontogenetic stages (Figure 15.3). At this stage the dorsal exoskeleton was a two-part, hinge-like structure made up of two units each comprising conjoined segments: the cephalon and the trunk. New segments accreted anamorphically near the rear of the trunk. New articulations between segments resulted in the progressive construction of the thorax, via the sequential release into it of trunk segments previously conjoined in the meraspid pygidium. Hence the construction of the trilobite thorax was a gradual and prolonged process with the thorax recognisable as a distinct region several instars after the segments that came to constitute it first appeared near the rear of the trunk.

Traditionally the trilobite pygidium has been considered homologous to the abdomen of other (eu)arthropods (Burmeister 1846, Cisne *et al.* 1980), but this pattern of the exchange of segments between the 'abdomen' and thorax is unlike that in almost all other (eu)arthropods

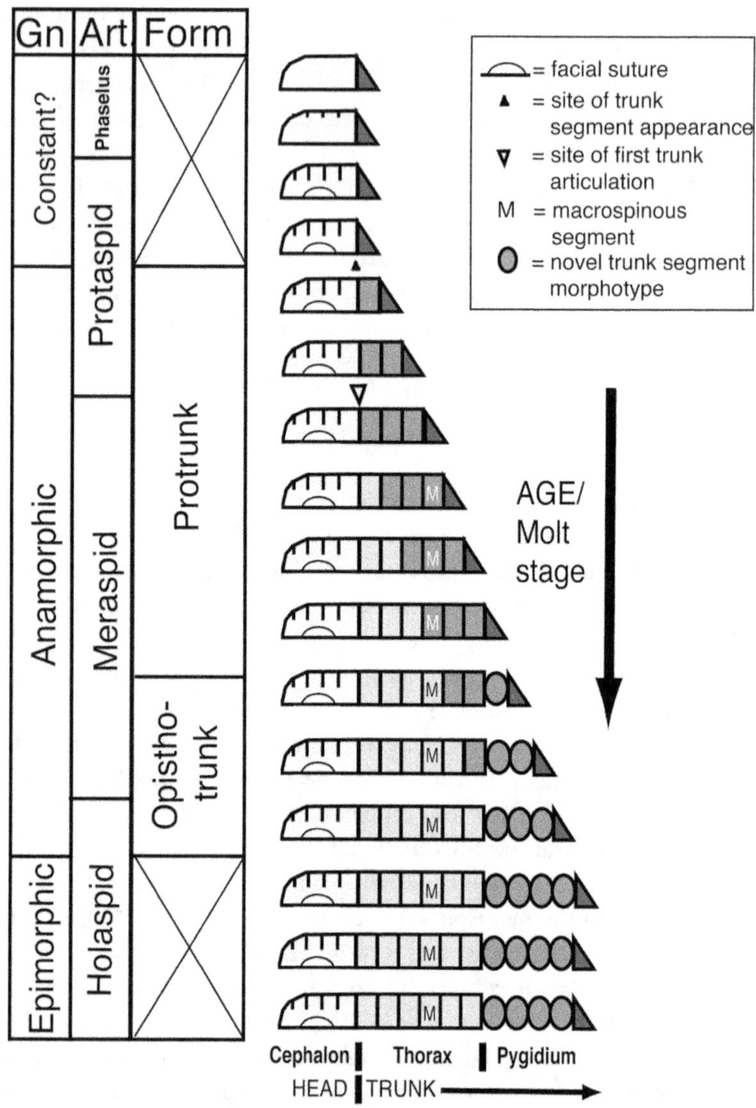

Figure 15.3 Generalised trilobite ontogeny showing the boundaries of ontogenetic stages based on three aspects of the development of segments: generation, articulation and morphology (see Hughes *et al.* 2006, fig. 3). 'Gn' refers to stages based on segment generation and contains a poorly known initial stage that may have had a constant set of cephalic segments, the anamorphic phase during which new segments appeared in the trunk, and the epimorphic phase after which the segment number was constant despite continued moulting. 'Form' refers to the morphology of newly generated trunk segments that in some trilobites are divided into discrete

(see below, and for distinction between limb-bearing thorax and limb-less abdomen, see Walossek and Müller 1998). All trilobites in which ontogenies are well known reached an instar in which segments ceased to be released into the thorax. Attainment of that state marked the transition from the meraspid stage to the holaspid stage. Like the anamorphic generation of trunk segments, the construction of the trilobite thorax was a gradual process, although it could depart from the regular release of one segment per instar. Cases in which more than one segment may have been released simultaneously, or two conjoined segments were released into the thorax together, are known but are rare. Other cases in which there may have been two or more instars for every segment released are also known but are uncommon.

THE FORM OF TRUNK SEGMENTS

Recognition of distinct body regions in arthropods is generally based on the morphology of segments, rather than their pattern of articulation. What little is known of trilobite trunk appendages does not indicate major differences in appendage morphology along the trunk: changes, where they do occur, tend to be mainly size differences, or relatively minor morphological modifications (Hughes 2003a). Nevertheless, the trilobite trunk exoskeleton did show marked differences in segment morphology in some cases (Hughes 2005). Examples include individual segments marked by unique features such as axial or pleural spines, or the grouping of segments into batches, each of which had a broadly uniform segment morphology that was distinct from that of other batches. The latter case is more comparable to the traditional tagmatic

---

Fig. 15.3 (cont.) batches of anterior (protrunk) and posterior (opisthotrunk) segments. 'Art' refers to developmental stages based on dorsal sclerite articulation pattern and includes the stages previously applied in studies of trilobite ontogeny. Site of the appearance of new trunk segments is shown for the first trunk segment only. Solid grey triangle is the terminal piece, darker grey segments are conjoined and part of the pygidium. Lighter grey segments are thoracic. Individualised segments, such as those that bore unusually large axial or pleural spines (i.e. a 'macrospinous' condition), retained the same position relative to the cephalic margin following first appearance, indicating that the site of appearance of new segments was subterminal, and that the boundary between articulating and conjoined segments migrated posteriorly during the meraspid phase (Stubblefield 1926). The functional significance of this segment was considered by Hughes (2003a).

boundaries recognised along the anterior–posterior axis of other arthropods. Indeed, trilobites present an interesting case in which the evolution of a morphologically distinct posterior tagma apparently occurred independently several times within the group. In trilobites in which all trunk segments, whether thoracic or pygidial, are homonomous, arguments for recognising the holaspid pygidium as a distinct tagma are weak (Minelli *et al.* 2003). However, where the trunk is heteronomous, and divided into batches of distinct segments, it is more reasonable to consider the trunk to comprise two tagmata, particularly where the boundary in segment morphotype coincides with the mature thoracic/pygidial boundary (Hughes 2003a,b, 2005).

The freedom to vary segment morphology within the trunk was strongly constrained in the trilobite thorax because in that region segments were required to articulate with one another. This constraint also apparently applied to the segments in the juvenile trunk that would eventually become thoracic: ultimately they had to achieve a form that permitted articulation. Trilobite trunk segments did not apparently change radically in shape as they were released from the pygidium into the trunk. This is evident in the early ontogenies of those trilobites in which segments that would ultimately become thoracic and those ultimately part of the mature pygidium are differentiated within the meraspid pygidium (Chatterton 1971, Hughes 2003a). Interestingly, peripheral features, such as marginal spines, seem to vary ontogenetically independently of their ultimate identity as thoracic or pygidial segments. However, variation in axial features, such as the courses of furrows marking segment boundaries, and in the relationship between axial and pleural segmentation, could be markedly different in the two regions, with tight covariation in these features in those segments that would become thoracic, and more independent variation in those that would remain pygidial throughout life.

DIVERSITY IN THE SEGMENTATION PROCESS AMONG TRILOBITES

Each of the three attributes of trunk segmentation discussed above reached a mature phase after which, although growth via moulting continued, each state remained invariant (Figure 15.3) (Hughes *et al.* 2006). In the case of segment generation this was the onset of a stable number of segments (epimorphic phase); for articulation, it was the attainment of a constant number of thoracic segments (holaspid stage); and with respect to segment form, it was the onset of production of the distinct set of posterior trunk segments (the opisthotrunk) in those trilobites with a heteronomous, two-batch trunk (Figures 15.2 and 15.3). Given the rather regular and progressive development of the trunk segmentation, we might expect that transitions to the mature phases of each of these attributes would be strongly coordinated. In some cases this was so, and particularly in those trilobites in which the mature thorax and pygidium represent distinctly different structures, each presumably with a distinctly different function (Figure 15.4). Nevertheless, trilobites showed a surprising diversity in the relative developmental timing of the transitions to the mature phases of these attributes. Cases in which transition to a stable number of segments coincided with onset of the holaspid phase are termed *synarthromeric*, those in which segment generation was completed prior to the termination of trunk articulation are *protomeric*, and those in which articulation was completed prior to onset of a stable number of segments are *protarthrous* (Hughes *et al.* 2006).

A review of such cases shows a variety of patterns (Hughes *et al.* 2006). For example, the numbers of trunk segments in the Silurian aulacopleurid trilobite *Aulacopleura konincki* varied, apparently intraspecifically, over a range of five segments, but each of the five morphs apparently showed synarthromeric growth (Fusco *et al.* 2004). On the other hand, two putative intraspecific morphs of the early Cambrian eodiscid trilobite *Neocobboldia chinlinica* ultimately apparently achieved the same total number of trunk segments. A suggested explanation for this is that one may have developed synarthromerically, the other protomerically (Hughes *et al.* 2006). A comprehensive review of trunk trilobite development has yet to be attempted, but it appears that variation in developmental mode commonly occurred at low taxonomic levels, although some taxa may be characterized by a particular mode. One case is known in which the epimorphic stage was achieved prior to the onset of trunk articulation: the late Cambrian *Schmalenseeia fusilis* never developed joints in the trunk and thus remained a permanent protaspid (Peng *et al.* 2005, Hughes *et al.* 2006).

The principal value of the developmental record of trilobites is the ability to explore how these different aspects of the segmentation process map onto the phylogeny of the group, and to consider the evolutionary trade-offs between the flexibility to vary aspects of the trunk independently, and the advantages of increasingly integrated covariation. Key to exploring this is the delay during development between the appearance of segments and the attainment of their final functional role in the thorax. Thorough exploration of the developmental record of

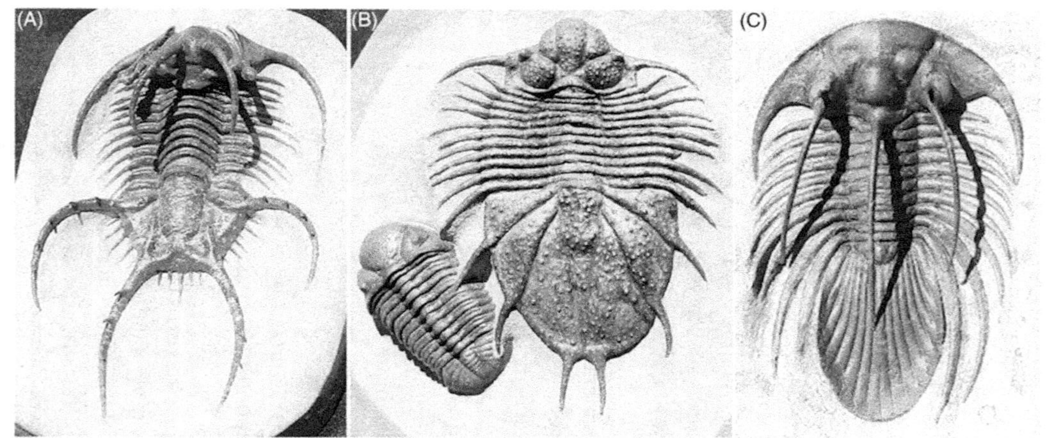

Figure 15.4 The 'two-batch' heteronomous trunk condition in which the mature thoracic and pygidial segments had very different morphologies occurred in a variety of major trilobite taxa, as illustrated by the large specimens in each of the three figures. All specimens are from the AM Limestone of the Devonian of Morocco. A, *Ceratarges* sp. (Order Lichida), length 8.4 cm. B, The larger specimen is the lichiid trilobite *Acanthopyge (Acanthopyge)* cf. *haueri* (Order Lichida), length 8.4 cm. The smaller trilobite (Order Phacopida, species undescribed) shows the homonomous trunk condition. C, *Kolihapeltis* sp. (Order Corynexochida), length about 5 cm. Note that the boundary in segment morphotype coincides with the mature thoracic–pygidial boundary in each of these cases. Modified from Hughes (2007).

trilobites will require a combination of careful phylogenetic and functional analyses and detailed individual studies of growth mechanics, and firm conclusions are not yet warranted. Nevertheless, those trilobites that varied trunk segment numbers at low taxonomic levels do appear to be those in which the trunk consisted of homonomous segments, while those with a highly tagmatised trunk were invariant in segment numbers (Hughes 2005). The fact that the tendency toward a more highly integrated, segment invariant trunk occurred independently among several trilobite taxa (and possibly also some allied taxa often referred to as 'trilobitomorphs') offers promise for exploring the functional context of such a transition.

DEVELOPMENTAL PATTERNS AMONG 'ORSTEN' ARTHROPODS

Several 'Orsten' (eu)arthropods are preserved with multiple developmental stages, the ontogeny of the eucrustacean *Rehbachiella kinnekullensis* being by far the best known (Walossek 1993) (Figure 15.5B). The earliest larva of this possible stem branchiopod was a true nauplius, a short-headed larva that is autapomorphic for Eucrustacea. This larva bore only three pairs of appendages, one pair of uniramous antennulae and two pairs of biramous limbs, the so-called antenna and mandible. Altogether, four 'naupliar' and 26 'post-naupliar' stages have been distinguished in *R. kinnekullensis*. During this phase, the trunk segments were added progressively at the rate of one segment for every two moults. This sequence led to a trunk with 13 limb-bearing segments, while the posteriormost trunk

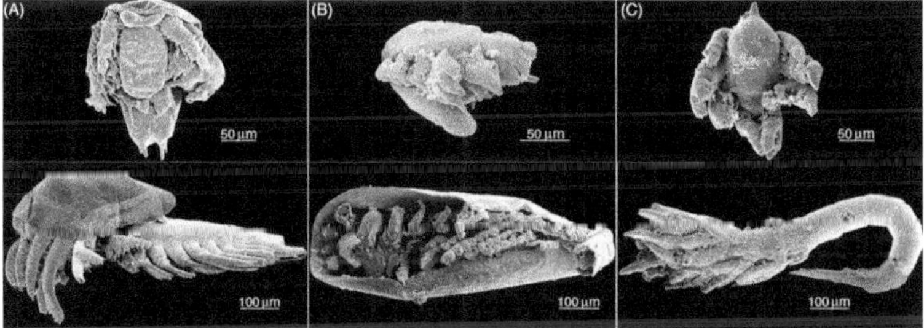

Figure 15.5 Three crustaceans from the 'Orsten' fauna. Top row shows early larvae, bottom row latest known stages. A, *Bredocaris admirabilis* (lower picture composed of different specimens). B, *Rehbachiella kinnekullensis*. C, *Martinssonia elongata*. All specimens illustrate the exceptional three-dimensional preservation.

limbs remained of larval shape and the hind body remained unsegmented (Walossek 1993).

The extant brine shrimp *Artemia salina* adds one trunk segment per moult, and after that phase the abdominal segments are added successively, while the appendages are modified into adult shape. Comparison with *Rehbachiella* indicates that even the latest stages of the fossil (size- and developmentally correlated with that of the latest larva of the first phase of *Artemia*) were still immature. Likewise, *Rehbachiella* would have required additional moults to develop the segmentation of the abdomen. Isolated limbs and body fragments also hint at older developmental stages, some being twice the size of the largest complete specimen. The 30-stage sequence of *Rehbachiella* is, therefore, far from being complete.

The strictly anamorphic developmental pattern exhibited by *Rehbachiella* has been used as a reference for the developmental patterns of other Crustacea (Figure 15.6A, B), serving to highlight deviations from such a regular system. It can even enable us to predict the size of a larva of a taxon based on its segmental stage via the correlation between the number of clearly expressed segments and overall size. This applies between moults that show large developmental 'jumps' between stages, with addition of several segments in one step (Figure 15.6C for penaeid decapod malacostracans).

As compared to the regular developmental pattern in *R. kinnekullensis*, the 'Orsten' eucrustacean, *Bredocaris admirabilis* (Figure 15.5A), exhibited a derived mode of development. The first larva was already a

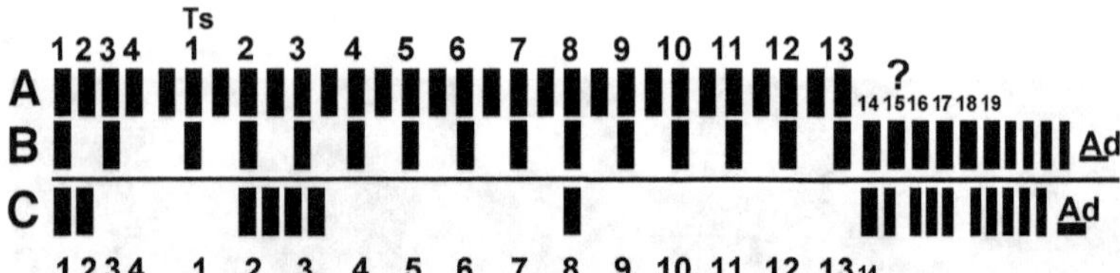

Figure 15.6 Schematic diagram of the developmental stages of various fossil and recent crustaceans. Numbers denote successive stages: 1–4 for early larvae, Ts 1 to *n* are for stages with developed postcephalic segments. Bars indicate stages expressed in the ontogenies of each taxon. Ad: adult, Ts: trunk segment. A, *Rehbachiella kinnekullensis*. B, *Artemia salina*. C, Example of a penaeid decapod malacostracan. *Rehbachiella* shows the most strictly anamorphic pattern. Deviation from the pattern is exemplified by large 'jumps' (gaps comprising missing stages) in the development of penaeid decapod malacostracans (cf. Walossek 1993).

metanauplius having a vestigial maxillula in the form of a bifid lobe and corresponding to a L3 *Rehbachiella* larva. In the four next stages, the segments of the second maxillae and the thorax were added anamorphically, but were not separated on the undivided hind body and can be recognised only through five ventral and progressively appearing limb buds, which remain as such during the entire phase (in a strict anamorphic sequence, limbs develop progressively and also add setae). With the next known stage, presumably a single moult further, the now fully developed trunk bore the developed maxilla and seven thoracopods. Hence *B. admirabilis* deviated from the scheme of development seen in *R. kinnekullensis*, still adding segments at each moult, but delaying trunk and limb development until a final step toward another instar, which might have been either a juvenile or the adult. This specific ontogenetic pattern is represented by extant barnacles, identifying *B. admirabilis* as a stem-lineage member of the taxon Thecostraca (cf. characterisation by Müller and Walossek 1988), which embraces cirripeds and a set of small-sized crustaceans such as ascothoracids and facetotectans (Høeg and Kolbasov 2002).

'Orsten' larval evidence is also available for a few non-eucrustacean taxa, including phosphatocopines, the sister taxon of Eucrustacea, derivatives of the stem lineage of Labrophora = Eucrustacea and Phosphatocopina, taxa like *Agnostus pisiformis* and a stem representative of the sea spiders, Pycnogonida (Müller and Walossek 1986a, Waloszek and Dunlop 2002). Investigations of the developmental patterns of more derivatives of the stem crustaceans are still in progress, but we can state that all of them started their ontogenies with an early larva with antennae and three pairs of functional head limbs, a condition named the 'head larva' by Walossek and Müller (1990, 1998) that characterised the ground pattern of the Euarthropoda.

Martinssonia elongata (Figure 15.5C) is the only species of the labrophoran stem-lineage derivatives with a larval series described in detail although this series is incomplete (Müller and Walossek 1986b). The first three stages were all head larvae, bearing three post-antennular biramous limbs and not adding further segments but changing proportions and developing a mouth and anus (they hatched as lecithotrophic larvae). The next known stage already had a head with an additional pair of limbs, a trunk with five ring-shaped segments and an elongate caudal end bearing the anus ventral-subterminally. Further addition of segments in the trunk region did not increase the number of limbs, nor change their developmental status – at least in the next stage, which bore seven trunk portions plus the caudal end.

The ontogeny of *Agnostus pisiformis* is also well known from the 'Orsten' (Müller and Walossek 1987). The phylogenetic relationships of *A. pisiformis* and the other agnostids traditionally assigned to trilobites are still controversial (Walossek and Müller 1990, Stein *et al.* 2005; but see Cotton and Fortey 2005). Agnostids lack several trilobite autapomorphies, such as the dorsal location of their compound eyes (*A. pisiformis* may have had remains of ventral compound eyes in the form of small soft humps in front of the hypostome; Müller and Walossek 1987, their Plate 11), or a multi-annulated, flagellum-like and possibly sensorial antenna (antennula in crustaceans) (*A. pisiformis* bore a leg-like antenna/antennula for food gathering; Müller and Walossek 1987). Furthermore, *A. pisiformis* had two pairs of post-antennal/antennular limbs that were specialised for swimming (Stein *et al.* 2005), a feature known otherwise only in Crustacea. Yet, the dorsal shield features, such as the glabellar lobes, may rather link agnostids with trilobites, and agnostids showed a developmental pattern that is similar to trilobites – the progressive release of trunk segments (Figure 15.7; see paragraphs above).

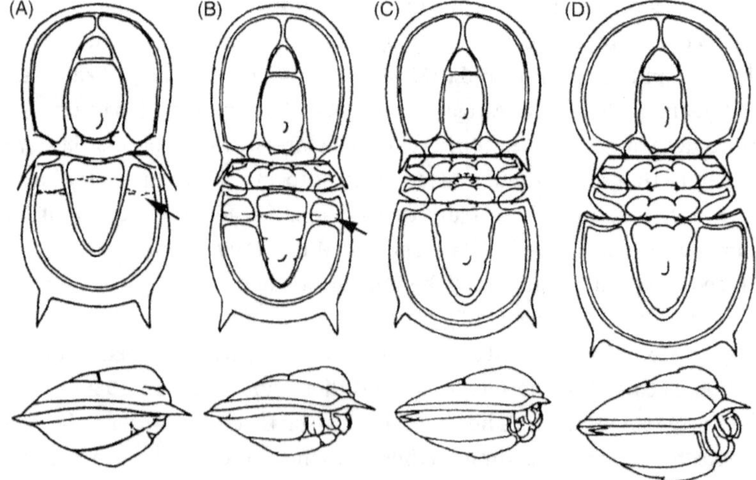

Figure 15.7 Developmental stages of *Agnostus pisiformis*. Top row: dorsal, outstretched view (probably not living position), bottom row: enrolled lateral view. A, First meraspid stage 0.327–0.43 mm; B, second meraspid stage 0.52–0.82 mm; C, first holaspid stage 0.9 mm; D, late holaspid stage *c.* 2 mm (modified from Müller and Walossek 1987). Arrows indicate future thoracic segments before being released. This developmental pattern is similar to the one known from trilobites.

TRILOBITE AND 'ORSTEN' ARTHROPOD DEVELOPMENT COMPARED,
AND THAT OF OTHER ARTHROPODS

The records of trilobite and 'Orsten' arthropod ontogenies contrast markedly in preservational style. The value of phylogenetic and temporal richness of the record of trilobite ontogeny is tempered by its limitation to exoskeletal information alone. 'Orsten' preservation is incomparably richer in morphological detail but, although occurring in several sites worldwide, is currently restricted to a handful of taxa. Given these differences, the fact that both data sources show broadly consistent patterns is likely to be of some significance. Although the appendages of unequivocal juvenile trilobites are unknown, those of the cephalon of mature trilobites, which bore a pair of uniramous antennae followed by three pairs of biramous appendages (Hughes 2003a) support the idea that the 'head larva' was the basal euarthropod condition. Furthermore, Waloszek's hypothesis (Walossek 1993) that crustaceans, as exemplified by *Rehbachiella*, basically show an extended phase of anamorphic development, with morphological change at each moult limited in scope, is consistent with the pattern seen in the development of the trilobite exoskeleton, in which change was generally progressive, track-like and incremental (Hughes *et al.* 2006). The degree of allometric shape change between instars was comparatively modest and generally gradual, especially at later phases of ontogeny. Even when so-called 'metamorphoses' occurred in trilobites they were not such radical reorganisations of the body as seen in the development of some derived arthropods. It is notable that hemianamorphic development is also basal condition in myriapods (Fusco 2005), raising the possibility that gradual, anamorphic development characterised basal mandibulates and their closest relatives (Hughes *et al.* 2006).

Interpretation of fossil ontogeny in functional terms is often difficult, but the fact that each trilobite instar had to function effectively in the external environment (as opposed to those arthropods that attained the mature form within the egg) may have placed a functional constraint on the ability to achieve dramatic morphological transitions between instars (Hughes 2003a). However, other free-living arthropods have been able to modify their life cycles (and accompanying morphologies) radically and examples of this kind of development were already present among the 'Orsten' arthropods discussed above, so this cannot account completely for the relative conservatism of trilobite ontogeny. Nevertheless, gradual, anamorphic development apparently occurred widely among early euarthropods and so this pattern need not require a trilobite-specific explanation. Perhaps food sources in the early Palaeozoic marine ecosystems were not partitioned into discrete, heterogeneous size classes, and this permitted early consumers to maintain a similar form, and presumably also similar feeding mode, across a wide range of absolute size.

ACKNOWLEDGEMENTS

We thank Extinctions.com for providing access to the images used in Figure 15.4. N.C.H.'s contribution is supported by National Science Foundation EAR grant number 0616574. J.H.'s and D.W.'s studies are funded by German Research Foundation grant number WA 754/11-1. We thank the reviewers and editors for helpful suggestions. The editors also hosted the delightful meeting in Venice and invited this contribution.

REFERENCES

Burmeister, H. 1846. *The Organization of Trilobites*. London: The Ray Society.

Chatterton, B. D. E. 1971. Taxonomy and ontogeny of Siluro-Devonian trilobites from near Yass, New South Wales. *Palaeontographica, Abteilung A* **137**, 1–108.

Chatterton, B. D. E. 1994. Ordovician proetide trilobite *Dimeropyge*, with a new species from northwestern Canada. *Journal of Paleontology* **68**, 541–556.

Chatterton, B. D. E. & Speyer, S. E. 1997. Ontogeny. In H. B. Whittington (ed.) *Treatise on Invertebrate Paleontology, Part O, Arthropoda 1. Trilobita, revised*. Boulder, CO: The Geological Society of America and Lawrence, KS: The University of Kansas, pp. 173–247.

Cisne, J. L., Chandlee, G. O., Rabe, B. D. & Cohen, J. A. 1980. Geographic variation and episodic evolution in an Ordovician trilobite. *Science* **209**, 925–927.

Cotton, T. J. & Fortey, R. A. 2005. Comparative morphology and relationships of the Agnostida. In S. Koenemann & R. A. Jenner (eds.) *Crustacea and Arthropod Phylogeny. Crustacean Issues 16*, Boca Raton, Florida: CRC Press, pp. 95–136.

Enghoff, H., Dohle, W. & Blower, J. G. 1993. Anamorphosis in millipedes (Diplopoda). The present state of knowledge and phylogenetic considerations. *Zoological Journal of the Linnean Society* **109**, 103–234.

Evitt, W. R. 1961. Early ontogeny in the trilobite family Asaphidae. *Journal of Paleontology* **35**, 986–995.

Fusco, G. 2005. Trunk segment numbers and sequential segmentation in myriapods. *Evolution & Development* **7**, 608–617.

Fusco, G., Hughes, N. C., Webster, M. & Minelli, A. 2004. Exploring developmental modes in a fossil arthropod: growth and trunk segmentation of the trilobite *Aulacopleura konincki*. *American Naturalist* **163**, 167–183.

Høeg, J. T. & Kolbasov, G. A. 2002. Lattice organs in Y-cyprids of the Facetotecta and their significance in the phylogeny of the Crustacea Thecostraca. *Acta Zoologica* **83**, 67–79.

Hughes, N. C. 2003a. Trilobite tagmosis and body patterning from morphological and developmental perspectives. *Integrative and Comparative Biology* **41**, 185–206.

Hughes, N. C. 2003b. Trilobite body patterning and the evolution of arthropod tagmosis. *BioEssays* **25**, 386–395.

Hughes, N. C. 2005. Trilobite construction: building a bridge across the micro and macroevolutionary divide. In D. E. G. Briggs (ed.) *Evolving Form and Function: Fossils and Development*. New Haven: Peabody Museum of Natural History, Yale University, pp. 138–158.

Hughes, N. C. 2007 The evolution of trilobite body patterning. *Annual Reviews of Earth and Planetary Sciences* **35**, 401–434.

Hughes, N. C. & Chapman, R. E. 1995. Growth and variation in the Silurian proetide trilobite *Aulacopleura konincki* and its implications for trilobite palaeobiology. *Lethaia* **28**, 333–353.

Hughes, N. C., Minelli, A. & Fusco, G. 2006. The ontogeny of trilobite segmentation: a comparative approach. *Paleobiology* **32**, 602–627.

Maas, A. & Waloszek, D. 2005. Phosphatocopina – ostracode-like sister group of Eucrustacea. *Hydrobiologia*, **538**, 139–152.

McNamara, K. J., Yu, F. & Zhou, Z. 2003. Ontogeny and heterochrony in the oryctocephalid trilobite *Arthricocephalus* from the Early Cambrian of China. *Special Papers in Palaeontology* **70**, 103–126.

Minelli, A., Fusco, G. & Hughes, N. C. 2003. Tagmata and segment specification in trilobites. *Special Papers in Palaeontology* **70**, 31–43.

Müller, K. J. 1985. Exceptional preservation in calcareous nodules. *Philosophical Transactions of the Royal Society of London B* **311**, 67–73.

Müller, K. J. & Walossek, D. 1986a. Arthropod larvae from the Upper Cambrian of Sweden. *Transactions of the Royal Society of Edinburgh: Earth Sciences* **77**, 157–179.

Müller, K. J. & Walossek, D. 1986b. *Martinssonia elongata* gen. et. sp. n., a crustacean-like euarthropod from the Upper Cambrian 'Orsten' of Sweden. *Zoologica Scripta* **15**, 73–92.

Müller, K. J. & Walossek, D. 1987. Morphology, ontogeny, and life habit of *Agnostus pisiformis* from the Upper Cambrian of Sweden. *Fossils and Strata* **19**, 1–124.

Müller, K. J. & Walossek, D. 1988. External morphology and larval development of the Upper Cambrian maxillopod *Bredocaris admirabilis*. *Fossils and Strata* **23**, 1–70.

Paterson, J. R. & Edgecombe, G. D. 2006. The Early Cambrian trilobite family Emuellidae Pocock, 1970: Systematic position and revision of Australian species. *Journal of Paleontology* **80**, 496–513.

Peng, S., Babcock, L. E. & Lin, H. 2005. *Polymerid Trilobites from the Cambrian of Northwestern Hunan, China*, Vol. 2. *Ptychopariida, Eodiscida, and Undetermined Forms*. Beijing: Science Press.

Stein, M., Waloszek, D. & Maas, A. 2005. *Oelandocaris oelandica* and the stem lineage of Crustacea. In S. Koenemann & R. A. Jenner (eds.) *Crustacea and Arthropod Phylogeny, Crustacean Issues 16*, Boca Raton, Florida: CRC Press, pp. 55–71.

Stubblefield, C. J. 1926. Notes on the development of a trilobite, *Shumardia pusilla* (Sars). *Zoological Journal of the Linnean Society of London* **35** 345–372.

Walossek, D. 1993. The Upper Cambrian *Rehbachiella* and the phylogeny of Branchiopoda and Crustacea. *Fossils and Strata* **32**, 1–202.

Walossek, D. & Müller, K. J. 1990. Upper Cambrian stem-lineage crustaceans and their bearing on the monophyletic origin of Crustacea and the position of *Agnostus*. *Lethaia* **23**, 409–427.

Walossek, D. & Müller, K. J. 1998. Cambrian 'Orsten'-type arthropods and the phylogeny of Crustacea. In R. A. Fortey & R. Thomas (eds.) *Arthropod Relationships (Systematics Association Special Volume 55)*. London: Chapman & Hall, pp. 139–153.

Waloszek, D., Chen, J., Maas, A. & Wang, X. 2005. Early Cambrian arthropods –
new insights into arthropod head and structural evolution. *Arthropod Structure & Development* **34**, 189–205.

Waloszek, D. & Dunlop, J. A. 2002. A larval sea spider (Arthropoda: Pycnogonida)
from the Upper Cambrian 'Orsten' of Sweden, and the phylogenetic position
of pycnogonids. *Palaeontology* **45**, 421–446.

Waloszek, D. & Maas, A. 2005. The evolutionary history of crustacean segmentation: a fossil-based perspective. *Evolution & Development* **7**, 515–527.

16

Developmental transitions during the evolution of plant form

JANE A. LANGDALE AND C. JILL HARRISON

THE INVASION OF LAND

Land plants evolved from aquatic algal ancestors. The algae are a poly-phyletic group from which the transition to land, and acquisition of developmental features associated with land plants, have occurred many times (Lewis and McCourt 2004). Recent phylogenetic evidence points to the charophyte algal lineage as the sister group to the land plants (Figure 16.1). Developmental features shared by charophytes and land plants are cell cleavage by phragmoplasts, plasmodesmatal connections between cells, and a placental link between haploid and diploid phases of growth (Marchant and Pickett-Heaps 1973). These and other features of derived charophytes, in particular growth from an apical cell in the gametophyte (Graham 1996, Graham *et al.* 2000), suggest that many of the cellular characteristics required for the development of land plants may have evolved within their common stem group.

FROM HAPLOID TO DIPLOID

The major character that distinguishes land plants from charophyte algae is the development of a diploid embryo. In charophytes, the majority of the life cycle is represented by the haploid gametophyte and only the unicellular zygote is diploid, undergoing meiosis immediately after formation. Embryo development represents a major growth transition in that meiotic division of the zygote is delayed and diploid sporophytic cells divide by mitosis, giving rise to a multicellular body. Although in the earliest land plants the gametophyte generation

Evolving Pathways: Key Themes in Evolutionary Developmental Biology, ed. Alessandro Minelli and Giuseppe Fusco. Published by Cambridge University Press.

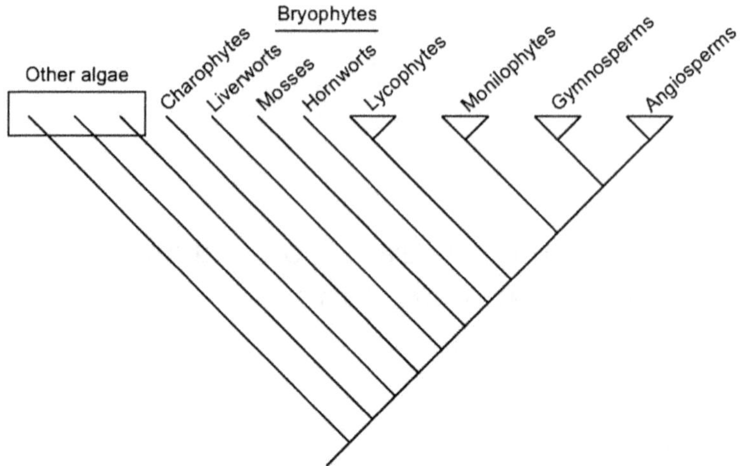

Figure 16.1 Phylogenetic relationships between extant land plants. The diagram has been compiled from phylogenies aimed at resolving particular nodes of the plant tree (Pryer *et al.* 2001, Lewis and McCourt 2004), although there is still conflict between topologies retrieved by different researchers. Angiosperm species mentioned in the text are *Arabidopsis thaliana, Populus tremuloides, Zea mays, Antirrhinum majus, Nicotiana sylvestris* and *Pisum sativum.*

remained dominant, this transition in growth pattern led to a dramatic change in life history such that the sporophyte generation gradually became the dominant form (Graham 1985, Kenrick 1994, Graham and Wilcox 2000). Consequently, the dominant part of the life cycle became diploid and thus was better protected against deleterious mutations.

The transition from growth in water to growth on land required innovations that aided reproductive success in a new, drier environment (see Figure 16.2 for a schematic sequence of character acquisitions). Most land plants have waxy cuticles enclosing vegetative tissue and sporopollenin-coated spores, both of which protect against desiccation. Cuticle formation subsequently necessitated the development of stomata to permit gaseous exchange with the environment. Cuticle formation, and the re-localisation of sporopollenin deposition from the zygote in charophytes to the spores in land plants, are shared ancestral character states (plesiomorphies) of land plants. Possession of stomata may also be a land-plant plesiomorphy but because liverworts do not possess true stomata, this scenario would imply a loss in this group. The alternative explanation is that stomatal formation is homoplastic.

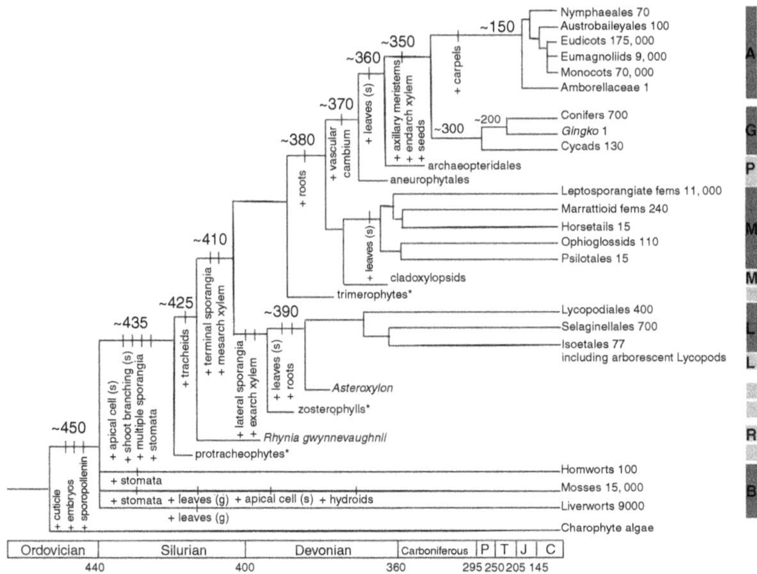

Figure 16.2 Phylogeny of selected extant and fossil land-plant taxa,
showing predicted character acquisitions on each branch (in grey, s: sporophyte; g: gametophyte). Approximate timing of acquisition is given and is
reflected by the bar at the base of the figure (P: Permian; T: Triassic;
J: Jurassic; C: Cretaceous) using the time scales of Gradstein *et al.* (2005).
Numbers after taxon names indicate the number of extant species within
the taxon, taken from Palmer *et al.* (2004). Asterisks indicate groups that
are putatively paraphyletic. Protracheophytes refers to non-vascular polysporangiates including *Aglaophyton major*, *Horneophyton lignieri* and some
non-vascular *Cooksonia*-like fossils; zosterophylls refers to Zosterophyllop-
sida and Lycophytina; trimerophytes refers to taxa such as *Psilophyton* and
Pertica (Crane *et al.* 2004). Cladoxylopsids refers to *Cladoxylon*, *Hyenia*, *Pseudo-
sporochnus* and *Lorophyton goense* (Fairon-Demaret and Li 1993). On the bar at
the right, dark boxes indicate extant taxa and light bars indicate fossil taxa.
A: angiosperms; P: progymnosperms (paraphyletic grade); G: gymnosperms; M: monilophytes; L: lycophytes; R: Rhyniaceae; B: bryophytes.
Bryophytes are depicted as paraphyletic in accordance with Qiu *et al.*
(2006); lycophyte relationships are in accordance with Bateman (1994) and
DiMichele and Bateman (1996); monilophyte and seed plant relationships
are in accordance with Palmer *et al.* (2004).

Either way, cuticle, sporopollenin-coated spores and stomata are considered to be key innovations that conferred desiccation resistance on
vegetative and reproductive structures, and so assisted in the colonisation of land (reviewed in Cronk 2001).

The earliest land plants were bryophytes (mosses, liverworts and hornworts), and extant bryophytes form a paraphyletic grade at the base of land-plant phylogenies (Wellman *et al.* 2003, Qiu *et al.* 2006). Recent evidence points to the hornworts as the vascular plant sister group (Qiu *et al.* 2006). In bryophytes, as in the charophytes, the conspicuous phase of growth is gametophytic (haploid), and plants have filamentous, thalloid and/or leafy form and bear egg-producing archegonia and sperm-producing antheridia. The diploid embryo remains attached to the gametophyte and develops into a single naked sporophytic axis that subtends a sporangium. In contrast, tracheophytes have conspicuous sporophytes and reduced gametophytes. Such sporophyte dominance is most pronounced in seed plants where the sporophyte can be a large ancient tree, yet the gametophyte lives for at most several months and can consist of just a few cells.

In the context of land-plant evolution, the significance of the transition from unicellular to multicellular growth in the sporophyte cannot be disputed. One of the most important questions to address in plant evo-devo is therefore how cell division is regulated following fertilisation of the zygote. The plane of the first cell division in bryophytes is variable; cleavage in mosses and liverworts is transverse, whereas cleavage in most hornworts is longitudinal (Kato and Akiyama 2005). The importance of such early divisions has recently been demonstrated by work with the angiosperm *Arabidopsis thaliana*, where the zygotic cell is usually cleaved unequally and transversely to give a small apical and larger basal cell. In *gnom* mutant zygotes, division is symmetrical, and as a consequence, embryos fail to develop a shoot or root. The GNOM protein functions as an auxin response factor (Geldner *et al.* 2003, 2004) suggesting that recruitment of an auxin pathway may have been pivotal in the transition to multicellular development in the sporophyte (Cooke *et al.* 2003). However, without studying GNOM function in species other than *A. thaliana*, this is wild speculation – particularly since early cell division patterns in the *A. thaliana* embryo represent only one of six patterns recognised in angiosperms (Johri 1984) and may be unrelated to patterns in other plants. It may be more relevant to study the basis of multicellularity in bryophytes, particularly as the bryophyte sporophyte presents a relatively simple morphology.

GROWING FROM THE TOP

Although bryophyte sporophytes are united by having a single axis of growth that terminates in sporangium formation, distinct growth patterns are seen in each of the main groups. Liverworts grow from cell divisions throughout the structure, mosses grow from both an apical cell and an intercalary meristem at the base of the sporophyte, and hornworts grow from the base (Kato and Akiyama 2005). In contrast, all tracheophytes grow from one or more apical cells, and this mode of sporophytic growth is a plesiomorphy of vascular plants that had evolved by the mid Silurian (~420 million years ago [mya]).

Apical growth in most lycophytes, monilophytes and gymnosperms is thought to be restricted to one or few cells in the surface layer of the meristem, but how these meristems function is poorly understood. In contrast, angiosperm meristems are layered and have distinct zones of specialised function, and their development is reasonably well understood. In the layered meristem of *A. thaliana*, apical growth is sustained through activity of the WUSCHEL (WUS)/CLAVATA (CLV) pathway (Mayer *et al.* 1998, Brand *et al.* 2000, Schoof *et al.* 2000). The homeobox transcription factor WUS acts non-cell autonomously to promote stem-cell activity in cells overlying the WUS expression domain. In addition, WUS activity indirectly induces expression of the CLV3 ligand in cells at the surface of the meristem. CLV3 itself acts non-cell autonomously to activate the presumed CLV1 receptor in cells surrounding the WUS domain. CLV activity then restricts WUS activity to a small region of the meristem; in this way, a regulated population of stem cells is maintained. Recent reports suggest that a WUS/CLV-like feedback pathway also operates in meristems of other angiosperms (Taguchi-Shiobara *et al.* 2001, Bommert *et al.* 2005, Kieffer *et al.* 2006).

Meristems in early diverging land-plant lineages are structurally different from angiosperm meristems and arguably less complex. They are more likely to resemble the ancestral growth form that existed in the late Silurian (~400 mya). In this context, it will be interesting to see whether components of the WUS/CLV pathway function to maintain stem-cell activity in non-layered meristems that facilitate apical growth or whether this pathway is a synapomorphy (acquired shared character state) of angiosperms.

BRANCHING OUT

Whereas bryophytes have unbranched sporophytes, protracheophyte fossil sporophytes from the Silurian have multiple branches that terminate in sporangia, and are therefore named polysporangiates (reviewed in Crane *et al.* 2004). Unlike extant polysporangiates, the protracheophyte fossils *Aglaophyton major* and *Horneophyton lignieri* are non-vascular.

From a developmental perspective, this observation is important because in combination with the phylogenetic position of these fossil taxa, it implies that axial branching evolved before vascularisation. The switch from unbranched to branched growth habit required at least a transient switch from determinate to indeterminate shoot development. In the absence of extant leafless polysporangiates, it is not possible to determine exactly how this was achieved at the molecular level. Presumably, either the activity of apical cells was altered, or cells in the stem were partitioned to form two distinct axes. Regardless of the exact process, such sporophyte branching mechanisms evolved before vascular tissue.

Branching occurs by different developmental mechanisms in different plant groups. In lycophytes and some monilophytes, sporophytes branch as a consequence of equal or unequal bifurcation of the shoot apex. In contrast, angiosperms branch from axillary meristems that are formed as part of what can be visualised as the phytomer unit that also includes a leaf, node and internode (Galinat 1959). These distinct mechanisms of branching suggest that the developmental toolkit adopted for branching may have been different in each case.

At least in *A. thaliana*, formation of the axillary meristem is associated with axis formation in the leaf. Transcription factors such as PHABULOSA (PHB) and KANADI, which function to distinguish and maintain adaxial (upper) and abaxial (lower) leaf domains thus play a role in axillary meristem formation (Eshed *et al.* 2001, McConnell *et al.* 2001). Specifically, adaxialisation promoted by PHB activity is required for axillary meristem formation. Once formed, outgrowth from axillary meristems is inhibited by the plant growth regulator auxin, and this inhibition is mediated by a novel plant hormone that is regulated by the MAX pathway (Booker *et al.* 2005, Bennett *et al.* 2006).

In contrast to our understanding of branching processes in angiosperms, the mechanism of bifurcation in lycophyte meristems is poorly understood with respect to the underlying genetic and physiological mechanisms. In monilophytes the actual mechanism of generating new axes of growth is disputed, with evidence of meristem bifurcation in some species and branching from rhizomes in other species (Bierhorst 1977). Clearly, both developmental and genetic analyses need to be carried out with carefully selected lycophyte and monilophyte species to determine whether similar pathways are used to promote branching in distinct plant lineages, and to provide insight into the developmental module that was first recruited to facilitate branching.

VASCULAR HIGHWAYS

Branching led to increasingly complex body plans and competition for light, which then led to the evolution of taller plants. As a consequence, the need for mechanical support and for transport of water intensified. Coupled with the origin of lignin in vascular plants, these drivers led to the evolution of lignified conducting cells (tracheids). Whereas moss and some liverwort gametophytes transport water through conducting hydroids consisting of elongated cells with just primary cell walls, and protracheophyte fossils show evidence of a similar system (Boyce *et al.* 2003), true tracheophytes transport water through xylem cells that have lignified secondary walls.

Tracheophyte fossils indicate that tracheids first evolved in the mid-Silurian (~420 mya) (Edwards *et al.* 1983). In fossils such as *Rhynia gwynnevaughnii*, tracheids possess two layers of secondary cell wall – an inner degradation-prone layer and an outer degradation-resistant layer (Kenrick and Crane 1991). Because tracheids in well-studied extant seed plants have a single lignified secondary cell wall, the singularity of tracheid evolution has been questioned. However, tracheid secondary walls in at least one lycophyte and at least one monilophyte are bilayered as in tracheophyte fossils, supporting the argument that tracheids per se have a single origin (Cook and Friedman 1998, Friedman and Cook 2000).

The developmental mechanisms that specifically contribute to tracheid (xylem) formation in extant plants include cell-wall pitting and lignin deposition. Recent work in the angiosperms *A. thaliana* and *Populus tremuloides* has shown that the activity of two transcription factors – VASCULAR-RELATED NAC-DOMAIN6 (VND6) and VND7 – is sufficient to induce trans-differentiation of various cells into protoxylem or metaxylem vessel elements (Kubo *et al.* 2005). The relative simplicity of this switching mechanism is consistent with the reported single origin of xylem formation. However, developmental patterns vary in different plant groups. Xylem differentiation in lycophytes is exarch, in that the earliest maturing xylem (protoxylem) is at the periphery of the vascular bundle and the later maturing metaxylem is at the centre. In contrast, xylem formation in monilophytes is mesarch, with metaxylem developing both inward and outward of protoxylem relative to the centre of the bundle; and is endarch in seed plants, with protoxylem at the centre of the bundle and metaxylem at the periphery. Whether the VND system ubiquitously regulates xylem formation in land plants remains to be determined.

Although lignification is a defining characteristic of tracheids and other water-conducting cells in seed plants, biochemical analyses suggest that xylem lignification in late Silurian fossils (~400 mya) is minimal compared with that seen in extant plants (Boyce *et al.* 2003). The inferred increase of lignin deposition in xylem cells through evolutionary time (Boyce *et al.* 2003) and the developmental mechanisms operating in extant plants (McCann 1997) both suggest that cellulose secondary thickenings are a prerequisite for lignin deposition. Thus, the evolution of xylem lignification most likely required the transfer of a pre-existing lignin biosynthesis pathway to cellulose-thickened conducting cells and the development of a mechanism for targeted cell wall deposition.

Lignified vasculature paved the way for plants to increase in height. However, without an accompanying capacity for increased girth, and hence support, an increase in height is biomechanically untenable. Expansion in girth is facilitated by growth from a secondary meristem known as the vascular cambium. The most significant cambial development is seen in the trees of extant seed plants and in fossil lyco-phytes that evolved tree form independently (Phillips and DiMichele 1992, Bateman 1994). The developmental processes that regulate cambial development are poorly understood, but genes that regulate shoot meristem activity also play a role in regulating cambial activity and lignin biosynthesis (Mele *et al.* 2003, Oh *et al.* 2003, McHale and Koning 2004b, Groover and Robischon 2006). This suggests that although cambial meristems arise de novo in growing tissue, the developmental mechanism used may be the same as that recruited during embryogenesis to form the shoot apical meristem.

DIGGING DEEP

Whereas tracheids provided a conduit for water transport in aerial parts of the plant, increasingly complex body plans and greater sporophyte surface areas demanded greater water absorption from the environment and additional mechanical support at the base of the plant. The ecological driver was thus in place for the evolution of sporophyte roots.

Charophyte algae and free-living gametophytes develop root-like structures, but 'true' roots are only found in tracheophyte sporophytes. Gametophyte 'roots' are single-celled rhizoid-like structures that are analogous to root hairs, multicellular rhizoid-like projections from a subterranean rhizome, or mycorrhizas formed following a fungal association (Raven and Edwards 2001). In these cases, the rhizoids provide conduits for water and nutrient transport but only the subterranean rhizomes enhance mechanical support. In contrast, true roots provide both transport capability and mechanical support. By definition they are multicellular, have a defined endodermis (with the exception of extant Lycopodiaceae roots), produce lateral organs endogenously and have a root cap.

Sporophyte roots first appear in the fossil record in the early Devonian (~390 mya). Earlier rootless fossils are classified into two groups, the zosterophylls and trimerophytes (Banks 1975), the xylem arrangement and sporangial position of which are reflected in extant lycophytes and euphyllophytes, respectively. These observations suggest that the lycophyte and euphyllophyte lineages may have diverged before roots evolved (Raven and Edwards 2001). This suggestion is further supported by evidence of lycophyte roots in *Asteroxylon* fossils found in the early Devonian Rhynie Chert in Aberdeenshire (Kidston and Lang 1920), in contrast to the earliest evidence of euphyllophyte roots from the mid-Devonian cladoxylopsid fossil *Lorophyton goense* (Fairon-Demaret and Li 1993).

To date, our understanding of genetic mechanisms regulating root patterning is restricted to seed plants. In this case, work carried out mainly with *A. thaliana* has shown that a stem-cell niche is established during embryogenesis and that differentiation of cells produced from that niche depends on positional signals (van den Berg *et al.* 1997). The stem-cell niche consists of a small group of slowly dividing cells known as the quiescent centre (QC) plus four sets of initials that give rise to the cell types of the mature root (Dolan *et al.* 1993, Scheres *et al.* 1994). Stem-cell activity is regulated by the activity of two related transcription factors – SCARECROW (SCR) and SHORTROOT (SHR) (Sabatini *et al.* 2003). SCR acts cell autonomously within the QC to regulate QC identity. The QC then acts in a non-cell autonomous manner to maintain stem-cell activity of the surrounding initials. Derivatives of the initials are physically separated from the stem-cell-inducing QC and thus differentiate. SCR and SHR also play a role in regulating cell-type differentiation in these derivatives and thus contribute to radial patterning in the root (Helariutta *et al.* 2000, Nakajima *et al.* 2001). Super-imposed upon the SCR/SHR system is the PLETHORA/auxin system which regulates stem-cell activity and longitudinal patterning in the root (Aida *et al.* 2004), and the WOX system (Kamiya *et al.* 2003). Significantly with respect to understanding the evolution of root structures, SCR and SHR play a role in radial patterning of the shoot (Wysocka-Diller *et al.* 2000). WOX genes are related to WUSCHEL, and components of a CLV-like pathway are involved in maintaining root meristem activity

(Casamitjana *et al.* 2003). Root evolution in the euphyllophytes may therefore have involved co-option of mechanisms already functioning in the shoot. Whether similar genetic mechanisms operate in lycophytes awaits investigation but the developmental anatomy of lycophyte roots suggests that they are derived from shoots.

HARVESTING ENERGY

The next significant phase of land-plant evolution could have been driven by a large drop in atmospheric CO_2 and temperature, or by competition for light (Beerling *et al.* 2001, Beerling 2005). Whatever the cause, lateral flattened photosynthetic organs (leaves) evolved. The evolution of leaves followed a similar trajectory to that of roots in that small leaf-like structures are present on some extant free-living gametophytes, but 'true' leaves, exhibiting lignified vasculature and multiple cell layers, are found only on tracheophyte sporophytes. Like roots, sporophyte leaves have multiple origins.

Leaves evolved on at least two separate occasions in the Devonian, once in the lycophytes and once in the euphyllophytes (monilophytes, progymnosperms and seed plants). The independent origins and morphological differences between extant lycophyte and euphyllophyte leaves have led to several theories to explain how leaves evolved. Lycophyte and euphyllophyte leaves are traditionally termed microphylls and megaphylls, respectively. Most microphylls have a single vascular trace and no leaf gap at the point of insertion in the stem whereas megaphylls have complex venation and stem leaf gaps. The utility of the microphyll/megaphyll dichotomy is questionable, particularly since microphylls are seen in some euphyllophytes and some lycophytes have leaves with complex venation. However, the distinction fuelled leaf evolution theories for many years. Bower (1935) first proposed that microphyllous leaves arose as enations from the stem that later became vascularised. Kenrick and Crane (1997) challenged this theory, arguing instead that microphylls were derived from bracts subtending sporangia and that sterilisation of one member of a sporangial pair resulted in the leaf. Megaphylls, however, were proposed to derive as a consequence of modified stem branching patterns. The proposed mechanism was first put forward in Zimmermann's telome theory, which invoked the idea that one branch in a dichotomously branched axis became overtopped, and then went through a progressive series of planation, webbing and fusion events to form a complex flattened lateral organ (Zimmermann 1965, Kenrick 2002). Notably, there is fossil evidence to support all of the intermediate stages but there are several criticisms of the theory (see Stein and Boyer 2006 for discussion). Despite conflicting theories of evolutionary mechanism, the separate origin of lycophyte and euphyllophyte leaves has rarely been questioned.

As with most other developmental processes, leaf development is best understood in a few model angiosperms. In this case, however, the mechanism has been examined in several different species and a shared process has been revealed. Determinate leaves are formed on the flanks of indeterminate shoot apical meristems (SAM). Indeterminacy in the SAM is maintained through the activity of KNOTTED1-like homeobox (KNOX) transcription factors, whereas determinacy in the leaf is maintained by repressing KNOX activity (Smith *et al.* 1992, Jackson *et al.* 1994). In *A. thaliana, Zea mays* and *Antirrhinum majus*, the orthologous myb transcription factors ASYMMETRIC LEAVES 1, ROUGH SHEATH 2 and PHANTASTICA (ARP) were found to mediate KNOX repression in the leaf (Waites and Hudson 1995, Schneeberger *et al.* 1998, Timmermans *et al.* 1999, Tsiantis *et al.* 1999, Byrne *et al.* 2000). Later studies showed that similar mechanisms operate in *Nicotiana sylvestris* (McHale and Koning 2004a) and *Pisum sativum* (Tattersall *et al.* 2005), leading to the suggestion that the KNOX/ARP module may be a fundamental requirement for leaf development, at least in angiosperms.

Unlike many other developmental processes, the genetic basis of leaf development has also been studied in non-seed plants, most notably in the lycophyte *Selaginella kraussiana*. Despite *S. kraussiana* having meristems and leaves that are morphologically distinct from those seen in angiosperms, the KNOX/ARP pathway also regulates the switch from indeterminate shoot to determinate leaf growth (Harrison *et al.* 2005). As lycophyte and seed-plant leaves are thought to have evolved independently, this observation implies that the same mechanism was recruited twice in parallel, and thus suggests that there was a developmental constraint on leaf evolution. In this context, it will be interesting to determine the role of the KNOX/ARP pathway in monilophytes. However, the observation that the same developmental mechanism underpins leaf development in at least one lycophyte and in various seed plants also supports the view that leaves per se, or at least the potential to produce leaves, is a plesiomorphy of tracheophytes.

REPRODUCTIVE TRANSITION

Before the late Devonian, plants reproduced via spores, with most lineages being homosporous and some heterosporous (Bateman and

DiMichele 1994). Fertilisation required water, or at least moist conditions, because sperm swim. The evolution of seed plants, the oldest fossils of which date from the late Devonian (~360 mya) (Gillespie *et al.* 1981), changed both the developmental and environmental landscape. In seed plants the female gametophyte develops inside a modified and protected megasporangium (the nucellus). The male gametophyte is the pollen grain that develops within the microsporangium (the pollen sac). Female and male reproductive organs are borne on condensed and modified branching systems that are flowers or cones in gymnosperms, and flowers in angiosperms (reviewed in Bateman *et al.* 2006). These modifications permitted radiation of the seed plants by allowing dispersal of pollen and seed, and specialisation to attract animal pollinators. Genetic events that led to specialised functions on reproductive shoots in angiosperms, such as reproductive fate, modified organ fate and enclosure of floral organs, are reasonably well characterised in model angiosperm species. This work has been extensively reviewed in recent literature and will not be discussed here (Endress 2001, Albert *et al.* 2002, Theißen *et al.* 2002, Frohlich 2003, Irish 2003, Irish and Litt 2005, Bateman *et al.* 2006). Suffice to say that the data obtained have provided a foundation for exciting advances in species that represent transitional evolutionary stages, and have provided a framework for addressing questions about floral evolution and development.

FUTURE CHALLENGES

As can be seen from the overview provided, the field of plant evo-devo currently offers more questions than answers. This is particularly true in relation to macro-evolutionary questions. So what are the major challenges for the future? Most noteworthy is the conflict between phylogenetic hypotheses suggesting which bryophyte lineage is sister group to tracheophytes. Until this is known, hypotheses of character state transitions within the bryophytes, and between bryophytes and vascular plants, cannot be critically examined. Elsewhere, the phylogeny is well resolved at deep nodes and the major limitations are those concerning the ease with which non-flowering plant species can be grown in the laboratory and can be experimentally manipulated. Owing to the amount of time needed to establish different species as experimental organisms, this is an area that needs wide consultation and agreement. Despite these limitations, the field is expanding, not only to ask questions of how developmental mechanisms evolved but also to look at the variety of mechanisms operating in extant plant lineages, particularly in response to changing environments. With genomes currently being sequenced in selected mosses, lycophytes and gymnosperms, and RNAi technologies being developed in several relevant species, the future is both exciting and unpredictable. Furthermore, there are plenty of empty niches available for young investigators who are not afraid to work with non-model systems.

ACKNOWLEDGEMENTS

We thank Liam Dolan for helpful comments on the manuscript and Richard Bateman for an extremely thorough review. Work in J.A.L.'s group is supported by grants from the BBSRC and the Gatsby Charitable Foundation.

REFERENCES

Aida, M., Beis, D., Heidstra, R. *et al.* 2004. The *PLETHORA* genes mediate patterning of the *Arabidopsis* root stem cell niche. *Cell* **119**, 109–120.

Albert, V. A., Oppenheimer, D. G. & Lindqvist, C. 2002. Pleiotropy, redundancy and the evolution of flowers. *Trends in Plant Science* **7**, 297–301.

Banks, H. P. 1975. Reclassification of *Psilophyta*. *Taxon* **24**, 401–413.

Bateman, R. M. 1994. Evolutionary developmental change in the growth architecture of fossil rhizomorphic lycopsids: scenarios conducted on cladistic foundations. *Biological Reviews* **69**, 527–597.

Bateman, R. M. & DiMichele, W. A. 1994. Heterospory: the most iterative key innovation in the evolutionary history of the plant kingdom. *Biological Reviews* **69**, 345–417.

Bateman, R. M., Hilton, J. & Rudall, P. J. 2006. Morphological and molecular phylogenetic context of the angiosperms: contrasting the 'top-down' and 'bottom-up' approaches used to infer the likely characteristics of the first flowers. *Journal of Experimental Botany* **57**, 3471–3503.

Beerling, D. J. 2005. Leaf evolution: gases, genes and geochemistry. *Annals of Botany* **96**, 345–352.

Beerling, D. J., Osborne, C. P. & Chaloner, W. G. 2001. Evolution of leaf-form in land plants linked to atmospheric CO_2 decline in the Late Palaeozoic era. *Nature* **410**, 352–354.

Bennett, T., Sieberer, T., Willett, B., Booker, J., Luschnig, C. & Leyser, O. 2006. The *Arabidopsis* MAX pathway controls shoot branching by regulating auxin transport. *Current Biology* **16**, 553–563.

Bierhorst, D. W. 1977. On the stem apex, leaf initiation and early leaf ontogeny in filicalean ferns. *American Journal of Botany* **64**, 125–152.

Bommert, P., Lunde, C., Nardmann, J. *et al.* 2005. *Thick tassel dwarf1* encodes a putative maize ortholog of the *Arabidopsis* CLAVATA1 leucine-rich repeat receptor-like kinase. *Development* **132**, 1235–1245.

Booker, J., Sieberer, T., Wright, W. *et al.* 2005. MAX1 encodes a cytochrome P450 family member that acts downstream of MAX3/4 to produce a carotenoid-derived branch-inhibiting hormone. *Developmental Cell* **8**, 443–449.

Bower, F. 1935. *Primitive Land Plants*. London: Macmillan.

Boyce, C. K., Cody, G. D., Fogel, M. L. *et al.* 2003. Chemical evidence for cell wall lignification and the evolution of tracheids in Early Devonian plants. *International Journal of Plant Science* **164**, 691–702.

Brand, U., Fletcher, J. C., Hobe, M., Meyerowitz, E. M. & Simon, R. 2000. Dependence of stem cell fate in *Arabidopsis* on a feedback loop regulated by CLV3 activity. *Science* **289**, 617–619.

Byrne, M. E., Barley, R., Curtis, M. *et al.* 2000. *ASYMMETRIC LEAVES1* mediates leaf patterning and stem cell function in *Arabidopsis*. *Nature* **408**, 967–971.

Casamitjana, M., Nez, E., Hofhuis, H. F. *et al.* 2003. Suppressors implicate a CLV-like pathway in the control of *Arabidopsis* root meristem maintenance. *Current Biology* **13**, 1435–1441.

Cook, M. E. & Friedman, W. E. 1998. Tracheid structure in a primitive extant plant provides an evolutionary link to earliest fossil tracheids. *International Journal of Plant Science* **159**, 881–890.

Cooke, T. J., Poli, D. & Cohen, J. D. 2003. Did auxin play a crucial role in the evolution of novel body plans during the Late Silurian–Early Devonian radiation of land plants? In I. Poole & A. R. Helmsley (eds.) *The Evolution of Plant Physiology*. London: Elsevier Academic Press, pp. 85–107.

Crane, P. R., Herendeen, P. & Friis, E. M. 2004. Fossils and plant phylogeny. *American Journal of Botany* **91**, 1683–1699.

Cronk, Q. C. B. 2001. Plant evolution and development in a post-genomic context. *Nature Reviews Genetics* **2**, 607–619.

DiMichele, W. A. & Bateman, R. M. 1996. Plant paleoecology and evolutionary inference: two examples from the Paleozoic. *Review of Paleobotany and Palynology* **90**, 223–247.

Dolan, L., Janmaat, K., Willemsen, V. *et al.* 1993. Cellular organization of the *Arabidopsis thaliana* root. *Development* **119**, 71–84.

Edwards, D., Feehan, J. & Smith, D. G. 1983. A late Wenlock flora from County Tipperrary, Ireland. *Botanical Journal of the Linnean Society* **86**, 19–36.

Endress, P. K. 2001. Origins of flower morphology. *Journal of Experimental Zoology B (Molecular and Developmental Evolution)* **291**, 105–115.

Eshed, Y., Baum, S. F., Perea, J. V. & Bowman, J. L. 2001. Establishment of polarity in lateral organs of plants. *Current Biology* **11**, 1251–1260.

Fairon-Demaret, M. & Li, C.-S. 1993. *Lorophyton goense* gen. et sp. nov. from the Lower Givetian of Belgium and a discussion of the Middle Devonian Cladoxylopsida. *Review of Palaeobotany and Palynology* **77**, 1–22.

Friedman, W. E. & Cook, M. E. 2000. The origin and early evolution of tracheids in vascular plants: integration of paleobotanical and neobotanical data. *Philosophical Transactions of the Royal Society of London B* **355**, 857–868.

Friedman, W. E., Moore, R. C. & Purugganan, M. D. 2004. The evolution of plant development. *American Journal of Botany* **91**, 1726–1741.

Frohlich, M. W. 2003. An evolutionary scenario for the origin of flowers. *Nature Reviews Genetics* **4**, 559–566.

Galinat, W. C. 1959. The phytomer in relation to floral homologies in the American *Maydea*. *Botanical Museum Leaflets Harvard University* **19**, 1–32.

Geldner, N., Anders, N., Wolters, H. *et al.* 2003. The *Arabidopsis* GNOM ARF-GEF mediates endosomal recycling, auxin transport, and auxin-dependent plant growth. *Cell* **112**, 219–230.

Geldner, N., Richter, S., Vieten, A. *et al.* 2004. Partial loss-of-function alleles reveal a role for GNOM in auxin transport-related, post-embryonic development of *Arabidopsis*. *Development* **131**, 389–400.

Gillespie, W., Rothwell, G. & Schlecker, S. 1981. The earliest seeds. *Nature* **293**, 462–464.

Gradstein, F., Ogg, J. & Smith, A. 2005. *A Geological Time Scale 2004*. Cambridge: Cambridge University Press.

Graham, L. E. 1985. The origin of the life cycle of land plants. *American Scientist* **73**, 178–186.

Graham, L. E. 1996. Green algae to land plants: an evolutionary transition. *Journal of Plant Research* **109**, 241–251.

Graham, L. E., Cook, M. E. & Busse, J. S. 2000. The origin of plants: body plan changes contributing to a major evolutionary radiation. *Proceedings of the National Academy of Sciences USA* **97**, 4535–4540.

Graham, L. E. & Wilcox, L. W. 2000. The origin of alternation of generations in land plants: a focus on matrotrophy and hexose transport. *Philosophical Transactions of the Royal Society of London B* **355**, 757–766.

Groover, A. & Robischon, M. 2006. Developmental mechanisms regulating secondary growth in woody plants. *Current Opinion in Plant Biology* **9**, 55–58.

Harrison, C. J., Corley, S. B., Moylan, E. C. *et al.* 2005. Independent recruitment of a conserved developmental mechanism during leaf evolution. *Nature* **434**, 509–514.

Helariutta, Y., Fukaki, H., Wysocka-Diller, J. *et al.* 2000. The *SHORT-ROOT* gene controls radial patterning of the *Arabidopsis* root through radial signaling. *Cell* **101**, 555–567.

Irish, V. F. 2003. The evolution of floral homeotic gene function. *BioEssays* **25**, 637–646.

Irish, V. F. & Litt, A. 2005. Flower development and evolution: gene duplication, diversification and redeployment. *Current Opinion in Genetics and Development* **15**, 454–460.

Jackson, D., Veit, B. & Hake, S. 1994. Expression of maize *Knotted1* related homeobox genes in the shoot apical meristem predicts patterns of morphogenesis in the vegetative shoot. *Development* **120**, 404–413.

Johri, B. M. 1984. *Embryology of Angiosperms*. Berlin: Springer.

Kamiya, N., Nagasaki, H., Morikami, A., Sato, Y. & Matsuoka, M. 2003. Isolation and characterization of a rice *WUSCHEL*-type homeobox gene that is specifically expressed in the central cells of a quiescent center in the root apical meristem. *Plant Journal* **35**, 429–441.

Kato, M. & Akiyama, H. 2005. Interpolation hypothesis for origin of the vegetative sporophyte of land plants. *Taxon* **54**, 443–450.

Kenrick, P. 1994. Alternation of generations in land plants: new phylogenetic and paleobotanical evidence. *Biological Reviews* **69**, 293–330.

Kenrick, P. & Crane, P. R. 1991. Water conducting cells in early fossil land plants: implications for the early evolution of tracheophytes. *Botanical Gazette* **152**, 335–356.

Kenrick, P. & Crane, P. R. 1997. *The Origin and Early Diversification of Land Plants: A Cladistic Study*. London: Smithsonian Institution Press.

Kenrick, P. 2002. The telome theory. In Q. C. B. Cronk, R. M. Bateman & J. A. Hawkins (eds.) *Developmental Genetics and Plant Evolution*. London: Taylor and Francis, pp. 365–387.

Kidston, R. & Lang, W. H. 1920. On Old Red Sandstone plants showing structure from the Rhynie Chert bed, Aberdeenshire. Part III. *Asteroxylon mackiei*. *Transactions of the Royal Society of Edinburgh* **52**, 643–680.

Kieffer, M., Stern, Y., Cook, H. *et al.* 2006. Analysis of the transcription factor WUSCHEL and its functional homologue in *Antirrhinum* reveals a potential mechanism for their roles in meristem maintenance. *Plant Cell* **18**, 560–573.

Kubo, M., Udagawa, M., Nishikubo, N. *et al.* 2005. Transcription switches for protoxylem and metaxylem vessel formation. *Genes & Development* **19**, 1855–1860.

Lewis, I. A. & McCourt, R. M. 2004. Green algae and the origin of land plants. *American Journal of Botany* **91**, 1535–1556.

Marchant, H. J. & Pickett-Heaps, J. D. 1973. Mitosis and cytokinesis in *Coleochaete scutata*. *Journal of Phycology* **9**, 461–471.

Mayer, K. F. X., Schoof, H., Haecker, A. *et al.* 1998. Role of *WUSCHEL* in regulating stem cell fate in the *Arabidopsis* shoot meristem. *Cell* **95**, 805–815.

McCann, M. C. 1997. Tracheary element formation: building up to a dead end. *Trends in Plant Science* **2**, 333–338.

McConnell, J. R., Emery, J., Eshed, Y. *et al.* 2001. Role of *PHABULOSA* and *PHAVO-LUTA* in determining radial patterning in shoots. *Nature* **411**, 709–713.

McHale, N. A. & Koning, R. E. 2004a. PHANTASTICA regulates development of the adaxial mesophyll in *Nicotiana* leaves. *Plant Cell* **16**, 1251–1262.

McHale, N. A. & Koning, R. E. 2004b. MicroRNA-directed cleavage of *Nicotiana sylvestris PHAVOLUTA* mRNA regulates the vascular cambium and structure of apical meristems. *Plant Cell* **16**, 1730–1740.

Mele, G., Ori, N., Sato, Y. & Hake, S. 2003. The *knotted1*-like homeobox gene *BREVIPEDICELLUS* regulates cell differentiation by modulating metabolic pathways. *Genes & Development* **17**, 2088–2093.

Nakajima, K., Sena, G., Nawy, T. & Benfey, P. N. 2001. Intercellular movement of the putative transcription factor SHR in root patterning. *Nature* **413**, 307–311.

Oh, S. H., Park, S. & Han, K.-Y. 2003. Transcriptional regulation of secondary growth in *Arabidopsis thaliana*. *Journal of Experimental Botany* **54**, 2709–2722.

Palmer, J. D., Soltis, D. E. & Chase, M. W. 2004. The plant tree of life: an overview and some points of view. *American Journal of Botany* **91**, 1437–1445.

Phillips, T. L. & DiMichele, W. A. 1992. Comparative ecology and life-history biology of arborescent lycopsids in Late Carboniferous swamps of Euroamerica. *Annals of the Missouri Botanical Garden* **79**, 560–588.

Pryer, K. M., Schneider, H., Smith, A. R. *et al.* 2001. Horsetails and ferns are a monophyletic group and the closest living relatives to seed plants. *Nature* **409**, 618–622.

Qiu, Y. L., Li, L., Wang, B. *et al.* 2006. The deepest divergences in land plants inferred from phylogenomic evidence. *Proceedings of the National Academy of Sciences of the USA* **103**, 15511–15516.

Raven, J. A. & Edwards, D. 2001. Roots: evolutionary origins and biogeochemical significance. *Journal of Experimental Botany* **52**, 381–401.

Sabatini, S., Heidstra, R., Wildwater, M. & Scheres, B. 2003. SCARECROW is involved in positioning the stem cell niche in the *Arabidopsis* root meristem. *Genes & Development* **17**, 354–358.

Scheres, B., Wolkenfeit, H., Willemsen, V. *et al.* 1994. Embryonic origin of the *Arabidopsis* primary root and root meristem initials. *Development* **120**, 2475–2487.

Schneeberger, R., Tsiantis, M., Freeling, M. & Langdale, J. A. 1998. The *rough sheath2* gene negatively regulates homeobox gene expression during maize leaf development. *Development* **125**, 2857–2865.

Schoof, H., Lenhard, M., Haecker, A. *et al.* 2000. The stem cell population of *Arabidopsis* shoot meristems is maintained by a regulatory loop between the *CLAVATA* and *WUSCHEL* genes. *Cell* **100**, 635–644.

Smith, L. G., Greene, B., Veit, B. & Hake, S. 1992. A dominant mutation in the maize homeobox gene, *Knotted-1*, causes its ectopic expression in leaf cells with altered fates. *Development* **116**, 21–30.

Stein, W. E. & Boyer, J. S. 2006. Evolution of land plant architecture: beyond the telome theory. *Paleobiology* **32**, 450–482.

Taguchi-Shiobara, F., Yuan, Z., Hake, S. & Jackson, D. 2001. The *fasciated ear2* gene encodes a leucine-rich repeat receptor-like protein that regulates shoot meristem proliferation in maize. *Genes & Development* **15**, 2755–2766.

Tattersall, A. D., Turner, L., Knox, M. R. *et al.* 2005. The mutant *crispa* reveals multiple roles for *PHANTASTICA* in pea compound leaf development. *Plant Cell* **17**, 1046–1060.

Theißen, G., Becker, A., Winter, K.-U. *et al.* 2002. How land plants learned their floral ABCs: the role of MADS box genes in the evolutionary origin of flowers. In Q. C. B. Cronk, R. M. Bateman & J. A. Hawkins (eds.) *Developmental Genetics and Plant Evolution*. London: Taylor and Francis, pp. 173–205.

Timmermans, M. C. P., Hudson, A., Becraft, P. W. & Nelson, T. 1999. ROUGH SHEATH2: A myb protein that represses *knox* homeobox genes in maize lateral organ primordia. *Science* **284**, 151–153.

Tsiantis, M., Schneeberger, R., Golz, J. F., Freeling, M. & Langdale, J. A. 1999. The maize *rough sheath2* gene and leaf development programs in monocot and dicot plants. *Science* **284**, 154–156.

van den Berg, C., Willemsen, V., Hendricks, G., Weisbeek, P. & Scheres, B. 1997. Short-range control of cell differentiation in the *Arabidopsis* root meristem. *Nature* **390**, 287–289.

Waites, R. & Hudson, A. 1995. *Phantastica*: a gene required for dorsoventrality of leaves in *Antirrhinum*. *Development* **121**, 2143–2154.

Wellman, C. H., Osterloff, P. L. & Mohiuddin, U. 2003. Fragments of the earliest land plants. *Nature* **425**, 282–285.

Wysocka-Diller, J. W., Helariutta, Y., Fukaki, H., Malamy, J. E. & Benfey, P. N. 2000. Molecular analysis of SCARECROW function reveals a radial patterning mechanism common to root and shoot. *Development* **127**, 595–603.

Zimmermann, W. 1965. *Die Telomtheorie*. Stuttgart: Fischer.

Part IV   Evolving body features

'Distinctive features', 'key apomorphies', 'novelties', 'innovations'. All these terms point to the fact that some characters in an organism are 'special' in some respect, and these have attracted the most interest from biologists and challenged the explanatory capacity of evolutionary theories. These are characters that emerged as new features during the evolution of a lineage, or that are believed to be the principal cause of a successful phyletic radiation. 'Evolution is tinkering' and in any such feature there are both new and conserved components, even if it is not always obvious what is new and what is not. By taking development into the picture, can evo-devo provide deeper insight into the origin, evolution and diversification of such characters? The answer seems to be yes.

An initial example is offered by Cassandra Extavour (Chapter 17), who addresses the evolution of bilaterian reproductive systems by examining comparative data on somatic gonad and germ cell specification during development. Surprisingly enough, although reproduction is possibly the most important quality for a living being, the evolution of bilaterian reproductive systems and strategies has received comparably little attention. Was the last common ancestor of bilaterians (the so-called Urbilateria) hermaphroditic? Did it possess a sequestered germ line, or distinctive gonads? The data presented here can tell us what kinds of general features, or basic pattern, Urbilateria's reproductive system was likely to have had, thus accounting for the systems found in extant bilaterian lineages.

Although segmentation is not an apomorphy for the (eu)arthropods, as it was inherited (at least) from a pan-arthropodan ancestor, it is certainly one of the most distinctive features of the arthropod body plan, and a key element of arthropod evolution and diversification.

Ariel Chipman (Chapter 18) performs a comparative analysis of various aspects of the arthropod trunk segmentation process, searching for those components of the process that may have been present in the last common ancestor of all modern arthropods. The author elaborates on the hypothesis that the ancestral arthropod segmentation process consisted in the patterning of a posterior population of undifferentiated cells through a periodic signal during axis elongation. In particular, he examines the evidence for a mechanism that generates a segmental pattern through the transduction of a temporal pattern of cell state oscillation into a spatially periodic pattern of gene expression.

Digging deeper into the arthropod body, Angelika Stollewerk (Chapter 19) addresses a fundamental question on arthropod segmentation, i.e. the origin and evolution of a segmented central nervous system. The author reviews data on early neurogenesis in all main arthropod lineages to uncover ancestral states and possibly derived homologies. Considering both the morphological processes of neural precursor formation and the genetic network involved in neural precursor specification, she attempts a reconstruction of the ground pattern of neurogenesis in arthropods, and speculates on how neurogenesis evolved in this group.

Moving from segment production to segment differentiation, Nikola Prpric and Wim Damen (Chapter 20) confront the problem of evolution of arthropod appendage diversity. Which was the 'ground state' of the appendage that served as the basis for the evolution of this unique diversity? What are the underlying genetic mechanisms and how did they change during evolution? Searching for an answer to these fundamental questions, they observe that what is valid for an animal as a whole seems also to be valid for individual parts of it. The same 'hour-glass model' used to describe the high diversity of early embryonic and adult body plans with respect to the limited diversity of phenotypes manifested at the so-called 'phylotypic stage' might also be applied for the walking legs of arthropods. These in fact present a high diversity in early development and adult morphology that contrasts with an intermediate 'podotypic stage' characterised by conserved developmental gene interactions.

Abandoning the ecdysozoans, Claus Nielsen (Chapter 21) brings us to the evolution of development of the brain of some lophotrochozoans. Several lophotrochozoan groups with spiral cleavage are sometimes grouped within a taxon Spiralia, although the composition and the monophyletic status of this assembly are still matters for debate. The author reviews available evidence from descriptive embryology, corroborated by new data from studies of Hox gene expression. Considering cleavage pattern, cell lineage, developmental origin of ganglia and nerve cords, the ontogeny of the brain would support a clade including Annelida, Mollusca, Sipuncula, Entoprocta, Nemertini and Platyhelminthes s.str., deriving from a common ancestor whose post-embryonic development included a trochophora larva.

17

Urbisexuality: the evolution of bilaterian germ cell specification and reproductive systems

CASSANDRA G. M. EXTAVOUR

A key focus of evolutionary developmental biology (evo-devo) in recent years has been to elucidate the evolution of developmental mechanisms as a means to reconstructing the hypothetical last common ancestors of various clades. Prominent among such reconstructions have been proposals as to the nature of the mysterious Urbilateria, originally defined as the last common ancestor (LCA) of the extant Bilateria (Ecdysozoa, Lophotrochozoa and Deuterostomia) (De Robertis and Sasai 1996, Kimmel 1996). Indeed, drawings of this animal can now be found, as well as detailed information on the genetics and morphological processes that it used to construct its gut, heart, eyes, appendages, segments and body region identities (Gilbert and Singer 2006). Perhaps surprisingly, however, no explanations have yet been offered of how it might have achieved the successful reproduction that must have been necessary for it to give rise to still surviving lineages. This chapter will examine the comparative data available on the specification of bilaterian reproductive systems during development, with special emphasis on the cells containing the genetic hereditary material, the germ cells, and speculate on the possible gonad structure and reproductive strategy of Urbilateria.

Before proceeding, we should clarify our expectations as to what the study of extant species can tell us about Urbilateria. In this chapter, I wish to avoid suggesting that extant reproductive systems are simply variations on a defined metazoan reproductive 'Bauplan' theme; the great weakness of the current evo-devo approach stems

Evolving Pathways: Key Themes in Evolutionary Developmental Biology, ed. Alessandro Minelli and Giuseppe Fusco. Published by Cambridge University Press.
© Cambridge University Press 2008.

from dilution of explanatory force with inappropriate fixations on strict, confining definitions of this kind (Scholtz 2004, 2005, Hübner 2005). I will review the current and historical literature on germ-cell and somatic gonad anatomy, embryonic specification and development, studies obviously all carried out on extant species, but will not infer from these data that Urbilateria must have had specific, archetypical genetic or developmental characteristics of its reproductive system anatomy or reproductive strategies; rather, I will suggest that these data can tell us what kinds of general features, or basic pattern, its reproductive system was likely to have had, in order for it to have given rise to these systems as manifest in extant bilaterian lineages.

Over the past couple of decades, comparative gene expression patterns, and, to a lesser extent, comparative morphology, have been used as tools in the dig for LCAs. The result has been a rather detailed description of the genetic networks, or at least major genetic players, which are proposed to have been active in Urbilateria to give it various features, including axial polarity, body regionalisation, light-sensing cells, a heart or circulatory system, and a regionalised nervous system. However, no suggestions have been forthcoming as to how this animal might have made gametes, ensured their fertilisation if necessary, and given rise to the first generation of bilaterian LCAs. Several questions about this aspect of Urbilateria come to mind. Was it hermaphroditic or parthenogenetic, or did separate sexes exist? Did it have a dedicated germ-cell population? If so, how was it specified? Did it have a discrete gonad? If so, from which germ layer did it originate? How was fertilisation achieved? To begin to examine some of these questions, we first need to define the components of functional reproductive systems.

COMPONENTS OF BILATERIAN REPRODUCTIVE SYSTEMS

There is a minimum of two aspects to successful sexual reproduction: (1) cells to make gametes, and (2) a fertilisation strategy. Most bilaterian reproductive systems possess a third critical element, which is a dedicated group of somatic cells to enclose, support, and extrude the gametogenic cells.

The germ line

Our starting point is the bilaterian LCA, a multicellular animal with multiple cell types and a division of labour, albeit of unknown extent, among different cell populations. Bilaterian outgroups do show a distinction between germline and soma: although a dedicated and exclusive gametogenic cell population may not exist (reviewed in Extavour and Akam 2003), most of the cells of these animals are *not* capable of producing gametes. The true innovation in the evolution of the germ line was not therefore the generation of a gametogenic lineage, but rather the loss of gametogenic potential from the majority of cells of the organism. Here, I do not consider this evolutionary innovation in detail; such explanation lies beyond the scope of this paper, and has been dealt with extensively by several researchers. Nonetheless, it is appropriate to briefly review current ideas as to the evolution of a germ-cell lineage.

Even general developmental biology textbooks that do not explicitly include evolutionary biology in their remit often recognise that 'development from more than one cell presents problems, as mutations could occur in some of the cells' (Wolpert *et al.* 2007: 521). More explicitly, 'The only way for the genome to be fully tested is to have only one line of germ cells' (Gerhart and Kirschner 1997: 249). Sequestration of a dedicated germ line early in development circumvents this problem, as the organism can thus develop from only one cell, but in its final form be composed of millions. We could reasonably expect that, in order to effectively confer the advantage of protection from somatic mutation, such a lineage might show reduction of mitotic activity (since more rounds of DNA replication give more opportunity for mutation through copy error; Sweasy *et al.* 2006), reduced transcriptional activity (because genes may be more subject to mutation when actively transcribed; Medvedev 1981) and reduced transposable element mobility (which, although it can be a 'positive' force in adaptive evolution, indisputably leads to increased mutation rates; McDonald 1993, Fedoroff 1999, Deragon and Capy 2000). In fact, the germ line displays all of these features. Germ cells are typically mitotically quiescent from the time of their specification during embryogenesis, until the time that gametogenesis begins, usually during larval or adult life. They are relatively transcriptionally quiescent during most of embryonic development, as revealed by diagnostic histone modifications and single-cell transcription analysis (Schaner *et al.* 2003). Finally, RNA-mediated silencing of transposable elements has recently been documented in the germ lines of *Caenorhabditis elegans* and *Drosophila melanogaster* (Aravin *et al.* 2004, Robert *et al.* 2004, Vagin *et al.* 2006).

It has further been suggested that the invention of a gametogenic lineage was not just an added bonus, but in fact a *sine qua non* of the evolution of multicellular organisms that acted, and were acted on by natural selection, as true individuals (Michod 1999). This is because as long as all cells retain the possibility to contribute to future generations, intra-individual competition among cell lineages is predicted to prevent the fitness gains of the group (that is, the multicellular organism) from exceeding the fitness gains of the component cells. In summary, Urbilateria, as a *bona fide* metazoan, can be assumed to have possessed at least a majority of truly somatic cells, so that it depended for its reproductive success on the successful specification and protection throughout development of a germ line.

The soma

What all somatic reproductive systems have in common is that they comprise a network of non-gametogenic cells whose role is to support, enclose, transport and expel the gametic products of the individual. Beklemishev (1969) defined five components of the somatic reproductive system as follows: (1) gonads (where gametogenesis takes place); (2) genital ducts (for storing, transporting or extruding gametic products); (3) copulatory organs (for transferring gametes between individuals of the opposite sex); (4) adaptations for creating envelopes for ova; (5) adaptations for bearing live young. We shall use these five categories to characterise the reproductive systems of the metazoan phyla, and as will become evident, a successful reproductive strategy may involve all or none of these elements.

Fertilisation strategies

Urbilateria, by definition, must have used some kind of reproductive strategy, but we have no way of knowing what it was. Once gametes have been made, if fertilisation is necessary then this needs to take place. Fertilisation can be wholly external (gametes of both sexes released without copulation), wholly internal (gametes of one sex deposited into the individual of the opposite sex, via copulation) or external–internal (gametes of one sex are released without copulation, then taken up by the opposite sex, so that fertilisation is internal). The type of fertilisation strategy used depends on the anatomy of the somatic reproductive system. For example, genital ducts and copulatory organs are prerequisites for wholly internal fertilisation. For this reason, we will only be able to begin speculation on an urbilaterian reproductive strategy once we have identified some patterns of comparative metazoan somatic gonad structures.

COMPARATIVE DATA ON SOMATIC GONAD SPECIFICATION

Anatomical studies of members of most extant bilaterian phyla provide data on the structure of the somatic reproductive system. More difficult to obtain are data on the developmental origin of the system, and on its functioning during reproduction, as these depend on availability of reliably staged developmental intermediates, and direct observations of copulation and/or fertilisation, respectively. What is immediately apparent even from the data available, however, is that on a bilaterian scale, a strictly phylogenetic consideration of reproductive system anatomy makes no sense without also considering life history and environmental factors.

We will use Beklemishev's (1969) five categories of reproductive system components to characterise the complexity of these systems across the Bilateria. We observe here the full range of complexities of reproductive systems, from free-floating gametes within the body cavity, which are extruded by epidermal rupture to engage in external fertilisation, to gametes confined within elaborate gonads, which can only be exposed to gametes of the opposite sex through copulation, and eventually travel through dedicated ducts to uteri specialised for viviparity (Table 17.1).

Among the protostomes, reproductive system structure can vary not only between phyla, but also within a single phylum. For example, within the Annelida, leeches have true gonads and gonoducts, as do oligochaetes and some polychaetes. However, many polychaete species lack discrete gonads; instead, their gametes mature in coelomic cavities from free-floating gametogonia, are released by body wall rupture and undergo external fertilisation in the water column (Beklemishev 1969). Some onychophorans have not only complex gonad structures but also uteri; fertilisation is internal, embryos develop in uteri, and animals give birth to live young (Manton 1949). As in many other segmented protostomes, somatic gonad components are formed from mesodermal cells of the splanchnic dorsal coelomic wall (Manton 1949, Anderson 1973).

Among the deuterostomes, *Xenoturbella* has the simplest known reproductive system: as in many sponges, cnidarians and flatworms, gametes develop freely in the coelom and are extruded through the mouth upon maturity (Beklemishev 1969). Many marine invertebrate deuterostomes have discrete gonads and gonoducts, but lack copulatory organs, and fertilisation takes place in the water column. Mammals have of course developed specialised copulatory organs, as well as adaptations

Table 17.1 *Fertilisation strategies and somatic reproductive systems across the Metazoa.*

	External–external (*No copulation: both gametes released externally*)					Internal–internal (*Copulation; neither gamete released externally*)					External–internal (*External sperm release & internal fertilisation*)					Self-fertile hermaphrodite				
Porifera	–	–	–	–	–						–	–	–	–	–	–	–	–	–	–
Cnidaria	–	–	–	–	–															
Ctenophora																1	2	–	4	–
Gastrotricha						1	2	3	4	–										
Rotifera						1	2	3	4	–						1	2	–	4	–
Annelida	1	2	–	4	–						(1)	(2)	(3)	(4)	–					
Echiura	1	2	–	–	–															
Sipuncula	1	–	–	–	–															
Nemertea	1	2	–	4	–															
Platyhelminthes											1	2	–	4	–					
Mollusca	1	–	(3)	4	–	1	–	3	4	–										
Gnathostomulida																1	2	3	4	–
Entoprocta											1	2	–	4	–					
Nematoda						1	2	3	4	–						1	2	–	4	–
Priapulida	1	2	–	4	–															

Kinorhyncha						1	2	(3)	4	–					
Tardigrada						1	2	3	4	–	1	2	3	4	–
Onychophora											1	2	3	4	(5)
Arthropoda						1	2	3	4	–	1	2	3	4	–
Chaetognatha											(1)	(2)	–	(4)	–
Xenoturbella	–	–	–	–	–										
Echinodermata	1	2	–	–	–						1	2	–	–	–
Hemichordata	1	2	–	–	–										
Cephalochordata	1	2	–	–	–										
Urochordata	1	2	–	–	–						1	2	3	–	–
Vertebrata	1	2	–	–	–	1	2	3	4	5					

Note: Numbers 1 through 5 indicate the following components: (1) gonads or genital glands (envelope of somatic cells enclosing gametogenic cells); (2) genital ducts (leading from gonads to gonopores for extrusion of gametes); (3) copulatory organs (may be used either for internal fertilisation or external gamete deposition, e.g. spermatophores); (4) adaptations for creating eggshells for ova (including vitelline membranes and chorions, but excluding extra-embryonic membranes of zygotic origin); (5) adaptations for viviparity (including uteri and placentae). Hyphens indicate phyla using the indicated fertilisation strategy but possessing none of the somatic reproductive system components. Parentheses indicate that some phylum members may not always possess the given feature. Data complied from references cited in Extavour and Akam (2003) and from Giese and Pearse (1974–1989) and Brusca and Brusca (2003).

for internal fertilisation, embryonic development and viviparity. The mammalian somatic gonad probably derives from the mesonephros and the adjacent coelomic epithelium (McLaren 2000).

For many studied metazoans, it is clear that the somatic and germ-line components of the reproductive system are specified or 'sequestered' separately during development; that is, they share limited or no lineage. The huge diversity in somatic reproductive systems should therefore not be surprising, given that independently sequestered lineages may display a certain modular independence in morphological evolution. West-Eberhard summarises this by saying that 'an increase in modularity . . . sometimes appears to have contributed to increased diversification of *that aspect of the phenotype* during the history of a taxon' (West-Eberhard 2003: 87 [italics original]).

Jury still out on urbilaterian gonads

Beyond a mesodermal origin for the somatic structures of the reproductive system, no general pattern emerges from a phylogenetic consideration of these systems across the Bilateria. Convergent evolution of every aspect of the system is apparent not only between phyla, but also within phyla. Most bilaterian outgroups lack true gonads, but while some acoels similarly lack gonads, others display compact, paired, ovaries, and many have male copulatory organs. Data on the molecular mechanisms specifying somatic gonad fate are largely limited to mice (McLaren 2000), nematodes (Hubbard and Greenstein 2000) and fruit flies (Moore *et al.* 1998, DeFalco *et al.* 2004). To date, the evidence for conservation of gene function in somatic gonad cells is limited to the protein product of a single gene (Li *et al.* 2003). We therefore cannot suggest homology of molecular pathways involved, consistent with repeated convergent evolution. In summary, while it is likely that Urbilateria lacked a complex somatic reproductive system, it is at present impossible to speculate on whether it possessed a true gonad, let alone any other somatic adaptations for reproduction.

COMPARATIVE DATA ON GERM-CELL SPECIFICATION

Germ cells are one of the most extensively studied metazoan cell lineages. They represent a crucial link between developmental biology and evolutionary biology, being responsible for both reproduction of the individual and genetic continuity of the species. Although germ-cell migration, polarity and differentiation are all fascinating developmental problems in their own right, I propose that the most crucial aspect of germ-cell development for understanding the evolution of the germ line is the first specification event of the lineage, that is, the mechanism that separates germ line from soma.

Over the past two centuries, a battery of tools for germ-cell identification and study has become available to researchers (reviewed in Extavour and Akam 2003). Germ cells can almost always be unambiguously distinguished from somatic cells by one or a combination of the following four criteria: (1) characteristic morphology under transmitted white light, including organelle-free cytoplasm, large nuclear:cytoplasmic ratio, rounded nuclei with prominent nucleoli and diffuse chromatin, and granular cytoplasmic inclusions usually localised in the perinuclear cytoplasm associated with nuclear pores; (2) electron-dense cytoplasmic granules (nuage) identifiable by transmission electron miscroscopy; (3) high levels of alkaline phosphatase activity (this criterion has been useful only in vertebrates); (4) localisation of mRNA or protein products of germ-cell-specific genes, notably the *vasa* and *nanos* gene family products. Some combination of these criteria always holds for germ cells at all stages of development, from their initial embryonic specification as primordial germ cells (PGCs), until their differentiation as male and female gametes.

Identifying germ cells at some stage of development is therefore feasible for any animal one wishes to study, given access to embryos or adults or both. Much more difficult, however, is discerning the time, place and mechanism responsible for the initial specification event giving rise to the germ line. This is because, as Balfour (1885) correctly noted, 'Since it is usually only possible to recognise generative elements after they have advanced considerably in development, the mere position of a generative cell, when first observed, can afford . . . no absolute proof of its origin'.

Specification and origin of extant metazoan PGCs: epigenesis and preformation

In 1979 and 1981, Nieuwkoop and Sutasurya published two excellent volumes summarising all available literature on PGCs across the metazoans, including, but not limited to, their initial specification (Nieuwkoop and Sutasurya 1979, 1981). More focused survey studies dealing specifically with the first embryological sequestration of the germline in both vertebrates and invertebrates are limited to three: two classic monographs of the last century (Bounoure 1939, Wolff 1964), and a modern review incorporating the last quarter of a century of genetic and experimental data (Extavour and Akam 2003). The results of these studies will be briefly summarised here.

Modern developmental genetic model systems have indicated that two basic types of molecular mechanisms are responsible for germ-cell specification; I will call these two types 'preformation' and 'epigenesis' (Extavour and Akam 2003). It is important to note that the two mechanisms are not necessarily mutually exclusive, but rather are better viewed as two extremes of the continuum along which germ-cell development can be mapped, since at some stage of germ-cell development, both types of mechanism are inevitably used.

Preformation refers to cell-autonomous acquisition of germ-cell fate through localised, inherited cytoplasmic determinants, which are both necessary and sufficient to confer germ-cell fate upon the cell containing them. The molecules composing these determinants are both mRNA and protein products of genes that are widely conserved across all metazoans. Dipterans and nematodes are well-known, long-standing examples of animals showing this mode of PGC specification.

Epigenesis refers to acquisition of germ-cell fate by reception of cell non-autonomous signals from germ layers adjacent to future PGCs. In this case, the signals are themselves necessary and sufficient to induce receiving cells to adopt PGC fate. Mice and axolotls clearly exhibit this mode of PGC specification, and while in the axolotl the inductive signals have not yet been identified (but see Johnson *et al.* 2003), in mice they are members of the BMP2/4 and 8b families.

Until very recently, it was widely held among most developmental biologists that since preformation was prevalent among model laboratory organisms, it was probably the most widespread and ancestral mechanism of PGC formation (contrast the second edition of the influential text Wolpert *et al.* 2002, with the most recent edition, Wolpert *et al.* 2007). However, closer examination of the available data demonstrates that this is unlikely to be the case (for details and comprehensive reference lists, see Extavour and Akam 2003).

For most ecdysozoans and lophotrochozoans, all studied members of a given phylum appear to use epigenesis to specify PGCs, while a few phyla (Platyhelminthes, Annelida, Mollusca and Arthropoda) contain both members showing epigenesis, and members displaying preformation (Figure 17.1). Only in the Nematoda, Rotifera and Chaetognatha do all studied members exhibit preformation. In other words, across both the Ecdysozoa and the Lophotrochozoa, epigenesis is the most common mechanism of PGC specification.

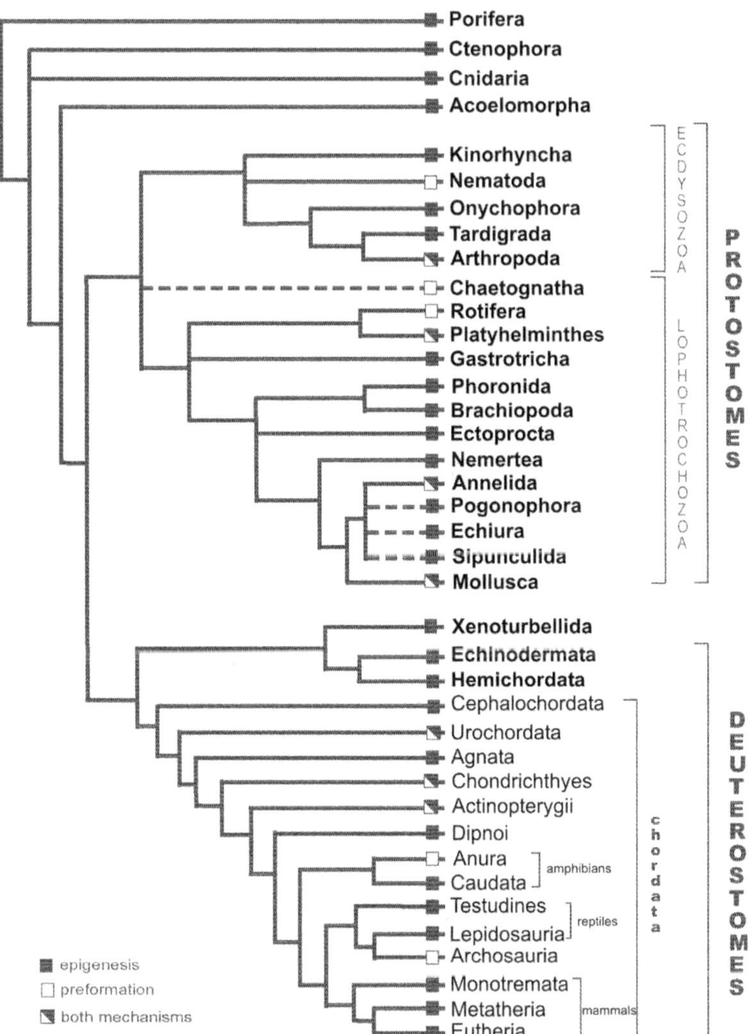

Figure 17.1 Distribution of PGC specification mechanisms across the
Metazoa. Epigenesis (black boxes), preformation (white boxes), or both
mechanisms (black and white boxes) are indicated only in phyla for which
at least two independent primary data sources provide morphological, cell
lineage, experimental or molecular evidence; phyla for which the data on
germ-cell specification mechanisms are insufficient have been omitted.
Dashed lines indicate phyla for which phylogenetic relationships are still
unclear. Details of source data are as described in Extavour and Akam
(2003). Adapted from Extavour and Akam (2003) with modifications as
follows: assignation of *Xenoturbella* to its own phylum within the
deuterostomes (Bourlat *et al.* 2003, 2006); evidence for epigenetic PGC

Within the deuterostomes, all studied members of these phyla, including all non-chordates, probably use epigenesis to specify PGCs (Figure 17.1). Of the chordates, only Urochordata, Chondrichthyes and Actinopterygii contain some members that use epigenesis and others that use preformation as a PGC specification mode. Finally, in only two clades (anuran amphibians and archosaurs) do all studied members exhibit preformation. To summarise, with the exception of some elasmobranchs, the only deuterostome clades containing preformistic members are those containing chordate model laboratory organisms other than mice: the solitary ascidians *Ciona intestinalis* and *Halocynthia roretzi* (but note that recent data on colonial ascidians are consistent with epigenesis; Sunanaga *et al.* 2006a,b); the frog *Xenopus laevis*; the teleost *Danio rerio*; and the chicken *Gallus gallus*. All other studied deuterostomes, including the Ambulacraria and *Xenoturbella*, show evidence for epigenesis as the mode of PGC specification.

A stem-cell origin of urbilaterian PGCs

The acoelomorph, protostome and deuterostome data summarised above, taken together with the observation that there are no data supporting preformation of the germ line in any of the bilaterian outgroups (Extavour and Akam 2003, Figure 17.1), strongly suggest that epigenetic establishment of the germ line was present in Urbilateria. Sponges, cnidarians and acoel flatworms use very similar strategies to obtain gametogenic cells. They all contain a population of endodermally derived pluripotent stem cells (sponge archaeocytes, cnidarian interstitial cells and acoel neoblasts) that acquire their fate in early to mid-embryogenesis, and can give rise to both somatic cell types and gametes. These cells are scattered throughout the gastral cavity and/or intercalated between other somatic cells. As we will see below from the basic patterns of somatic gonad structure, Urbilateria was unlikely to have had all of its gametogenic cells clustered together in one region, but rather might have had them scattered throughout the body. These potential PGCs would have been pluripotent stem cells: some of them would have been capable of creating or regenerating adult somatic tissue as well, throughout the lifetime of the animal.

Fig. 17.1 (cont.) specification in a colonial ascidian (Sunanaga *et al.* 2006, 2007); changed phylogenetic relationship of Urochordata and Cephalochordata within the Chordata (Bourlat *et al.* 2006, Delsuc *et al.* 2006, Vienne and Pontarotti 2006) and affiliation of Chaetognatha with the protostomes (Marletaz *et al.* 2006, Matus *et al.* 2006).

As well as using the general pattern of metazoan germ-cell specification modes to infer that Urbilateria's germ cells were a subpopulation of stem cells, we can also obtain evidence from modern molecular and functional comparisons between stem cells and germ cells. The electron-dense nuage material invariably found in germ cells using transmission electron microscopy has also been found in stem-cell lineages (Eddy 1975). Pluripotent cells often display all of the morphological features commonly used to identify germ cells, such as a large round nucleus with diffuse chromatin and a prominent nucleolus. This can lead to an inability to distinguish between germ cells and other types of stem cells (see for example Potswald 1969, 1972). Similarly, when using molecular markers to identify germ-cells, unless careful phylogenetic analysis of the gene homologues is carried out, researchers have run the risk of isolating genes that will not distinguish between germ cells and other pluripotent cells. For example, the products of *vasa* gene family members are nearly always exclusive to the germ-cell lineage (Raz 2000, Extavour and Akam 2003). The *vasa* gene family is thought to have evolved from the *PL10* family of helicases, which share significant structural similarity with *vasa* genes (Mochizuki *et al.* 2001). *PL10* products are usually localised to both germ cells and other pluripotent cell types. If *PL10* homologues are isolated and incorrectly assigned *vasa* homology because of insufficient analysis, using them to identify germ cells can give rise to ambiguous or inaccurate lineage assignation (see for example Shibata *et al.* 1999). On the morphological and gene expression levels, then, germ cells and stem cells are very similar.

Another level of similarity between germ cells and stem cells has been revealed by functional analysis in both vertebrate and invertebrate systems. Mammalian germ cells grown in culture and treated with fibroblast growth factor (FGF) can be induced to become pluripotent stem cells, called embryonic germ (EG) cells, that are very similar in differentiation potential to embryonic stem (ES) cells derived from the inner cell mass (ICM) of the blastocyst (Matsui *et al.* 1992, Resnick *et al.* 1992, Rohwedel *et al.* 1996, Shamblott *et al.* 1998). *Drosophila* germ cells already en route towards oogenic differentiation can be induced to revert back to a germline stem-cell state (Kai and Spradling 2004). Similar dedifferentiation and redifferentiation is seen in cells from teratocarcinomas. These are malignant tumours probably formed from ectopic or aberrant primordial germ cells, which contain multiple differentiated tissues as well as undifferentiated stem cells called embryonal carcinoma (EC) cells. Cultures of EC cells, used as in vitro models of mammalian differentiation and development, have demonstrated that PGCs may be able, after 'dedifferentiation' into EC cells, to 'redifferentiate' as multiple somatic cell types (Kleinsmith and Pierce 1964, Kahan and Ephrussi 1970). Even more strikingly, when transplanted into blastocysts, which are then implanted into host female uteri, mouse teratocarcinoma cells can contribute not only to many somatic tissues, but also to the germ line, of the resulting progeny (Stewart and Mintz 1981).

Because ES cells are usually derived from blastocyst ICM cells, they are generally assumed to be equivalent to ICM cells. Observed differences between ES cells and ICM cells might simply be the result of ES culture conditions. However, Zwaka and Thomson (2005) have hypothesised that EG, ES and EC cells may all have their closest in vivo equivalent not in ICM cells but rather in germ cells. This hypothesis is sufficient to explain the developmental origins of ES cells, but to explain the evolutionary origins of germ cells, we need to invert the hypothesis. I propose that PGCs may have their closest evolutionary equivalent in the pluripotent stem cells that are found in extant non-bilateria and basal bilaterians, and that almost certainly existed in Urbilateria.

Convergent evolution of preformation

If epigenesis was used by Urbilateria to specify the germ line, then preformation must have evolved convergently several times during the bilaterian radiation. We therefore require a feasible framework for conceiving the following: urbilaterian germ cells were a subpopulation of somatic cells, and repeatedly, in several descendant lineages of Urbilateria, germ cells acquired a cell-autonomous specification mechanism, and became a lineage independent of somatic cells, with the obvious caveat that somatic support structures are almost always required for successful gamete production, even in preformistic species. To demonstrate how this proposal represents a modification of previous models of germline continuity, I will compare it with the three major previous models of pangenesis, continuity, and modified continuity with somatic selection.

Darwin's (1859) pangenesis theory provided a biological explanation for Lamarck's ideas about inheritance of acquired characteristics (Lamarck 1809): all somatic cells produce invisible particles called gemmules, which travel through the body and lodge in the germ cells. Since germ cells do not initially contain all of the information necessary to reproduce the adult form in successive generations, including acquired characteristics, they need to receive this information from the gemmules. The germ line is neither immortal nor continuous, as it produces only the soma of the next generation, and that soma would produce the next germ line (Figure 17.2A). Weismann, on the other hand, was sure that germ cells are autonomously totipotent from the moment of their formation, and that their nuclear information is both impervious to somatic influence and sufficient for reproduction of the adult form (Weismann 1892). In other words, the germ line is both immortal and continuous, and the source of both soma and germ line of subsequent generations (Figure 17.2B). Since at least the 1920s, however, it has become increasingly clear that Weismann's hypothesis is in need of serious revision, given the existence of epigenesis in germ-line specification in many species (Hargitt 1919, Heys 1931, Berrill and Liu 1948). Buss (1983) has proposed an elegant revision to Weismann's hypothesis that takes into account both epigenetic germline origin and intra-individual cellular selection. In this model, while germline continuity may exist in some species (Figure 17.2C, bottom series), somatic

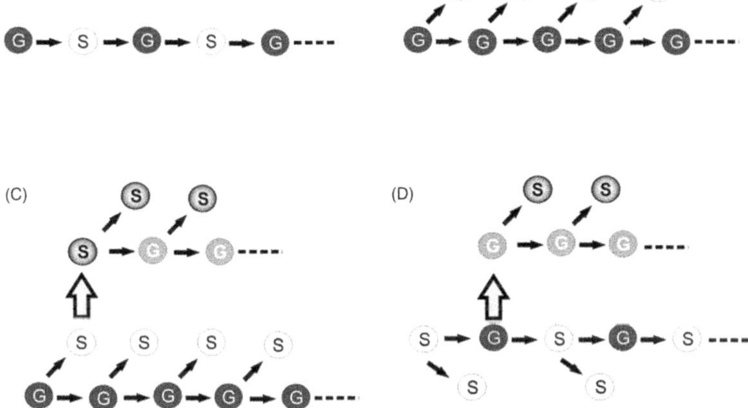

Figure 17.2 Models for the evolution of the relationship between germ line and soma. A, Pangenesis: the soma (white) informs and specifies the germ line (black), which in turn gives rise only to the soma. B, Immortality/continuity: the germ line is the sole progenitor of both germ line and soma, receiving no somatic input. C, Continuity allowing for somatic selection: somatic mutation (gradient) may allow specification of germ line (grey) from somatic cells (top series), representing a deviation (large arrow) from the usual continuity of the germ line (bottom series). D, Evolution of preformation from epigenesis: germline mutation (grey) may confer continuity on the germ line (top series), representing a deviation (large arrow) from its usual somatic stem-cell origin (bottom series).

mutation may sometimes allow a subpopulation of the soma to produce gametes (Figure 17.2C, top series).

To explain repeated evolution of preformation from epigenesis, it suffices to invert Buss's model (Figure 17.2D). Urbilateria would have segregated germ cells epigenetically, as a subpopulation of somatic cells: soma therefore gave rise to both germ line and soma (Figure 17.2D, bottom series). Where Buss's model suggests that mutations affecting the soma could allow somatic cells to produce gametes, I suggest that mutations affecting the germ line could allow cell-autonomous segregation of germ cells in a subsequent generation (Figure 17.2D, top series). This mechanism of preformation would then be inherited in subsequent generations. In order to understand what kind of germline mutation could have had this effect, in the next section we will consider known examples of germ cells that segregate by preformation.

Evolving preformation from epigenesis: a transitional model

All known molecular mechanisms of preformation rely on localisation of germ-cell-specific molecules (germ plasm components) to a particular place in the oocyte, either before or after fertilisation (see for example Illmensee *et al.* 1976, Ressom and Dixon 1988, Carré *et al.* 2002). In several cases, notably the *vasa* and *nanos* gene families, the genes encoding these molecules, and their germline expression, are conserved across all bilaterian species for which data are available (Extavour and Akam 2003). Many germ plasm components are expressed and required not only in primordial germ cells but also during gametogenesis (see for example Styhler *et al.* 1998, Tanaka *et al.* 2000, Extavour *et al.* 2005). The major difference between epigenesis and preformation is thus the relative expression timing and gene product localisation of germ-cell-specific genes: in epigenesis, these genes are downregulated and/or their products are eliminated from the oocyte, after gametogenesis. Their products are not present in the cytoplasm of the fertilised egg and cannot therefore be inherited cell-autonomously by PGCs; instead the genes must be zygotically activated in PGCs through epigenetic signalling (Figure 17.3A). In preformation, germ-cell-specific gene products persist through completion of oogenesis in the zygotic cytoplasm, and are therefore available for inclusion into PGCs before the initiation of zygotic transcription (Figure 17.3B). In this context, we can now see that in order to make the transition from epigenesis to preformation, only two things are necessary: (1) persistence (and

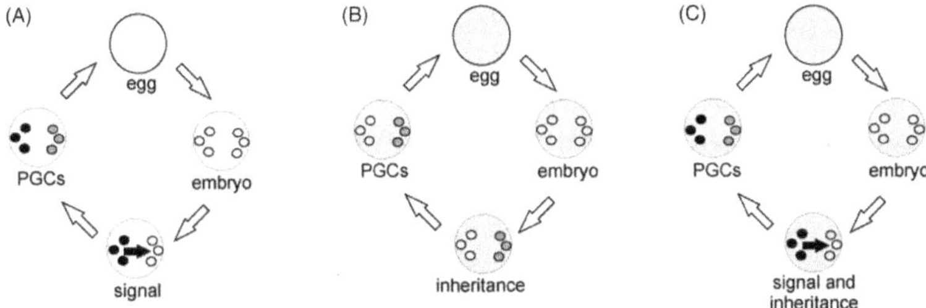

Figure 17.3 A transitional model for the evolution of preformation from epigenesis. A, Epigenesis: germ-cell-specific molecules expressed during gametogenesis are not present in oocytes at the time of fertilisation. During embryogenesis, inductive signals (black) specify PGCs, which begin zygotic expression of germ-cell-specific molecules (dark grey). Germ cells produce gametes to complete the cycle. B, Preformation: maternal germ-cell determinants (light grey) are localised to oocyte cytoplasm and inherited cell-autonomously by PGCs forming in early embryogenesis. Germ cells produce gametes to complete the cycle. C, Transition from epigenesis to preformation: germ-cell-specific molecules expressed during gametogenesis are retained in oocytes through to the time of fertilisation. They are inherited cell-autonomously by PGCs forming in early embryogenesis. Inductive signals (black) produced during embryogenesis are now redundant with respect to PGC formation. Germ cells produce gametes to complete the cycle. Loss of inductive signals is predicted over evolutionary time, so that this system comes to be like that shown in B.

possibly, through further refinement of the mechanism, cytoplasmic localisation within the oocyte) of germ-cell-specific gene products through the end of gametogenesis; and (2) inheritance of these products, which would now constitute germ plasm components, by future PGCs (Figure 17.3C).

Mutations arising in the germ line that affected oocyte cytoskeletal dynamics or mRNA or protein localisation of germ cell molecules could allow persistence and/or localisation of these molecules in mature oocytes. Once preformation had arisen in a heritable way through such mutation(s), signals from somatic tissues to induce germline fate would no longer be necessary to ensure species survival. We would therefore expect gradual loss of these signalling mechanisms, since 'unnecessary but costly structures or activities should be lost in evolution' (Michod 1999: 55). This model can explain why we see the repeated evolution of autonomous germline determinants in several groups (Figure 17.1), but never observe examples of epigenesis in phyla where preformation is plesiomorphic (e.g. Rotifera, Chaetognatha, Nematoda).

One prediction of the model is the existence at some time of species in which both preformation and epigenesis were operative, or at least operable. In most preformistic model organisms, however, when PGCs or their precursors are eliminated through physical ablation or genetic manipulation, the resulting animals are sterile, presumably unable to replace the ablated germ line through epigenetic mechanisms (reviewed in Saffman and Lasko 1999). These animals may belong to lineages in which preformation evolved so long ago that epigenetic signalling mechanisms have become unusable through lack of positive selection. Given that all currently used developmental genetic model organisms are derived with respect to many other aspects of embryogenesis, this explanation is not unreasonable. Alternatively, our failure thus far to observe widespread coexistence of both PGC specification mechanisms may simply be reflective of poor taxon sampling. Intriguingly, in the solitary ascidian *C. intestinalis*, although convincing embryological and molecular genetic data indicate that preformation specifies PGCs, when the PGCs are ablated in larval stages the resulting adults are still fertile (Takamura *et al.* 2002). The mechanism responsible for this germ line replacement is currently unknown. I suggest that as more species from the diversity of the Bilateria become amenable to molecular analysis of embryogenesis and development, further examples of species able to use both epigenetic and preformation to specify germ cells will emerge.

CONCLUSIONS

Urbilateria was unlikely to have had a complex somatic reproductive system, but whatever somatic support it did have for gametogenic cells was almost certainly of mesodermal origin. The changes in life histories undergone by urbilaterian descendant lineages, as they occupied different ecological niches, led to morphogenetic modification of these mesodermal derivatives, resulting in convergent evolution of different elements of somatic reproductive systems, including gonads, gonoducts and gonopores, copulatory organs and adaptations for viviparity. Urbilaterian germ cells were likely probably specified as a subpopulation of pre-existing somatic pluripotent stem cells, through inductive signals of unknown molecular identity. Its germ cells expressed *vasa* gene family members and possibly also *nanos* gene(s). Changes in the expression timing (heterochrony) and ooplasmic localisation (heterotopy/heterotypy) of germ-cell differentiation genes led to early embryonic cytoplasmic inheritance of germ-cell determinants that was both heritable and independent of somatic epigenetic signalling later in embryonic development, resulting in convergent evolution of preformation. In descendant lineages that had evolved preformation, epigenetic germ-cell specification mechanisms would have gradually deteriorated owing to lack of positive selection.

ACKNOWLEDGEMENTS

I am grateful to the BBSRC for providing project grant BBS/B/07586 for support during the development of these ideas.

REFERENCES

Anderson, D. T. 1973. *Embryology and Phylogeny in Annelids and Arthropods.* Oxford: Pergamon Press.

Aravin, A. A., Klenov, M. S., Vagin, V. V. *et al.* 2004. Dissection of a natural RNA silencing process in the *Drosophila melanogaster* germ line. *Molecular and Cellular Biology* **24**, 6742–6750.

Balfour, F. M. 1885. *A Treatise on Comparative Embryology.* London: MacMillan and Co.

Beklemishev, W. N. 1969. *Principles of Comparative Anatomy of Invertebrates: Organology.* Chicago: University of Chicago Press and Edinburgh: Oliver and Boyd.

Berrill, N. J. & Liu, C. K. 1948. Germplasm, Weismann, and Hydrozoa. *Quarterly Review of Biology* **23**, 124–132.

Bounoure, L. 1939. *L'origine des cellules reproductrices et le problème de la lignée germinale.* Paris: Gauthier-Villars.

Bourlat, S. J., Juliusdottir, T., Lowe, C. J. *et al.* 2006. Deuterostome phylogeny reveals monophyletic chordates and the new phylum Xenoturbellida. *Nature* **444**, 85–88.

Bourlat, S. J., Nielsen, C., Lockyer, A. E., Littlewood, D. T. & Telford, M. J. 2003. *Xenoturbella* is a deuterostome that eats molluscs. *Nature* **424**, 925–928.

Brusca, G. J. & Brusca, R. C. 2003. *Invertebrates.* Sunderland, MA: Sinauer Associates.

Buss, L. W. 1983. Evolution, development, and the units of selection. *Proceedings of the National Academy of Sciences of the USA* **80**, 1387–1391.

Carré, D., Djediat, C. & Sardet, C. 2002. Formation of a large Vasa-positive granule and its inheritance by germ cells in the enigmatic chaetognaths. *Development* **129**, 661–670.

Darwin, C. 1859. *On the Origin of Species by Means of Natural Selection, or the Preservation of Favoured Races in the Struggle for Life.* London: John Murray.

De Robertis, E. M. & Sasai, Y. 1996. A common plan for dorsoventral patterning in Bilateria. *Nature* **380**, 37–40.

DeFalco, T., Le Bras, S. & Van Doren, M. 2004. *Abdominal-B* is essential for proper sexually dimorphic development of the *Drosophila* gonad. *Mechanisms of Development* **121**, 1323–1333.

Delsuc, F., Brinkmann, H., Chourrout, D. & Philippe, H. 2006. Tunicates and not cephalochordates are the closest living relatives of vertebrates. *Nature* **439**, 965–968.

Deragon, J. M. & Capy, P. 2000. Impact of transposable elements on the human genome. *Annals of Medicine* **32**, 264–273.

Eddy, E. M. 1975. Germ plasm and the differentiation of the germ cell line. *International Review of Cytology* **43**, 229–280.

Extavour, C. & Akam, M. E. 2003. Mechanisms of germ cell specification across the metazoans: epigenesis and preformation. *Development* **130**, 5869–5884.

Extavour, C., Pang, K., Matus, D. Q. & Martindale, M. Q. 2005. *vasa* and *nanos* expression patterns in a sea anemone and the evolution of bilaterian germ cell specification mechanisms. *Evolution & Development* **7**, 201–215.

Fedoroff, N. V. 1999. Transposable elements as a molecular evolutionary force. *Annals of the New York Academy of Sciences* **870**, 251–264.

Gerhart, J. & Kirschner, M. 1997. *Cells, Embryos and Evolution: Toward a Cellular and Developmental Understanding of Phenotypic Variation and Evolutionary Adaptability.* Malden, MA: Blackwell Science.

Giese, A. C. & Pearse, J. S. (eds.) 1974–1989. *Reproduction of Marine Invertebrates.* New York: Academic Press.

Gilbert, S. F. & Singer, S. R. 2006. *Developmental Biology.* Sunderland, MA: Sinauer Associates.

Hargitt, G. T. 1919. Germ cells of Coelenterates. VI. General considerations, discussion, conclusions. *Journal of Morphology* **33**, 1–60.

Heys, F. 1931. The problem of the origin of germ cells. *Quarterly Review of Biology* **6**, 1–45.

Hubbard, E. J. & Greenstein, D. 2000. The *Caenorhabditis elegans* gonad: a test tube for cell and developmental biology. *Developmental Dynamics* **218**, 2–22.

Hübner, C. 2005. *Hox* genes, homology and axis formation: the application of morphological concepts to evolutionary developmental biology. *Theory in Biosciences* **124**, 371–396.

Illmensee, K., Mahowald, A. P. & Loomis, M. R. 1976. The ontogeny of germ plasm during oogenesis in *Drosophila*. *Developmental Biology* **49**, 40–65.

Johnson, A. D., Crother, B., White, M. E. *et al.* 2003. Regulative germ cell specification in axolotl embryos: a primitive trait conserved in the mammalian lineage. *Philosophical Transactions of the Royal Society of London. Series B: Biological Sciences* **358**, 1371–1379.

Kahan, B. W. & Ephrussi, B. 1970. Developmental potentialities of clonal in vitro cultures of mouse testicular teratoma. *Journal of the National Cancer Institute* **44**, 1015–1036.

Kai, T. & Spradling, A. 2004. Differentiating germ cells can revert into functional stem cells in *Drosophila melanogaster* ovaries. *Nature* **428**, 564–569.

Kimmel, C. B. 1996. Was Urbilateria segmented? *Trends in Genetics* **12**, 329–331.

Kleinsmith, L. J. & Pierce, G. B. Jr. 1964. Multipotentiality of single embryonal carcinoma cells. *Cancer Research* **24**, 1544–1551.

Lamarck, J.-B. 1809. *Philosophie Zoologique.* Paris.

Li, M. A., Alls, J. D., Avancini, R. M., Koo, K. & Godt, D. 2003. The large Maf factor Traffic Jam controls gonad morphogenesis in *Drosophila*. *Nature Cell Biology* **5**, 994–1000.

Manton, S. M. 1949. Studies on the Onychophora. VII. The early embryonic stages of *Peripatopsis*, and some general considerations concerning the morphology and phylogeny of the Arthropoda. *Philosophical Transactions of the Royal Society of London B* **233**, 483–580.

Marletaz, F., Martin, E., Perez, Y. *et al.* 2006. Chaetognath phylogenomics: a protostome with deuterostome-like development. *Current Biology* **16**, R577–578.

Matsui, Y., Zsebo, K. & Hogan, B. L. 1992. Derivation of pluripotential embryonic stem cells from murine primordial germ cells in culture. *Cell* **70**, 841–847.

Matus, D. Q., Copley, R. R., Dunn, C. W. *et al.* 2006. Broad taxon and gene sampling indicate that chaetognaths are protostomes. *Current Biology* **16**, R575–R576.

McDonald, J. F. 1993. Evolution and consequences of transposable elements. *Current Opinion in Genetics and Development* **3**, 855–864.

McLaren, A. 2000. Germ and somatic cell lineages in the developing gonad. *Molecular and Cellular Endocrinology* **163**, 3–9.

Medvedev, Z. A. 1981. On the immortality of the germ line: genetic and biochemical mechanism. A review. *Mechanisms of Ageing and Development* **17**, 331–359.

Michod, R. E. 1999. Individuality, immortality, and sex. In L. Keller, (ed.) *Levels of Selection in Evolution*. Princeton, NJ: Princeton University Press, pp. 53–74.

Mochizuki, K., Nishimiya-Fujisawa, C. & Fujisawa, T. 2001. Universal occurrence of the *vasa*-related genes among metazoans and their germline expression in *Hydra*. *Development Genes & Evolution* **211**, 299–308.

Moore, L. A., Broihier, H. T., Van Doren, M. & Lehmann, R. 1998. Gonadal mesoderm and fat body initially follow a common developmental path in *Drosophila*. *Development* **125**, 837–844.

Nieuwkoop, P. D. & Sutasurya, L. A. 1979. *Primordial Germ Cells in the Chordates*. Cambridge: Cambridge University Press.

Nieuwkoop, P. D. & Sutasurya, L. A. 1981. *Primordial Germ Cells in the Invertebrates: From Epigenesis to Preformation*. Cambridge: Cambridge University Press.

Potswald, H. E. 1969. Cytological observations on the so-called neoblasts in the serpulid *Spirorbis*. *Journal of Morphology* **128**, 241–260.

Potswald, H. E. 1972. The relationship of early oocytes to putative neoblasts in the serpulid *Spirorbis borealis*. *Journal of Morphology* **137**, 215–228.

Raz, E. 2000. The function and regulation of *vasa*-like genes in germ cell development. *Genome Biology* **1**, reviews 1017.1011–1017.1016.

Resnick, J. L., Bixler, L. S., Cheng, L. & Donovan, P. J. 1992. Long-term proliferation of mouse primordial germ cells in culture. *Nature* **359**, 550–551.

Ressom, R. E. & Dixon, K. E. 1988. Relocation and reorganization of germ plasm in *Xenopus* embryos after fertilization. *Development* **103**, 507–518.

Robert, V. J., Vastenhouw, N. L. & Plasterk, R. H. 2004. RNA interference, transposon silencing, and cosuppression in the *Caenorhabditis elegans* germ line: similarities and differences. *Cold Spring Harbour Symposia on Quantitative Biology* **69**, 397–402.

Rohwedel, J., Sehlmeyer, U., Shan, J., Meister, A. & Wobus, A. M. 1996. Primordial germ cell-derived mouse embryonic germ (EG) cells in vitro resemble undifferentiated stem cells with respect to differentiation capacity and cell cycle distribution. *Cell Biology International* **20**, 579–587.

Saffman, E. E. & Lasko, P. 1999. Germline development in vertebrates and invertebrates. *Cellular and Molecular Life Sciences* **55**, 1141–1163.

Schaner, C. E., Deshpande, G., Schedl, P. D. & Kelly, W. G. 2003. A conserved chromatin architecture marks and maintains the restricted germ cell lineage in worms and flies. *Developmental Cell* **5**, 747–757.

Scholtz, G. 2004. *Bauplâne versus* ground patterns, phyla versus monophyla: aspects of patterns and processes in evolutionary developmental biology. In G. Scholtz, (ed.) *Evolutionary Developmental Biology of Crustacea (Crustacean Issues)* 15. Lisse: Balkema, pp. 3–16.

Scholtz, G. 2005. Homology and ontogeny: pattern and process in comparative developmental biology. *Theory in Biosciences* **124**, 121–143.

Shamblott, M. J., Axelman, J., Wang, S. *et al.* 1998. Derivation of pluripotent stem cells from cultured human primordial germ cells. *Proceedings of the National Academy of Sciences of the USA* **95**, 13726–13731.

Shibata, N., Umesono, Y., Orii, H. *et al.* 1999. Expression of *vasa* (*vas*)-related genes in germline cells and totipotent somatic stem cells of planarians. *Developmental Biology* **206**, 73–87.

Stewart, T. A. & Mintz, B. 1981. Successive generations of mice produced from an established culture line of euploid teratocarcinoma cells. *Proceedings of the National Academy of Sciences of the USA* **78**, 6314–6318.

Styhler, S., Nakamura, A., Swan, A. & Suter, B. 1998. *vasa* is required for GURKEN accumulation in the oocyte, and is involved in oocyte differentiation and germline cyst development. *Development* **125**, 1569–1578.

Sunanaga, T., Saito, Y. & Kawamura, K. 2006. Postembryonic epigenesis of Vasa-positive germ cells from aggregated hemoblasts in the colonial ascidian, *Botryllus primigenus*. *Development Growth and Differentiation* **48**, 87–100.

Sunanaga, T., Watanabe, A. & Kawamura, K. 2007. Involvement of *vasa* homolog in germ line recruitment from coelomic stem cells in budding tunicates. *Development Genes & Evolution* **217**, 1–11.

Sweasy, J. B., Lauper, J. M. & Eckert, K. A. 2006. DNA polymerases and human diseases. *Radiation Research* **166**, 693–714.

Takamura, K., Fujimura, M. & Yamaguchi, Y. 2002. Primordial germ cells originate from the endodermal strand cells in the ascidian *Ciona intestinalis*. *Development Genes & Evolution* **212**, 11–18.

Tanaka, S. S., Toyooka, Y., Akasu, R. *et al.* 2000. The mouse homolog of *Drosophila* Vasa is required for the development of male germ cells. *Genes & Development* **14**, 841–853.

Vagin, V. V., Sigova, A., Li, C. *et al.* 2006. A distinct small RNA pathway silences selfish genetic elements in the germline. *Science* **313**, 320–324.

Vienne, A. & Pontarotti, P. 2006. Metaphylogeny of 82 gene families sheds a new light on chordate evolution. *International Journal of Biological Sciences* **2**, 32–37.

Weismann, A. 1892. *The Germ-Plasm: A Theory of Heredity*. London: Walter Scott.

West-Eberhard, M. J. 2003. *Developmental Plasticity and Evolution*. New York, NY: Oxford University Press.

Wolff, E. 1964. *L'origine de la lignée germinale chez les vertébrés et chez quelques groupes d'invertebrés*. Paris: Hermann.

Wolpert, L., Beddington, R., Jessell, T. *et al.* 2002. *Principles of Development*. Oxford: Oxford University Press.

Wolpert, L., Jessell, T., Lawrence, P. *et al.* 2007. *Principles of Development*. Bath: Oxford University Press.

Zwaka, T. P. & Thomson, J. A. 2005. A germ cell origin of embryonic stem cells? *Development* **132**, 227–233.

18

Thoughts and speculations on the ancestral arthropod segmentation pathway

ARIEL D. CHIPMAN

In the past decade or so, there has been a significant increase in the available data on the developmental mechanisms underlying the process of segmentation in a wide range of arthropod taxa. This large body of data makes it possible to attempt, albeit cautiously, a comparative analysis of the various aspects of the segmentation process, and to try to find which of its features and components may have been present in the arthropod common ancestor. A recent review (Peel *et al.* 2005) covers much of what is known about the diversity of segmentation processes in arthropods, although even at the time of this writing, less than a year later, there is already a substantial amount of newly published data not covered therein. My aim in this chapter is not to repeat the review and synthesis presented in Peel *et al.* (2005), but to build on it, adding the most recent data, and expand the discussion into the more speculative domain of evolutionary reconstructions. The reader is encouraged to refer to that review for more details of the currently available data and for a more complete bibliography.

When addressing a large-scale evolutionary question, such as that suggested in the title of this chapter, it is important to define the boundaries of the problem discussed. In this review, I will focus only on the mechanisms of trunk segmentation, ignoring the differentiation and segmentation of the head region, and the posterior unsegmented region. Although it is difficult to think of the trunk as a distinct compartment that stands on its own, and many of the relevant developmental processes are continuous between the trunk and the areas

Evolving Pathways: Key Themes in Evolutionary Developmental Biology, ed. Alessandro Minelli and Giuseppe Fusco. Published by Cambridge University Press.
© Cambridge University Press 2008.

immediately anterior and posterior to it, not enough is known about how these areas are defined in a range of arthropods. For the purpose of the discussion, trunk segmentation will be thought of as a continuing process, without any consideration of where it begins or ends.

The phylogenetic scope of the analysis will include the Euarthropoda only, excluding tardigrades and onychophorans. In fact, because of the almost complete lack of data on segmentation in pycnogonids, these are also excluded from the discussion, leaving us with an evolutionary reconstruction of the last common ancestor of the Cormogonida (arthropods excluding pycnogonids; Zrzavy *et al.* 1997, Dunlop and Arango 2005). For simplicity, I will continue talking about 'arthropods', even though I will, in fact, be covering a somewhat smaller group. For it to be possible to reconstruct the common ancestor of such a diverse clade, we need a sample that covers as much of its diversity as possible. There are good data about many species of insects; mostly, but not exclusively, holometabolous insects. Outside the insects, information is more limited, but segmentation has been studied extensively in the spider *Cupiennius salei*, as a representative chelicerate; and in two centipedes, *Strigamia maritima* and *Lithobius forficatus*, and a millipede *Glomeris marginata* representing the myriapods. Within the crustaceans most of the available data come from the Malacostraca, mainly *Parhyale hawaiiensis*, but also others.

As mentioned above, there is a wealth of data about segmentation mechanisms in diverse arthropods. One must be cautious when dealing with such a large amount of information to differentiate between characters that are specific to one system or taxon and characters that can be considered generalities and can be used for ancestral reconstruction. I will attempt to concentrate only on the latter, and try to avoid being bogged down with too many bits of specific information. It is worth pointing out that many arthropods do not generate all of the trunk segments during embryogenesis, but continue adding segments during larval stages, a phenomenon known as anamorphic development (see Fusco 2005 for a discussion). A further complication is that in a few cases, the segmentation process is not necessarily generating a single series of segments along the trunk, but a number of distinct serial structures as in the dissociation of dorsal and ventral segmentation in the pill millipede (Janssen *et al.* 2004). For simplicity, I will not consider these two exceptions, but will stay with the case of a single series of segmental structures formed as part of a single process. A final pitfall I hope to avoid is getting entrenched in the so-called '*Drosophila* paradigm'. Although I will use what is known about segmentation in *Drosophila melanogaster* as a reference, it will be made clear that *Drosophila* is but one example, and an unusual one at that, so no general conclusions can be drawn from information on *Drosophila* alone.

The reconstruction of the segmentation process in the arthropod common ancestor will use both available molecular data and cellular/morphological data. The latter data type is, in general, much poorer for diverse arthropods, and necessarily reconstruction of cellular and morphological processes relies more on deduction from basic principles and extrapolation from the little that is known. For molecular data, the approach I will take is essentially one of ancestral character reconstruction using parsimony, although the number of data points for any given molecular player or interaction is usually much too small for a formal or mathematical analysis. Of course, the ever-present spectre of convergence is a problem for parsimony-based reconstructions. Nonetheless, I will assume that the expression of homologous genes in comparable domains during similar processes is sufficient for assuming common ancestry. In many cases, homologous genes are apparently performing the same function in very different cellular environments, so gene expression patterns alone tell us little about cellular processes. Thus, molecular data and cellular/morphological data will be considered separately, to be joined only at the end when I go to the ancestral reconstruction itself.

Finally, what I will present is my own interpretation of the data. The data, despite their wealth, are still very partial, and are open to many alternative interpretations. Some readers may prefer different evolutionary scenarios, and I agree that many are possible. Nonetheless, the evolutionary scenario I will present is consistent with the data and addresses the key questions about the ancestral condition of the arthropod segmentation mechanism.

Making a segmented body involves defining a main body axis and generating a repeated pattern along that axis during ontogeny. The main source of information about segmentation in arthropods is the fruit fly *Drosophila melanogaster*. In *Drosophila*, the repeated pattern is generated by a progressive subdivision of the blastoderm – a thin layer of tissue surrounding a yolky egg, which will ultimately give rise to the embryo itself. This is possible in the *Drosophila* embryo because almost the entire anterior-posterior extent of the blastoderm will give rise to the segmented germ band (so-called 'long germ' embryos). In most arthropods, this is not the case, and the initial embryonic rudiment is much shorter than the ultimate extent of the germ band. In embryos with this kind of short rudiment (often referred to as 'short germ'), an additional step is required before a reiterated pattern can be generated, namely axial elongation, in which the initial short rudiment is extended to give the full length of the germ band. These three processes – axis determination, axial elongation (in non-long-germ embryos) and gener-ation of a repeated pattern – are central to the creation of segmented body plan during embryogenesis. I will start by going through these three and examining how they are manifested in diverse arthropod embryos. In what follows, I will use the term 'germ band' to refer to axially polarised tissue that has undergone a certain amount of differen-tiation and arrangement to distinguish it from the initial undifferen-tiated field of cells. The germ band includes both tissue that is overtly segmented, and tissue that is still unsegmented but may have already undergone some of the molecular processes that precede segmentation.

SETTING UP AN ANTERIOR–POSTERIOR AXIS

The anterior–posterior axis in the best-studied model, *Drosophila melano-gaster*, is set up by maternal determinants deposited in the egg during oogenesis. As with many *Drosophila* characters, it is difficult to extrap-olate from this to other arthropods, because the existence of nurse cells, which are responsible for the loading of these maternal determi-nants, is not universal within arthropods. However, many of the genes and gene products that are active in axis determination in *Drosophila* can be found in other arthropods as well. First and foremost among these is the homeobox-containing transcription factor Caudal. Expression of *caudal* in the posterior of early embryos has been found in all arthropods where it has been looked for (Peel *et al.* 2005, Olesnicky *et al.* 2006), and indeed in other metazoans as well (Epstein *et al.* 1997, Holland 2002, de Rosa *et al.* 2005, Shimizu *et al.* 2005). This posterior expression pattern, together with functional studies in *Drosophila*, suggests a role in determining the posterior pole. The widespread appearance of posterior *caudal* in arthropods and in outgroups indicates that it probably had a similar role in the arthropod common ancestor.

A second gene that may have a conserved ancestral role is *nanos*. In *Drosophila*, it is expressed posteriorly, and represses translation of the maternally deposited transcription factor *hunchback* (Irish *et al.* 1989). Homologues of *nanos* can be found in most metazoans, indicating that the gene itself existed in the arthropod ancestor. Most of what is known about its role is limited to insects. The translational repression of *hunchback* is conserved in dipterans (Curtis *et al.* 1995), as well as in grasshoppers (Lall *et al.* 2003) and possibly in the jewel wasp *Nasonia vitripennis* (Pultz *et al.* 2005). Nothing is known about *nanos* in other arthropods, but in the very distantly related Cnidaria, it also has a posterior expression pattern in the developing embryo (Torras *et al.* 2004, Torras and González-Crespo 2005). Another homeobox-containing transcription factor that is involved in axis formation in some insects is Orthodenticle. In *N. vitripennis*, *orthodenticle* patterns both the anterior and the posterior poles of the embryo (Lynch *et al.* 2006), whereas in the beetle *Tribolium castaneum* it has a role in defining the anterior pole (Schröder 2003). However, studies in non-insect arthropods do not support such a role outside of the insects (Browne *et al.* 2006), so its involvement in axis determination in an arthropod ancestor cannot be confirmed.

AXIAL ELONGATION

Drosophila melanogaster cannot provide any clues about the evolutionary history of axial elongation mechanisms in arthropods, since as a long-germ insect it already has the entire anterior–posterior extent of the germ band present at very early developmental stages. Looking at expression patterns and functional studies in other arthropods can give some hints as to the ancestral players in this process.

I have already mentioned *caudal* as a key player in the determination of the posterior pole during axial specification in many arthropods. In addition to this role, it also has a central role in axial elongation. Experimental knock-down of *caudal* expression results in a complete disruption of the segmentation process and a truncation of the growth of the embryo posterior to the gnathal segments (Copf *et al.* 2003, Shinmyo *et al.* 2005).

A similar phenotype to the *caudal* knock-down is seen when knocking down *even-skipped* in *Oncopeltus fasciatus* (Liu and Kaufman 2005a). In several insects *even-skipped* is expressed in a broad posterior domain (reviewed in Liu and Kaufman 2005b). In the brine shrimp *Artemia franciscana* it is also expressed broadly in the posterior (Copf *et al.* 2003), and in the centipede *Strigamia maritima* an *even-skipped* homologue, *eve2*, is one of a group of genes expressed very early in the segmentation process (A. Chipman and M. Akam, unpublished data). These results suggest an early role and possible involvement in axial elongation for *even-skipped*, a surprising suggestion, given that in *Drosophila*, *even-skipped* is generally thought of as a gene that is involved in the segmentation cascade at a relatively late stage. However, looking outside of arthropods, it has been suggested that *even-skipped* and *caudal* are jointly involved in axial elongation even in the annelid *Platynereis dumerilii* (de Rosa *et al.* 2005).

At the cellular level, there are two possible mechanisms of axial elongation. Elongation can be done either through the activity of a growth zone, in which cell proliferation contributes new tissue throughout the elongation process as seen in many malacostracan crustaceans (Scholtz *et al.* 1994, Wolff and Scholtz 2002), or through rearrangement and recruitment of existing tissue to the elongating germ band as seen in the centipede *Strigamia maritima* (Chipman *et al.* 2004a). These two possibilities represent two extremes of a continuum, and in most cases, axial elongation is probably achieved through an intermediate process incorporating contributions by both mechanisms. It should be pointed out that the details of axial elongation in most arthropods are very poorly known, and have not been studied extensively (Liu and Kaufman 2005b). Keller (2006) discusses, in a much wider phylogenetic scope, the many different types of cellular mechanisms that are involved in elongation processes in development. Although he does not give much information about arthropods, he provides possible clues to the types of processes we could look for in axial elongation. What seems clear is that in probably all cases, the elongation zone is subterminal, that is, there is a terminal zone, which remains constant and does not participate in the elongation process.

GENERATING A REPEATED PATTERN

Once again, the mechanism for generating a repeated pattern in *Drosophila melanogaster* cannot give us information about other arthropods, since in higher Diptera – unusually – all segments are formed almost simultaneously through a stepwise subdivision of the germ band. Generating a repeated pattern in an elongating embryo, as is more common in arthropods, can be accomplished in two ways, dependent in part on the mode of axial elongation. One possibility is that one end of the rudiment is made of a population of cells that is constantly proliferating. As new cells are formed, each is given an identity – based on the timing of its birth – that corresponds to a specific role or position within the forming segment, and thus a reiterated pattern is created simultaneously with the generation of new tissue. The alternative possibility is that tissue addition is independent of generating a repeated pattern, and the pattern is generated through a periodic input acting on unsegmented tissue that has already been recruited to the germ band. The first alternative is only possible in embryos where the process of elongation is tied to the generation of new cells, as seen in the extreme case of malacostracan crustaceans. Since this type of elongation is probably a malacostracan apomorphy (Scholtz *et al.* 1994, Scholtz 1998), the ancestral arthropod pattern is more likely to be one of patterning an undifferentiated population of cells through a periodic signal.

The models that best predict the appearance of a periodic pattern in an undifferentiated field of cells are models including a cellular oscillator or clock, in which cells oscillate between a series of cell states with a fixed periodicity. The cell states can be thought of as specific expression levels of a set of genes or their products. The first formulation of such a model was the clock and wavefront model (Cooke and Zeeman 1976), in which all the cells of the pre-segmented tissue oscillate with a linked phase (i.e. they are all simultaneously at the same phase of the cycle) at a relatively high frequency. A slow moving wave of cell states, or expression of a different set of genes and gene products, passes over the oscillating tissue in an anterior–posterior direction, with the front of the wave including a rapid change of cell state (i.e. a steep gradient of expression). Each cell in the pre-segmented tissue gets fixed in a specific state, depending on when it meets the wavefront. More recent models are based on a cell-autonomous oscillator, in which cells emerge from a growth zone or progress zone and continue oscillating between states. In this model the cells are not locked in the same phase, but each cell communicates with cells anterior and posterior to it, creating a travelling wave of oscillating states that moves in an anterior direction. These oscillations slow and eventually stop in a specific state based on the time since they emerged from the growth zone (Jaeger and Goodwin 2001).

The existence of a somitogenesis clock in vertebrates has been demonstrated by a number of workers on different vertebrate systems (Palmeirim *et al.* 1997, Pourquié 2003, Dubrulle and Pourquié 2004, Giudicelli and Lewis 2004), and vertebrate segmentation seems to conform to the aforementioned theoretical models (Aulehla and Herrmann 2004). The main components of the vertebrate clock are genes in the Notch signalling pathway that are expressed in a cycling fashion through a negative feedback loop (Collier *et al.* 1996, Rida *et al.* 2004).

The discovery that Notch and its ligand Delta are involved in segmentation in the spider *Cupiennius salei* (Stollewerk *et al.* 2003) set off a flurry of interest in the similarities between vertebrate somitogenesis and arthropod segmentation. Involvement of Notch pathway genes in segmentation has since been found in centipedes (Chipman *et al.* 2004b; A. Chipman and M. Akam, unpublished data) and in the cockroach *Periplaneta americana* (J.-P. Couso, unpublished data). Although several attempts have been made to find similar genes in other insects, there have been no additional conclusive data. In the case of the centipede *Strigamia maritima* the Notch-ligand gene *Delta* and the Notch target *odd-skipped* are expressed in what looks like an oscillating pattern of travelling waves (Chipman *et al.* 2004b; A. Chipman and M. Akam, unpublished data), as would be expected if the segmentation clock models were true for centipede development. There is no current evidence to indicate the existence of a gradient or wavefront that would interact with an oscillator in arthropod segmentation, nor a clear indication as to its identity. Jaeger and Goodwin's (2001) cellular oscillator model does not require a specific gradient, and in the case of *S. maritima* it may be sufficient for the oscillating signal to become fixed when it reaches the germ band after travelling through the undifferentiated posterior zone. However, if we were to speculate about such a gradient, a possible candidate would be *caudal*, which is present in a graded pattern at the correct place and time. Circumstantial support for this hypothesis is provided by the fact that *caudal* is expressed in periodic stripes at exactly the point where the oscillating pattern of *S. maritima* is fixed (Chipman *et al.* 2004b), possibly through some feedback from the oscillator.

TRANSLATING A REPEATED PATTERN INTO SEGMENTS

The segmented germ band is a highly conserved stage in arthropod development, and has even been called the arthropod 'phylotypic stage' (Raff 1996, Galis *et al.* 2002). This conservation of a morphological stage is also represented at the molecular level (Peel *et al.* 2005), and unlike earlier stages in the process, information from *Drosophila* segmentation is applicable to other arthropods as well. The involvement of a series of segment polarity genes, namely *engrailed, hedgehog, wingless* and others in generation of segmental boundaries (von Dassow *et al.* 2000, Larsen *et al.* 2003), is conserved in all arthropods where it has been studied (Peel *et al.* 2005). Indeed, *engrailed* is so ubiquitously conserved that it is the standard marker for segmental boundaries in almost all studies of arthropod segmentation. Directly upstream of the segment polarity genes are a group of genes that in *Drosophila* are referred to as pair-rule genes. This group includes *even-skipped, odd-skipped, hairy, runt* and several others. In *Drosophila*, they are initially expressed in a two-segment periodicity, which is then split to give the single segment periodicity of the segmented germ band. Homologues of pair-rule genes have also been found wherever they have been looked for within the arthropods (Peel *et al.* 2005, Choe *et al.* 2006). However, their expression in a two-segment periodicity is not universal. In most insects, at least some of these genes are involved in generating a two-segment repeat. In non-insect arthropods, with the exception of the geophilomorph centipede *S. maritima*, they are expressed segmentally (Peel *et al.* 2005). The highly conserved involvement of pair-rule gene homologues in the final stages of arthropod segmentation suggests they may represent the primary output of the oscillator. Furthermore, in many arthropods there is a functional division within the pair-rule genes between primary and secondary genes (Coulter and Wieschaus 1988, Damen *et al.* 2005, Choe and Brown 2007), in which the primary pair-rule genes are upstream of the secondary ones, and hence possibly the immediate output of the oscillator. The exact members of each of these subgroups vary from species to species. It may be that such a functional division existed in the common ancestor of arthropods, but that individual genes have moved between primary and secondary roles with relative ease throughout evolution (Choe and Brown 2007).

WHAT ABOUT GAP GENES?

Readers who are familiar with *Drosophila* segmentation will have noticed the conspicuous absence of gap genes in my discussion up to this point. In the conceptual sequence of events leading to a segmented body plan, such as I have outlined in this chapter, the gap gene phase, which forms the crucial early pattern in *Drosophila* segmentation, is not necessary. However, gap gene homologues have been found in insects other than *Drosophila* and they are claimed to have a role in the generation of the segmented body plan (Peel *et al.* 2005). The changing role of gap genes in different insects has been discussed in Peel *et al.* (2005), and I will not repeat that discussion here. Suffice it to say that the evidence suggests that the gap genes in sequentially segmenting insects are not involved in segment generation per se, but rather in generating segmental identity. The only data on gap gene homologues outside of insects come from the centipede *S. maritima*, and for *hunchback* only, from the brine shrimp *Artemia franciscana*. In the centipede, two of the gap genes, *Krüppel* and *hunchback*, are suggested to have a role in neural precursor identity, long after the segments have formed (Chipman and Stollewerk 2006) – a role that is conserved in *Drosophila* as well (Isshiki *et al.* 2001). Both of these genes, as well as *knirps* and *giant*, have no relevant expression during segmentation stages (A. Chipman and M. Akam, unpublished results). In the brine shrimp, *hunchback* is expressed in already segmented mesoderm (as is also seen transiently in the centipede) but has no gap-like pattern, or obvious role in the segmentation process (Kontarakis *et al.* 2006). With this rather sketchy evidence, it is difficult to draw firm conclusions about the ancestral role of gap genes. I would suggest, however, that whatever this role was, gap genes were not significant players in the sequential segmentation process itself, although gap genes could have had a role in patterning the anterior part of the embryo, including the head segments.

RECONSTRUCTING THE ANCESTOR

Having pointed out the conserved aspects in each of the stages of the segmentation process in arthropods, I now move on to reconstructing a series of hypothetical segmentation events that may be similar to the process in the common ancestor of all arthropods (Figure 18.1).

Virtually nothing is known about what embryos of early arthropods looked like, and the fossil record has been silent on this question to date. Many aspects of the earliest phases of the segmentation process are dependent on the size, shape and yolk content of the egg and on the extent of the embryonic rudiment relative to the egg. Leaving these considerations aside, for lack of information, I will stay with the most simplistic generalities based on what we do know.

Initially, the embryo would include a uniform, unpatterned field of cells. One end of the field would have to be slightly different, either through an inherent asymmetry in the egg, or because of the point of sperm entry, or by random gravitational orientation. This end would form one of the poles of the anterior–posterior axis. The posterior pole would be defined by the expression of *caudal* and possibly *nanos*, as is probably the case in all extant arthropods. The target of one or both of these genes might be *hunchback*, which may have been initially distributed uniformly.

Once the anterior–posterior axis had been set up, a germ band would form along this axis. At the time of the beginning of axial elongation, it is likely that a head or head rudiment would already be in existence, but as mentioned at the beginning of the chapter, I will not discuss how this could have been accomplished. Tissue would be recruited to the posterior of the embryo by cell rearrangements with cells moving from a pool of undifferentiated tissue, which would be

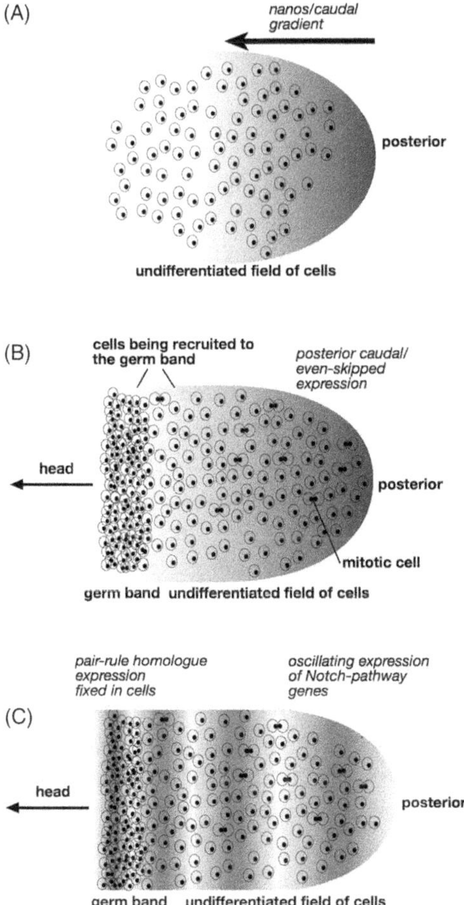

Figure 18.1 A schematic representation of the three main stages in the segmentation process of the hypothetical arthropod common ancestor. Cells are represented as ovals with a dot. Undifferentiated cells are drawn as larger than germ-band cells, but this is only for illustrative purposes and is not meant to imply that such a difference existed. Shades of grey represent gene expression levels. Captions referring to cellular events or domains are in bold while captions referring to molecular events are in italics. A, A gradient of *nanos* and/or *caudal* with a higher concentration at the posterior defines the posterior pole in an undifferentiated field of sparsely packed cells. B, The germ band has formed in the anterior (left) side of the embryo, and cells from the undifferentiated field are recruited for its elongation. Cell divisions are scattered throughout the undifferentiated area. Expression of *caudal* and/or *even-skipped* in the posterior is involved in the elongation of the germ band. C, Oscillating expression of *Notch* pathway genes and their downstream targets moves through the undifferentiated field, with the progress of the wave slowing as it moves towards the germ band, and eventually becoming fixed in the germ band.

replenished by unordered cell proliferation. The movement of cells and their addition to the growing germ band would be orchestrated by the activity of Caudal and Even-skipped. Exactly how these two would interact and what their targets would be remains unclear. It is possible that one or both of these would maintain posterior cells in an undifferentiated state, and as cells moved out of the control of these genes they would gain germ-band properties.

As the germ band is extending a series of genes would start expression in an oscillating pattern that overlays the process of germ-band extension. The primary oscillation would probably be through a Notch-Delta mediated negative feedback loop, with several other genes and gene-products oscillating with these two at different phases. These additional genes could either be direct components of the clock or immediate targets of some component of the clock. It is likely that the focus of the oscillating pattern would be in the very posterior of the embryo, which does not participate in the elongation process. The cycling expression patterns would move as a wave through the undifferentiated tissue, slowing as they entered the germ band and becoming fixed at different phases in different cells. The different phases of the cycle would be manifested by different combinations of pair-rule gene homologues – some activated directly by the main components of the Notch pathway and others secondarily by the earlier group of pair-rule homologues. The question of whether each cycle of the Notch-Delta oscillator and the downstream pair-rule homologues would represent one or two segments in the segmented germ band cannot be answered conclusively, since examples of both are found in different arthropod classes. I tend towards a single-segment periodicity as the output of the initial cycler, since a two-segment periodicity seems to be a derived feature, appearing convergently in specialised groups (the holometabolous insects and geophilomorph centipedes).

The combinations of pair-rule homologues would be read by segment polarity genes, such as *engrailed* and *wingless*, the products of which would activate the cellular components involved in morphological differentiation of the segments and the establishment of segmental boundaries.

Although the process has been described as including a series of discrete and separable steps, in reality all these steps are linked and difficult to tease apart. This is true both in the hypothetical ancestor and in real extant arthropods. Axis specification is closely tied to axial elongation (as is evident by the involvement of *caudal* in both processes). Axial elongation is closely linked to the generation of a repeated pattern.

Translating a repeated pattern into segments is a gradual process and not a series of leaps from stage to stage. Nonetheless, these are conceptually different process, and could probably vary independently throughout evolution.

It may be true that my description of the hypothetical arthropod ancestor reflects my own personal biases. The process presented above bears striking similarities to the segmentation process in the geophilomorph centipede *Strigamia maritima*. My favourite animal has much to recommend it, and was chosen as a research organism for unrelated reasons (Arthur and Chipman 2005). However, it turns out to have many features that make it useful for speculations about ancestry. The simplicity of its overall morphology – a large number of homonomous trunk segments, which are all generated during embryogenesis – makes it possible to draw generalities from the specifics of *S. maritima* development. Still, the general reconstruction I have presented is based on a wide comparison, and not just from my own work on centipede development. No feature of *S. maritima* development can be assumed (or has been assumed) to be ancestral, without corroborating comparisons with distantly related arthropods.

CONCLUSIONS

The ancestral reconstruction I have presented above might seem just an entertaining exercise in scenario building. However, it is more than that. First, by setting out the available data and building upon them, it clarifies exactly where there are gaps in our knowledge and points out interesting and potentially useful avenues of research. Second, it provides a testable hypothesis about what features in the arthropod segmentation process can be deemed as generalities. As the evo-devo community expands and we learn more about the development of different arthropod taxa, this scenario will be corrected and refined to give a more reliable picture of how the distant ancestors of the arthropods made segments.

ACKNOWLEDGEMENTS

This chapter has benefited from discussions over the past few years with many people working on arthropod segmentation. I would like to thank Nipam Patel, Diethard Tautz, Ernst Wimmer, Giuseppe Fusco, Graham Budd, Nick Monk, Juan-Pablo Couso and Nigel Hughes for sharing ideas and preliminary data. I am indebted to the members of the Museum Molecular Lab at the University Museum of Zoology,

Cambridge, UK, for creating a stimulating research environment and for challenging every piece of data or idea I have presented there. Specifically I would like to thank the laboratory head, Michael Akam, and my former colleagues, Johannes (Yogi) Jaeger, Andrew Peel and Joakim Eriksson. Wallace Arthur has been a collaborator on all the *Strigamia* work and always open to discussion and exchange of ideas. Andrew Peel and Johannes Jaeger commented on an earlier version of the manuscript. My work was funded by a grant from the BBSRC.

REFERENCES

Arthur, W. & Chipman, A. D. 2005. The centipede *Strigamia maritima*: what it can tell us about the development and evolution of segmentation. *BioEssays* **27**, 653–660.

Aulehla, A. & Herrmann, B. G. 2004. Segmentation in vertebrates: clock and gradient finally joined. *Genes & Development* **18**, 2060–2067.

Browne, W. E., Schmid, B. G. M., Wimmer, E. A. & Martindale, M. Q. 2006. Expression of *otd* orthologs in the amphipod crustacean, *Parhyale hawaiensis*. *Development Genes & Evolution* **216**, 581–595.

Chipman, A. D., Arthur, W. & Akam, M. 2004a. Early development and segment formation in the centipede *Strigamia maritima* (Geophilomorpha). *Evolution & Development* **6**, 78–89.

Chipman, A. D., Arthur, W. & Akam, M. 2004b. A double segment periodicity underlies segment generation in centipede development. *Current Biology* **14**, 1250–1255.

Chipman, A. D. & Stollewerk, A. 2006. Specification of neural precursor identity in the geophilomorph centipede *Strigamia maritima*. *Developmental Biology* **290**, 337–350.

Choe, C. P. & Brown, S. J. 2007. Evolutionary flexibility of pair-rule patterning revealed by functional analysis of secondary pair-rule genes, *paired* and *sloppy-paired* in the short-germ insect, *Tribolium castaneum*. *Developmental Biology* **302**, 281–294.

Choe, C. P., Miller, S. C. & Brown, S. J. 2006. A pair-rule gene circuit defines segments sequentially in the short-germ insect *Tribolium castaneum*. *Proceedings of the National Academy of Sciences of the USA* **103**, 6560–6564.

Collier, J. R., Monk, N. A. M., Maini, P. K. & Lewis, J. H. 1996. Pattern formation by lateral inhibition with feedback: a mathematical model of Delta-Notch intercellular signalling. *Journal of Theoretical Biology* **183**, 429–446.

Cooke, J. & Zeeman, E. C. 1976. Clock and wavefront model for control of number of repeated structures during animal morphogenesis. *Journal of Theoretical Biology* **58**, 455–476.

Copf, T., Rabet, N., Celniker, S. E. & Averof, M. 2003. Posterior patterning genes and the identification of a unique body region in the brine shrimp *Artemia franciscana*. *Development* **130**, 5915–5927.

Coulter, D. E. & Wieschaus, E. 1988. Gene activities and segmental patterning in *Drosophila*: analysis of *odd-skipped* and pair-rule double mutants. *Genes & Development* **2**, 1812–1823.

Curtis, D., Apfeld, J. & Lehmann, R. 1995. *nanos* is an evolutionarily conserved organizer of anterior–posterior polarity. *Development* **121**, 1899–1910.

Damen, W. G., Janssen, R. & Prpic, N. M. 2005. Pair rule gene orthologs in spider segmentation. *Evolution & Development* **7**, 618–628.

de Rosa, R., Prud'homme, B. & Balavoine, G. 2005. *caudal* and *even-skipped* in the annelid *Platynereis dumerilii* and the ancestry of posterior growth. *Evolution & Development* **7**, 574–587.

Dubrulle, J. & Pourquié, O. 2004. Coupling segmentation to axis formation. *Development* **131**, 5783–5793.

Dunlop, J. A. & Arango, C. P. 2005. Pycnogonid affinities: a review. *Journal of Zoological Systematics and Evolutionary Research* **43**, 8–21.

Epstein, M., Pillemer, G., Yelin, R., Yisraeli, J. K. & Fainsod, A. 1997. Patterning of the embryo along the anterior–posterior axis: the role of the *caudal* genes. *Development* **124**, 3805–3814.

Fusco, G. 2005. Trunk segment numbers and sequential segmentation in myriapods. *Evolution & Development* **7**, 608–617.

Galis, F., van Dooren, T. J. M. & Metz, J. A. 2002. Conservation of the segmented germband stage: robustness or pleiotropy? *Trends in Genetics* **18**, 504–509.

Giudicelli, F. & Lewis, J. 2004. The vertebrate segmentation clock. *Current Opinion in Genetics and Development* **14**, 407–414.

Holland, L. Z. 2002. Heads or tails? Amphioxus and the evolution of anterior–posterior patterning in deuterostomes. *Developmental Biology* **241**, 209–228.

Irish, V., Lehmann, R. & Akam, M. 1989. The *Drosophila* posterior-group gene *nanos* functions by repressing *hunchback* activity. *Nature* **338**, 646–648.

Isshiki, T., Pearson, B., Holbrook, S. & Doe, C. Q. 2001. *Drosophila* neuroblasts sequentially express transcription factors which specify the temporal identity of their neuronal progeny. *Cell* **106**, 511–521.

Jaeger, J. & Goodwin, B. C. 2001. A cellular oscillator model for periodic pattern formation. *Journal of Theoretical Biology* **213**, 171–181.

Janssen, R., Prpic, N. M. & Damen, W. G. M. 2004. Gene expression suggests decoupled dorsal and ventral segmentation in the millipede *Glomeris marginata* (Myriapoda: Diplopoda). *Developmental Biology* **268**, 89–104.

Keller, R. 2006. Mechanisms of elongation in embryogenesis. *Development* **133**, 2291–2302.

Kontarakis, Z., Copf, T. & Averof, M. 2006. Expression of *hunchback* during trunk segmentation in the branchiopod crustacean *Artemia franciscana*. *Development Genes & Evolution* **216**, 89–93.

Lall, S., Ludwig, M. Z. & Patel, N. H. 2003. *Nanos* plays a conserved role in axial patterning outside of the diptera. *Current Biology* **13**, 224–229.

Larsen, C. W., Hirst, E., Alexandre, C. & Vincent, J.-P. 2003. Segment boundary formation in *Drosophila* embryos. *Development* **130**, 5625–5635.

Liu, P. Z. & Kaufman, T. C. 2005a. *even-skipped* is not a pair-rule gene but has segmental and gap-like functions in *Oncopeltus fasciatus*, an intermediate germ-band insect. *Development* **132**, 2081–2092.

Liu, P. Z. & Kaufman, T. C. 2005b. Short and long germ segmentation: unanswered questions in the evolution of a developmental mode. *Evolution & Development* **7**, 629–646.

Lynch, J. A., Brent, A. E., Leaf, D. S., Pultz, M. A. & Desplan, C. 2006. Localized maternal *orthodenticle* patterns anterior and posterior in the long germ wasp *Nasonia*. *Nature* **439**, 728–732.

Olesnicky, E. C., Brent, A. E., Tonnes, L. *et al.* 2006. A *caudal* mRNA gradient controls posterior development in the wasp *Nasonia*. *Development* **133**, 3973–3982.

Palmeirim, I., Henrique, D., Ish-Horowicz, D. & Pourquié, O. 1997. Avian *hairy* gene expression identifies a molecular clock linked to vertebrate segmentation and somitogenesis. *Cell* **91**, 639–648.

Peel, A. D., Chipman, A. D. & Akam, M. 2005. Arthropod segmentation: Beyond the *Drosophila* paradigm. *Nature Reviews Genetics* **6**, 905–916.

Pourquié, O. 2003. The segmentation clock: converting embryonic time into spatial pattern. *Science* **301**, 328–330.

Pultz, M. A., Westendorf, L., Gale, S. D., *et al.* 2005. A major role for zygotic *hunchback* in patterning the *Nasonia* embryo. *Development* **132**, 3705–3715.

Raff, R. A. 1996. *The Shape of Life*. Chicago: The University of Chicago Press.

Rida, P. C., Le Minh, N. & Jiang, Y. J. 2004. A Notch feeling of somite segmentation and beyond. *Developmental Biology* **265**, 2–22.

Scholtz, G. 1998. Cleavage, germ band formation and head segmentation: the ground pattern of the Euarthropoda. In R. A. Fortey & R. A. Thomas (eds.) *Arthropod Relationships*. London: Chapman & Hall, pp. 317–332.

Scholtz, G., Patel, N. H. & Dohle, W. 1994. Serially homologous *engrailed* stripes are generated via different cell lineages in the germ band of amphipod crustaceans (Malacostraca, Peracarida). *International Journal of Developmental Biology* **38**, 471–478.

Schröder, R. 2003. The genes *orthodenticle* and *hunchback* substitute for *bicoid* in the beetle *Tribolium*. *Nature* **422**, 621–625.

Shimizu, T., Bae, Y. K., Muraoka, O. & Hibi, M. 2005. Interaction of *Wnt* and *caudal*-related genes in zebrafish posterior body formation. *Developmental Biology* **279**, 125–141.

Shinmyo, Y., Mito, T., Matsushita, T. *et al.* 2005. *caudal* is required for gnathal and thoracic patterning and for posterior elongation in the intermediate-germ-band cricket *Gryllus bimaculatus*. *Mechanisms of Development* **122**, 231–239.

Stollewerk, A., Schoppmeier, M. & Damen, W. G. M. 2003. Involvement of *Notch* and *Delta* genes in spider segmentation. *Nature* **423**, 863–865.

Torras, R. & González-Crespo, S. 2005. Posterior expression of *nanos* orthologs during embryonic and larval development of the anthozoan *Nematostella vectensis*. *International Journal of Developmental Biology* **49**, 895–899.

Torras, R., Yanze, N., Schmid, V. & Gonzalez-Crespo, S. 2004. *nanos* expression at the embryonic posterior pole and the medusa phase in the hydrozoan *Podocoryne carnea*. *Evolution & Development* **6**, 362–371.

von Dassow, G., Meir, E., Munro, E. M. & Odell, G. M. 2000. The segment polarity network is a robust developmental module. *Nature* **406**, 188–192.

Wolff, C. & Scholtz, G. 2002. Cell lineage, axis formation, and the origin of germ layers in the amphipod crustacean *Orchestia cavimana*. *Developmental Biology* **250**, 44–58.

Zrzavy, J., Hypsa, V. & Vlásková, M. 1998. Arthropod phylogeny: taxonomic congruence, total evidence and conditional combination approaches to morphological and molecular data sets. In R. A. Fortey & R. H. Thomas (eds.) *Arthropod Relationships*. London. Chapman & Hall, pp. 97–107.

19

Evolution of neurogenesis in arthropods

ANGELIKA STOLLEWERK

Several alternative hypotheses have been suggested that support various phylogenetic groupings of the individual euarthropod taxa, the chelicerates, myriapods, crustaceans and insects. The Tetraconata hypothesis suggests a sister-group relationship of insects and crustaceans in contrast to the traditional monophyletic grouping of myriapods and insects, the Tracheata or Atelocerata (see references in Stollewerk and Chipman 2006). The Mandibulata hypothesis suggests a clade consisting of insects, crustaceans and myriapods (see references in Harzsch *et al.* 2005). However, the relationships within this clade are being debated since this hypothesis excludes neither the Pancrustacea nor the Tracheata concept. The latest hypothesis suggests a sister-group relationship of chelicerates and myriapods. Although this theory was initially based on molecular phylogenetic analysis (Friedrich and Tautz 1995, Hwang *et al.* 2001, Kusche and Burmester 2001, Nardi *et al.* 2003, Mallatt *et al.* 2004, Pisani *et al.* 2004), recent morphological and molecular data on neurogenesis in these groups potentially support a close relationship (Stollewerk *et al.* 2001, 2003, Dove and Stollewerk 2003, Kadner and Stollewerk 2004, Stollewerk and Simpson 2005, Chipman and Stollewerk 2006, Stollewerk and Chipman 2006). On top of the various ideas on the relationships within the euarthropods, none of the groups except the insects is generally accepted as monophyletic.

New insights into the evolutionary relationships between the different taxa have been gained by comparing morphological features and expression patterns of genes involved in developmental processes. Besides the analyses of the expression domains of segmentation genes and Hox genes (see references in Stollewerk *et al.* 2001), morphological comparison of neurogenesis in insects and crustaceans has confirmed the molecular evidence of a sister-group relationship between these two groups (see references in Schachtner *et al.* 2005, Harzsch 2006, Strausfeld *et al.* 2006a,b). Furthermore, neurogenesis in myriapods is more similar to chelicerates than to insects and thus the data support a Myriochelata clade and contradict the Tracheata hypothesis (Dove and Stollewerk 2003, Kadner and Stollewerk 2004, Chipman and Stollewerk 2006, Pioro and Stollewerk 2006, Stollewerk and Chipman 2006). However, without knowing the ancestral state of neurogenesis (and any other developmental process for that matter) a correct interpretation of the molecular and morphological data is not possible. Since analysis of outgroups to the euarthropods has turned out to be difficult and time-consuming because of the microscopic size (tardigrades) or almost year-long embryogenesis (onychophorans), a thorough analysis of many representatives of each arthropod group can be used as an alternative method of reconstructing the ancestral pattern of developmental processes. It can be assumed that characters conserved in all arthropod groups are ancestral and thus were present in the last common ancestor. Characters that can only be found in subsets of arthropod groups can consecutively be analysed in outgroups to the arthropods to see if they are synapomorphies or rather reflect the ancestral pattern. With this approach, studies in difficult outgroups can be kept to a minimum since we have to focus only on specific questions using already established methods.

Neurogenesis is a perfect system to study since the complexity of this developmental process provides a pool of various characters that have to correspond in detail to be judged as homologous, reducing the risk of assessing superficial similarities as homologies. Here I re-evaluate published data on early neurogenesis in all arthropod groups to uncover ancestral and possibly derived homologies and speculate on the sequence of evolutionary changes that might have led to the different modes of neurogenesis in arthropods.

NEURAL PRECURSOR FORMATION: THE MORPHOLOGICAL PROCESSES

Individual stem-cell-like neuroblasts are formed in the ventral neuroectoderm of insects and crustaceans

Our detailed knowledge of the generation of neurons in insects is mainly based on studies of neurogenesis in *Drosophila melanogaster* and *Locusta*

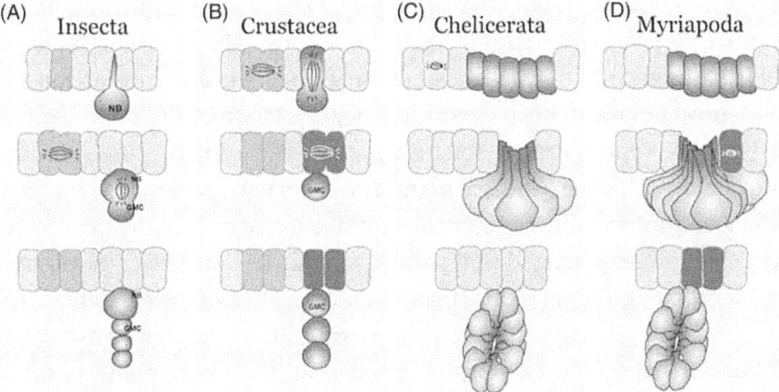

Figure 19.1 Comparison of the different modes of neurogenesis in the four euarthropod groups. Neuroectodermal cells are represented in light grey, neural precursors in multi-shaded grey, epidermal cells in middle grey. A, In insects, single neuroblasts (NBs) are recruited from the ventral neuroectoderm. Within minutes after their specification, the neuroblasts delaminate from the neuroectoderm into the interior of the embryo and divide to give rise to ganglion mother cells (GMC). Those cells that remain apical after five waves of neuroblast formation will give rise to epidermoblasts (middle grey) and divide in the plane of the neuroectoderm. B, In crustaceans, neuroblasts do not delaminate. After rotation of the mitotic spindle, the neuroblasts bud off ganglion mother cells into the interior of the embryo. Crustacean neuroblasts can switch from producing ganglion mother cells to generating epidermal cells (dark grey). C, In chelicerates the central region of the neuroectoderm gives rise exclusively to neural cells. Groups of mainly postmitotic neural precursors form an invagination site that persists in the neuroectoderm for several days. After formation of all invagination sites the groups eventually detach from the apical layer. D, A similar mode of neurogenesis is seen in myriapods, although in contrast to chelicerates single mitotic cells (dark grey) and groups of mitotic cells (not shown) are associated with forming invagination sites.

migratoria (Bate 1976, Bate and Grunewald 1981, Hartenstein and Campos-Ortega 1984). Individual stem cells – so-called neuroblasts – delaminate from a single-layered neuroectoderm to the interior of the embryo in five subsequent waves (Figure 19.1A, top). In this basal (interior) position, they divide asymmetrically to self-renew and to produce smaller ganglion mother cells that divide once to give rise to two neural cells (i.e. neurons or glia; Figure 19.1A). About 500 neuroblasts are generated in the ventral neuroectoderm forming a highly stereotyped temporal and spatial pattern (Hartenstein and

Campos-Ortega 1984, Goodman and Doe 1993). The cells remaining in the apical cell layer give rise to epidermal cells. The decision between epidermal and neural fate depends on direct cell–cell interactions of the ventral neuroectodermal cells (see below). This mode of neurogenesis seems to be representative for insects since consecutive studies on the flour beetle *Tribolium castaneum* and the silverfish *Ctenolepisma longicaudata* have confirmed the presence of stem-cell-like cells in the neuroectoderm which are arranged in a pattern similar to *Drosophila melanogaster* and *Locusta migratoria* (Truman and Ball 1998, Wheeler *et al.* 2003).

Neuroblasts have also been described in higher crustaceans (Malacostraca) and in two Branchiopoda, but their origin and position in the developing neuromeres is different from that of insects (see references in Whitington 2004, Stollewerk and Simpson 2005). In malacostracan crustaceans, amphipods excepted, neuroblasts arise from stereotyped divisions of ectoteloblasts. These are stem-cell-like cells that are located in the posterior region of the germ band anterior to the proctodeum. The asymmetric divisions of ectoteloblasts produce transverse rows of stereotyped individually identifiable cells that form the grid-like pattern of the postnaupliar germ band. These cells lack the typical morphology of neuroectodermal cells. They divide several times until the first neuroblasts can be identified by their typical mode of mitotic division. In contrast to insects, crustacean neuroblasts do not delaminate from the outer cell layer into the embryo but remain apical and divide perpendicular to the surface so that the daughter cells are pushed into the embryo (Figure 19.1B, top). After several rounds of division a dorso-ventral column of ganglion mother cells is visible (Dohle and Scholtz 1988; Figure 19.1B, bottom). Similar to insects, the ganglion mother cells divide once to give rise to two neural cells. In contrast to insects, crustacean neuroblasts can switch from the production of ganglion mother cells to the generation of epidermal cells (Figure 19.1B, middle). Interestingly, despite these differences the neuroblast pattern is similar in crustaceans and insects. In both groups, 25 to 30 neuroblasts are arranged in seven rows in each hemisegment (Dohle and Scholtz 1988, Scholtz 1992, Ungerer 2006).

Groups of neural precursors are recruited from the ventral neuroectoderm of chelicerates and myriapods

In a few classical accounts, neuroblasts have been described in three chelicerate species, but it is possible that the data were partly misinterpreted because of technical limitations at the time (Yoshikura 1955, Mathew 1956, Winter 1980). Apart from these studies, the literature suggests that neurogenesis occurs by a generalised inward proliferation of neuroectodermal cells to produce paired segmental thickenings both in chelicerates and myriapods (Anderson 1973). In some arachnids, the amblypygids and araneids, each neuromere is supposed to be formed by a number of invaginations (Weygoldt 1985). Groups of cells divide and form small clusters that invaginate. Recent analysis of neurogenesis in three chelicerates (spiders: *Cupiennius salei*, *Pholcus phalangioides*; Xiphosura: *Limulus polyphemus*) and four myriapods (diplopods: *Glomeris marginata*, *Archispirostreptus* sp.; chilopods: *Lithobius forficatus*, *Strigamia maritima*) showed that neural precursor formation is indeed significantly different from that of insects and crustaceans (Stollewerk *et al.* 2001, 2003, Dove and Stollewerk 2003, Kadner and Stollewerk 2004, Chipman and Stollewerk 2006, Pioro and Stollewerk 2006, Stollewerk and Chipman 2006). (Note the term 'neural precursor' refers to cells that are committed to the neural fate but have not yet developed into neurons and glial cells.) In the ventral neuroectoderm of chelicerates and myriapods, groups of precursors are specified for the neural fate (Figure 19.1C, D; Figure 19.2). The precursor groups form invagination sites at stereotyped positions which persist in the neuroectoderm for several days (Figure 19.1C, D, middle). After invagination, most of the neural precursors do not divide but directly differentiate into neurons and glia.

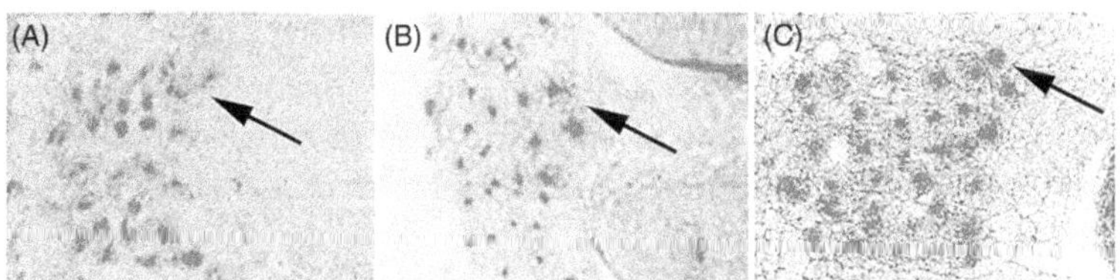

Figure 19.2 Comparison of the expression pattern of invaginating neural precursor groups in single hemisegments of three arthropod species. Confocal micrographs of embryos stained with phalloidin-rhodamine. Anterior is towards the top, the midline towards the left. The arrangement of invagination sites is similar in the chilopod *Lithobius forficatus* (A) the spider *Cupiennius salei* (B) and the diplopod *Glomeris marginata* (C). The arrows point to two lateral invagination sites that are located at similar positions in all three species.

Differences in the timing of neural precursor formation in individual chelicerate and myriapod species

Although neurogenesis follows the same general pattern in chelicerates and myriapods, differences in the timing of formation of invagination sites were observed that might coincide with distinct modes of embryogenesis in the individual species. In the spider *Cupiennius salei* the same numbers of invaginating cell groups arise simultaneously in the prosomal segments. Neural precursor groups are formed in four waves generating 5 to 13 invagination sites each. In the opisthosoma, invagination sites are formed in an anterior to posterior gradient, since new opisthosomal segments are generated by the posterior growth zone during the course of neurogenesis. In a similar way, in the diplopod *Glomeris marginata*, the same number of invaginating cell groups arises simultaneously in the five head segments and the first three leg segments, while the invagination sites are formed in an anterior to posterior gradient in the remaining leg segments. Similar to the spider, four waves of neural precursor group formation have been described. In the chilopod *Lithobius forficatus* and the diplopod *Archispirostreptus* sp., neurogenesis is less synchronised in the trunk segments as compared with *Glomeris marginata*. The formation of neural precursors seems to occur consecutively in each trunk segment in *Archispirostreptus* sp., while no more than two trunk segments show the same pattern of invagination sites in *Lithobius forficatus*. Although embryogenesis takes about the same time in these myriapods and the spider, neurogenesis lasts only two days in *Lithobius forficatus* and *Archispirostreptus* sp., while in the spider and *Glomeris marginata* this process takes five days to complete. The different timing of neural precursor formation might therefore be an adaptation to the acceleration of neurogenesis in these species. A distinct pattern of neurogenesis is also seen in the geophilomorph centipede *Strigamia maritima*. In contrast to *Lithobius forficatus* and *Glomeris marginata*, *Strigamia maritima* undergoes epimorphic development. Myriapods showing this kind of development generate all segments during embryogenesis, whereas in *Lithobius forficatus* and *Glomeris marginata* further segments are added during posthatching larval stages. In *Strigamia maritima*, about 50 segments are generated during embryogenesis and differentiate in quick succession. Expression studies and morphological analyses showed that each trunk segment exhibits a different differentiation state along the anterior–posterior axis during neurogenesis. In contrast to the other myriapods and the spider, neural precursor groups arise one-by-one in each hemisegment, rather than in several waves. This distinct mode of neural precursor formation might be an adaptation to the independent initiation of neurogenesis in each segment.

However, despite these differences the final pattern of invagination sites is strikingly similar in all chelicerate and myriapod species analysed. In each species about 30 invagination sites per hemisegment are arranged in a regular pattern of seven rows consisting of four to six invagination sites each.

Differences in the number and morphology of invaginating neural precursors

While the pattern of invagination sites is conserved between different chelicerate and myriapod species, the morphology of the invagination sites and the number of neural precursors forming an invagination group is not consistent. In the diplopod *Glomeris marginata* and the chilopod *Strigamia maritima* up to 12 cells contribute to an individual invagination site, while in the spider *Cupiennius salei* and in the chilopod *Lithobius forficatus* the invaginating neural precursor groups consist of only five to nine cells. Furthermore, in *Strigamia maritima*, the cell processes of the neural precursors are not attached to the apical surface but to a single cell of the precursor group. Initially the invagination groups form three rows per hemisegment which are rearranged to a final pattern of seven rows during the convergent extension movements that lead to an extension of the germ band. It can be speculated that the specific morphology of the invagination groups is necessary for the rearrangement of the neural precursor groups from three rows to seven rows during these medio-lateral movements in *Strigamia maritima*.

Neural stem cells comparable to insect and crustacean neuroblasts are missing in chelicerates and myriapods

Analysis of the mitotic pattern in the spider *Cupiennius salei* and the myriapod *Glomeris marginata* suggests that the neural precursors of the invaginating cell groups are not comparable to insect and crustacean neuroblasts. In contrast to the literature that suggests a connection between cell proliferation and invagination (see above), dividing cell groups or single mitotic cells, which prefigure regions where invagination sites arise, could not be detected in the neuroectoderm of the spider *Cupiennius salei*. In addition, most mitotic divisions occur in the apical cell layer. The neuroectodermal cells divide in the plane of the apical surface so that the daughter cells remain in the surface layer and are not pushed into the embryo. Since there are no cell divisions in the invaginating cell groups, it can be concluded that mainly postmitotic neuroectodermal cells are recruited for the neural fate. These results contrast with those for insects, since neuroblasts do not divide until they delaminate into the embryo and produce ganglion mother cells by asymmetric cell divisions (Hartenstein and Campos-Ortega 1984). They are also in contrast to crustaceans, since crustacean neuroblasts divide asymmetrically similar to insects, although without delamination. The absence of asymmetric cell divisions in the spider was confirmed by analysing the expression pattern of the neural cell fate determinant Prospero (Weller and Tautz 2003). In *Drosophila*, Prospero is asymmetrically distributed into ganglion mother cells during neuroblast division (Doe *et al.* 1991), while in the spider Prospero is equally distributed to both daughter cells in the few neural precursors that divide after invagination.

Studies of neurogenesis in different representatives of all myriapod groups have in most cases failed to reveal neural stem-cell-like cells with the characteristics of insect and crustacean neuroblasts (Heymons 1901, Tiegs 1940, 1947, Dohle 1964, Whitington *et al.* 1991). Knoll (1974) proposed that neuroblasts are present in the ventral neuroectoderm of the chilopod *Scutigera coleoptrata* generating vertical columns of neurons – a mode of neural precursor formation similar to the crustacean pattern. However, the cells that Knoll identified as neuroblasts are only insignificantly larger than the neural cells in the basal cell layers. Analysis of the mitotic pattern in the ventral neuroectoderm of *Glomeris marginata* revealed that single dividing cells are associated with invaginating neural precursors. Furthermore, groups of dividing cells seem to prefigure the regions where invagination sites arise. The single dividing cells are significantly larger in size than the surrounding cells. This pattern can be interpreted in two ways. The large dividing cells might be neural stem cells that divide asymmetrically in the plane of the neuroectoderm to produce a group of neural precursor cells that subsequently invaginates. On the other hand, the presence of groups of dividing cells suggests that a single cell divides giving rise to two daughter cells which divide again and so forth, until a group of about 12 cells is generated. However, it would be possible to distinguish between these two scenarios by dye labelling of individual mitotic cells. If the progeny of these cells give rise to clones of about 12 cells which subsequently invaginate, we can assume that neural stem cells are present in the ventral neuroectoderm of *Glomeris marginata*.

The ventral neuroectoderm of chelicerates and myriapods is comparable to the neural plate of vertebrates

Neurogenesis in chelicerates and myriapods shows an additional distinct feature compared with insects and crustaceans. In chelicerates and myriapods, the central region of the ventral neuroectoderm generates exclusively neural cells, while in the remaining arthropods both neural and epidermal cells arise from the ventral neurogenic region. This mode of neurogenesis is actually more similar to vertebrates. During primary neurulation in vertebrates the ectoderm becomes divided into the internally positioned neural plate, which will form the brain and the spinal cord and the externally positioned region from which the epidermis of the skin will arise. A similar division of the ectoderm into a medial neurogenic region and lateral epidermal precursors is visible in the ventral neuroectoderm of chelicerates and myriapods. In addition, most cell divisions occur in the apical neuroectoderm, while the neural precursors exit the cell cycle and differentiate in deeper cell layers. This mode of neurogenesis is also more similar to vertebrates than to insects and crustaceans. These data suggest that neurogenesis in chelicerates and myriapods reflects the ancestral pattern, while the formation of the nervous system in insects and crustaceans is derived.

The morphological processes of neural precursor formation

To summarise, several characters have been described in chelicerates and myriapods that cannot be found in equivalent form in the remaining arthropods. (1) Groups of neural precursors invaginate from the ventral neuroectoderm of chelicerates and myriapods, while single neuroblasts are specified in crustaceans and insects. (2) In contrast to insects and crustaceans, mainly postmitotic neural precursors are recruited for the neural fate. (3) The central region of the ventral neuroectoderm in chelicerates and myriapods generates exclusively neural cells, while in insects and crustaceans both neural and epidermal cells arise from the ventral neurogenic region. Despite these differences, the pattern of neural precursor groups/neuroblasts is strikingly similar in all arthropod groups. In all species analysed, about 30 neuroblasts/ neural precursor groups per hemisegment are arranged in seven transverse rows with four to six neural precursor groups/neuroblasts each indicating that this pattern is a conserved character of neurogenesis in arthropods.

Proneural and neurogenic genes are essential for the specification of neuroblasts

In *Drosophila*, early neurogenesis is controlled by proneural genes that encode transcription factors with a basic domain necessary for DNA binding and two helices that allow for the formation of heterodimers with other basic helix-loop-helix (bHLH) proteins (see references in Pioro and Stollewerk 2006). The proneural genes belong to two major subfamilies, the *achaete-scute* group and the *atonal* group. In the ventral neuroectoderm of *Drosophila*, members of the Achaete-Scute Complex (AS-C; *achaete*, *scute* and *lethal of scute*) are expressed in a stereotyped, partially overlapping pattern and are necessary for neuroblast formation. In loss of function mutants fewer neuroblasts are generated. Proneural proteins can only bind DNA as heterodimers with the ubiquitously expressed bHLH protein Daughterless. By recruiting proneural proteins to autoregulatory enhancer elements, Daughterless assists in an up-regulation of proneural gene expression in the precursor cell which is essential for activation of the neural programme.

Despite their function in selecting neural precursors, proneural genes specify neuronal subtype identity indicating that these genes activate both a common neural programme and neuronal subtype-specific target genes. In the *Drosophila* PNS, *achaete* and *scute* specify external sensory organ identity, while *atonal* mainly specifies chordotonal organ identity. Prior to delamination of the neuroblasts, the proneural genes are expressed in clusters of cells in the ventral neuroectoderm (Figure 19.3A). Because of the activity of a second group of genes, the neurogenic genes, the expression of the proneural genes becomes restricted to a single cell of the cluster, the future neuroblast. This process is called lateral inhibition and is mediated by the transmembrane proteins Notch and Delta. Binding of the ligand Delta to the Notch receptor eventually leads to the activation of the *Enhancer of split* gene complex. The gene products of this complex repress proneural gene expression. Since production of the ligand Delta is positively regulated by the proneural genes, activation of the Notch signalling pathway leads to a down-regulation of Delta. Because of this feedback loop that takes place between the cells of a proneural cluster, a slightly elevated level of proneural gene expression in one cell of the cluster, the future neuroblast, leads to a repression of proneural gene expression in the neighbouring cells. Mutations affecting the process of lateral inhibition lead to an overproduction of neurons – a neurogenic phenotype.

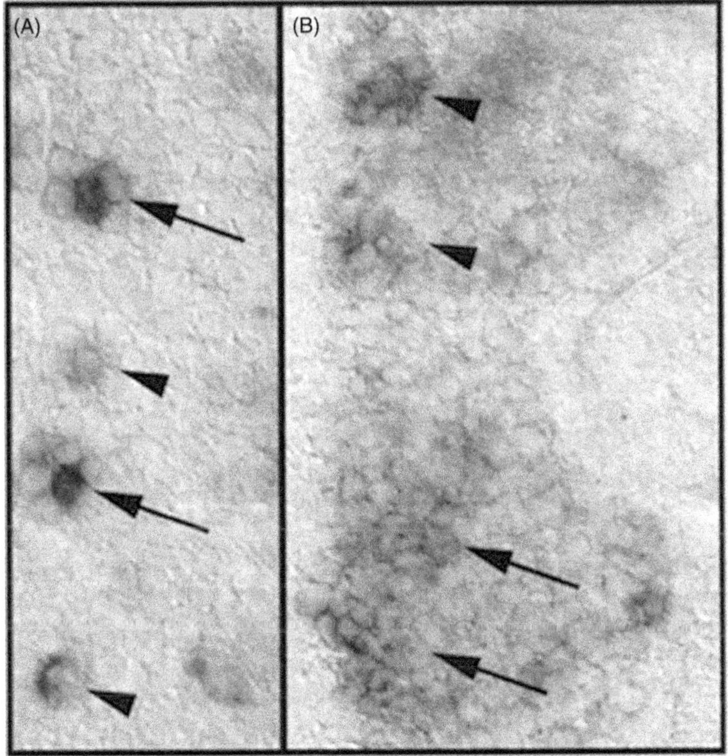

Figure 19.3 Expression of proneural genes in the insect *Drosophila melano-gaster* and the spider *Cupiennius salei*. Flat preparations of embryos stained with a DIG-labelled *Drosophila melanogaster achaete* probe (A) and a DIG-labelled *Cupiennius salei ASH1* probe (B). The midline is towards the left. A, In *Drosophila melanogaster*, the proneural gene *achaete* is first expressed in groups of cells (arrows). Because of lateral inhibition, expression becomes up-regulated in single cells (arrowheads) and down-regulated in the remaining cells of the proneural cluster. B, In the spider *Cupiennius salei*, *CsASH1* is expressed in fields of cells at the beginning of neurogenesis (arrows). In contrast to *Drosophila*, proneural gene expression becomes restricted to groups of neural precursors (arrowheads).

Within the insect group, proneural genes have been identified in several dipterans, a butterfly and the flour beetle *Tribolium castaneum* (Galant *et al.* 1998, Wülbeck and Simpson 2000, 2002, Pistillo *et al.* 2002, Skaer *et al.* 2002, Wheeler *et al.* 2003). A proneural function of the *achaete-scute* homologues in these dipteran species has been suggested by analysis of their expression in the peripheral nervous system. The butterfly *achaete-scute* homologue shows a restricted expression in small proneural clusters in the embryonic ventral neuroectoderm that cannot account for the formation of all neural precursors. However, the expression of the gene is down-regulated to single cells indicating a similar mode of neurogenesis to that in the remaining insects. The *Tribolium achaete-scute* homologue (*TcASH*) is expressed in proneural clusters in the ventral neuroectoderm and becomes restricted to single neuroblasts, in a pattern similar to *Drosophila*. Recently, two *achaete-scute* homologues have been identified in the branchiopod crustacean, *Triops longicaudatus*. *Tl-ASH1* seems to be expressed like a proneural gene, while *Tl-ASH2* is exclusively expressed in neuroblasts.

The function of achaete-scute homologues in neural precursor specification seems to be conserved in chelicerates and myriapods

Two *achaete-scute* homologues have been identified in the spider *Cupiennius salei* (Stollewerk *et al.* 2001), both of which are exclusively expressed in the developing central and peripheral nervous system. *CsASH1* shows a proneural expression pattern (Figure 19.3B). Transcripts prefigure the regions where invagination sites arise at each wave of neural precursor formation. However, in contrast to insect and crustacean proneural genes, expression of *CsASH1* becomes restricted not to single cells but to groups of neural precursors. The second *achaete-scute* homologue *CsASH2* is exclusively expressed in the invaginating neural precursor groups. The expression pattern of *CsASH2* can be compared with that of the fourth member of the *Drosophila* Achaete-Scute Complex, *asense*, which is exclusively expressed in neuroblasts. Analyses of *CsASH1* and *CsASH2* function during neurogenesis by RNA mediated interference revealed that *CsASH1* is required for the recruitment of all neural precursors, while *CsASH2* has a later function during differentiation.

One *achaete-scute* homologue each has been identified in the diplopod *Glomeris marginata* and the chilopod *Lithobius forficatus*. Both homologues show a proneural mode of expression and accumulate at higher levels in the neural precursors that are going to invaginate.

Furthermore, a member of the Atonal family has been identified in *Glomeris marginata* (Pioro and Stollewerk 2006). *Gm Atonal*, like its *Drosophila* homologue, is expressed exclusively in the PNS. However, in contrast to *Drosophila*, *Gm Atonal* seems to be expressed in external sensory organs. The expression domains at the tip of the appendages correspond to the positions of developing chemosensory organs (cone sensilla). *Gm Atonal* is not expressed in the sensory precursors of the body wall, lateral to the limb buds, while *GmASH* transcripts accumulate in this area. Similarly, the spider *achaete-scute* homologues *CsASH1* and *CsASH2* are partially expressed in non-overlapping domains in the developing PNS. These data suggest that the proneural genes of chelicerates and myriapods are involved in specification of neuronal subtype identity, similar to the *Drosophila* homologues.

In addition the identification of a *daughterless* homologue in *Glomeris marginata* suggests conserved interactions of the genetic network involved in neural precursor specification (Pioro and Stollewerk 2006). Similar to *Drosophila*, the heterodimerisation partner of the proneural genes is expressed ubiquitously during development, which might account for a function in a large number of developmental processes. However, in contrast to *Drosophila*, a higher accumulation of *daughterless* transcripts seems to correlate with regions where neural precursor groups form in the central and peripheral nervous system of *Glomeris marginata*. It has been speculated that up-regulation of *daughterless* in a proneural field might refine the precise position of proneural clusters in concert with the proneural genes (Pioro and Stollewerk 2006). The heterogenous expression of *daughterless* in the central and peripheral nervous system of *Glomeris marginata* supports this model.

Neurogenic genes mediate lateral inhibition in chelicerates and myriapods

Although groups of neural precursors, rather than single neuroblasts, are recruited for the neural fate from the ventral neuroectoderm of chelicerates and myriapods, the regular pattern and sequential generation of the invagination sites suggested that neurogenic genes might restrict the proportion of cells that arise at each wave of neural precursor formation. One *Notch* and two *Delta* homologues were identified in the spider *Cupiennius salei* (Stollewerk 2002). While *CsDelta1* is exclusively expressed in the invaginating cell groups, *CsDelta2* is expressed in all ventral neuroectodermal cells but shows a higher expression in the invaginating cells. Like *CsDelta2*, *CsNotch* transcripts are distributed over the entire ventral neuroectoderm, although there is heterogeneity in the expression level. *CsNotch* transcripts seem to accumulate at higher levels in the invagination sites after their formation suggesting a dual function of Notch in neural precursor formation and invagination. A similar pattern of expression of the single *Delta* and *Notch* homologues is seen in the ventral neuroectoderm of the myriapods *Glomeris marginata* and *Lithobius forficatus* (Dove and Stollewerk 2003, Kadner and Stollewerk

2004). Functional analysis by RNA mediated interference confirmed that the neurogenic genes of the spider mediate lateral inhibition. This is an interesting result, since the current model for singling out neural precursors from a group of initially equivalent cells via the Delta/Notch signalling pathway only applies to single cells rather than to groups of cells. The data suggest that the genetic interactions of components of the Notch signalling pathway must have changed during evolution to permit generation of single neuroblasts in insects and crustaceans on the one hand, but recruitment of groups of neural precursors in chelicerates and myriapods on the other hand. In this context, the expression pattern of the single *Delta* homologue in *Strigamia maritima* shows an interesting expression pattern (Chipman and Stollewerk 2006, Stollewerk and Chipman 2006). *StmDelta* seems to be expressed at higher levels in single cells of the neural precursor groups. However, it is also possible that *Delta* transcripts accumulate around individual cells, since the cell processes of all cells of an invagination group are attached to a single cell of the group (see above). In any case, the data suggest that individual cells of the precursor group are distinct. Although the whole precursor group will eventually invaginate and give rise to neural cells, Delta/Notch signalling might generate single cells with distinct properties within the precursor groups. These cells might have an important function during convergent extension movements in keeping individual cell groups together (see above). Therefore, neurogenesis in *Strigamia maritima* might represent an intermediate state between recruitment of groups of neural precursors in the remaining myriapods and chelicerates and singling out of individual neuroblasts in insects and crustaceans.

In the ventral neuroectoderm of *Drosophila melanogaster*, the decision between epidermal and neural fate depends on Delta/Notch signalling. Cells that eventually remain apical give rise to epidermis, while delaminating cells become neural precursors. Although this process has not been analysed in insects other than *Drosophila*, it can be assumed that Notch signalling is used in a similar way within this arthropod group, since the mode of neurogenesis is similar in all insects analysed. While in insects neuroblasts are singled out by cell–cell interactions between initially equivalent cells, neuroblasts of malacostracan crustaceans arise from stereotyped cell lineages. In addition, neuroblasts can switch from the production of ganglion mother cells to the production of epidermal precursors, indicating that the choice between two cell fates occurs within a single cell lineage rather than groups of equivalent cells as in *Drosophila*. This raises the question of whether

Notch signalling is required at all for the decision between epidermal versus neural fate in crustaceans. However, neurogenic genes have not been identified in crustaceans up to now. In chelicerates and myriapods, the ventral neurogenic region gives rise exclusively to neural cells (see above). The epidermal cells are derived from lateral regions of the neuroectoderm and overgrow the neurogenic region only after formation of all neural precursors. Therefore, Notch signalling is merely involved in the timing of neural precursor formation in the neurogenic region of chelicerates and myriapods, rather than in the decision between epidermal and neural fate.

Pan-neural genes switch on a common neural programme

Once neural precursors are selected, a group of genes referred to as pan-neural genes, such as *hunchback*, *deadpan* and *snail*, is expressed in most or all neuroblasts in *Drosophila*. These genes are either involved in asymmetric cell division or are part of a common neural programme and promote neural differentiation (see references in Stollewerk *et al.* 2003). In the *Drosophila* ventral neuroectoderm, two members of the Snail zinc finger family, *snail* and *worniu*, have a pan-neural mode of expression. Together with the third Snail family gene, *escargot*, they have partially redundant functions in the formation of the CNS and the mesoderm. Triple mutants show severe defects in the development of both the mesoderm and the nervous system. In these mutants, the neural determinants Prospero and Numb are no longer asymmetrically segregated into GMCs upon neuroblast division and the generation of GMCs is disrupted.

One *snail* homologue each has been identified in the myriapod *Glomeris marginata* and the spider *Cupiennius salei* which are both expressed in most or all neural precursor cells of the CNS, similar to *Drosophila* (Stollewerk *et al.* 2003, Pioro and Stollewerk 2006). In contrast to the spider, *Glomeris marginata snail* transcripts can be detected in the whole neuroectoderm at the beginning of neurogenesis and accumulate in groups of cells prior to formation of invagination sites. This expression is comparable to *Drosophila* where *snail* is expressed similar to the AS-C genes in proneural clusters in the ventral neuroectoderm. During specification of the neuroblasts, *snail* transcripts become restricted to all, or most, neural precursor cells. While the spider homologue is also expressed in the PNS, similar to *Drosophila*, *GmSnail* expression is restricted to the CNS. Additionally, *GmSnail* shows a strong expression in the ventral midline, which has been observed neither in *Drosophila* nor in the spider. It has been suggested that *Gm snail* is involved in cell shape changes in the ventral midline, since in *Drosophila* Snail induces cell shape changes in the wing imaginal disc and during ventral furrow formation. In addition, Snail might have a similar function in the ventral neuroectoderm of chelicerates and myriapods. In the *Drosophila* neuroectoderm, the *snail* genes are necessary for the asymmetric distribution of the cell fate determinants Prospero and Numb to ganglion mother cells. Since most of the neural precursors in chelicerates and myriapods do not proliferate after their specification and Prospero is not asymmetrically distributed to daughter cells (at least in the spider), Snail must have a different function in the invaginating neural precursors. The spider and millipede *snail* homologues might be involved in the maintenance of the cell shape changes that occur during formation of the invagination sites in the ventral neuroectoderm.

Within the arthropods, the expression pattern of the pan-neural protein Prospero has only been analysed in *Drosophila* and in the spider *Cupiennius salei*. In *Drosophila* Prospero is asymmetrically localised to the basal membrane of the neuroblasts. During mitosis Prospero is exclusively distributed into one daughter cell, the ganglion mother cell, where it translocates from the cytoplasm into the nucleus. It has been shown that Prospero inhibits expression of multiple cell cycle regulatory genes in ganglion mother cells entering their final mitotic division. In the spider Prospero is expressed in the nuclei of neural precursors which are located basally within the invagination groups (Weller and Tautz 2003). Most of these precursors do not divide after invagination but differentiate into neurons and glial cells. These data suggest a conserved role of Prospero in neural cell fate determination in the spider and the fly.

Generation of neural precursor diversity

It has been shown in the insect *Drosophila melanogaster* that once the neural precursors are selected they divide in a unique and invariant pattern generating a stereotyped sequential series of ganglion mother cells (GMC) (Doe 1992). Each GMC divides once to give rise to two neural cells. Neural precursor diversity in *Drosophila* is achieved by both spatial and temporal patterning mechanisms. During neurogenesis segment polarity and dorso-ventral patterning genes subdivide the ventral neuroectoderm into a grid-like structure (reviewed by Skeath 1999). Each proneural cluster thus expresses a unique set of genes giving rise to neuroblasts with spatial heterogeneity. The spatial cues change over time so that the identities of neuroblasts also correlate with their time of formation (Berger *et al.* 2001). After delamination from the ventral neuroectoderm, neuroblasts become independent of spatial patterning cues. Subsequently, temporal patterning mechanisms generate additional diversity among the cell lineages of individual neuroblasts (see references in Chipman and Stollewerk 2006). Temporal identity in neuroblasts is regulated by sequential expression of Hunchback, Krüppel, Pdm and Castor. The temporal expression profile is maintained in the progeny of the neuroblasts leading to expression of transcription factors in mutually exclusive cell layers in the ventral neuromeres. *Hunchback* is expressed in early-born neurons that are located in the deepest layer, while *Krüppel* is expressed at low levels in the *Hunchback* layer and in a distinct layer between *Hunchback* and *Pdm*. *Castor* transcripts accumulate in the late-born superficial layer neurons.

There are few comparative studies of the events that generate neural precursor diversity, following the recruitment of neural precursors, during early development of the ventral nerve cord in the different arthropod groups, and those studies are incomplete (Stollewerk and Simpson 2005). However, from the limited data available, the expression of the segment polarity genes does appear to have been conserved in arthropods. These genes are expressed during neurogenesis suggesting an additional (or even primary) function in neural precursor identity. Studies on the segment polarity gene *engrailed* have indeed revealed that this gene is specifically expressed in neuroblasts/neural precursor groups of rows 1, 6 and 7 in all arthropod groups (for references see Stollewerk and Chipman 2006). Within the arthropods, the expression pattern and function of the dorso-ventral patterning genes *ventral nerve cord defective*, *intermediate nerve cord defective* and *muscle segment homeodomain* have only been studied in *Drosophila melanogaster* and *Tribolium castaneum* (Skeath 1999, Wheeler *et al.* 2005). The overall expression of these genes in three longitudinal columns seems to be conserved, although slight differences in the spatiotemporal pattern were observed between the species.

However, it is obvious that spatial information from segment polarity genes and dorso-ventral patterning genes alone cannot account for the high complexity of cell types in the nervous system of arthropods. In *Drosophila* temporal identity genes like *hunchback*, *Krüppel*, *Pdm* and *castor* generate diversity within individual neuroblast lineages. But temporal identity mechanisms of the sort used by *Drosophila* cannot operate in a similar way in the remaining arthropod groups. In malacostracan crustaceans, neuroblasts do not delaminate and thus cannot escape the spatial cues of the neuroectoderm to initiate an independent temporal program. Furthermore, in chelicerates and myriapods stem-cell-like neuroblasts are absent and neural precursors are mainly postmitotic after invagination. Although this process has not been analysed in crustaceans, recent studies in chelicerates and myriapods suggest that the time-dependent expression of neural identity genes such as *hunchback* and *Krüppel* within the ventral neuroectoderm might generate additional diversity of neural precursor groups (Stollewerk *et al.* 2003, Chipman and Stollewerk 2006).

To summarise, although the components of the genetic network involved in specification of neural precursors are conserved in arthropods, the function of some of the genes might have changed, leading to different outcomes in the individual groups.

THE ANCESTRAL PATTERN OF NEUROGENESIS IN ARTHROPODS

The presented data indicate that some morphological and molecular aspects of neurogenesis are conserved in all arthropods and thus might have been present in their last common ancestor.

The ground pattern of neurogenesis in arthropods seems to be the successive formation of about 30 neuroblasts/groups of neural precursors that are arranged in seven rows in each hemisegment. The genetic network that controls the specification and identity of neural precursors is conserved, although the function of the genes is adapted to the specific modes of neurogenesis in the individual arthropod groups. One possible reason for the stereotyped arrangement of neural precursors is the connection of neural precursor identity with spatial cues that confer anterior–posterior and dorso-ventral identities within a segment. The neural precursor pattern might have been constrained along with the general patterning mechanisms. Hence, we would expect a similar pattern of neural precursors in all animals that show expression of segment polarity and dorso-ventral patterning genes comparable to the euarthropods. However, this theory has to be tested in the future.

How has neurogenesis evolved in arthropods? It is tempting to speculate that the ancestral pattern of neurogenesis is the formation of groups of neural precursors, since this mode of neurogenesis is present in the basal arthropod groups. Neurogenesis in myriapods might be the crosslink between formation of groups of neural precursors and single neuroblasts. In contrast to chelicerates, neural precursor formation in myriapods seems to be associated with a proliferation pattern that indicates a clonal relationship of the precursor groups. Furthermore, although in *Strigamia maritima* groups of cells are specified for the neural fate, one cell of the group is different since all the remaining cells of the group are attached to this cell during the convergent extension movements that lead to an elongation of the germ band. The next step in evolution could have been the appearance of single neuroblasts in the lineage leading to insects and crustaceans. However, another possibility is that the modifications in the mode of neurogenesis in the individual arthropod groups do not reflect the actual sequence of evolution but are merely adaptations to the specific modes of embryogenesis in the individual species.

Indeed, differences in neurogenesis seem to coincide with different modes of embryogenesis in arthropods. For example, the different timing and order of neural precursor formation in *Strigamia maritima* and *Archispirostreptus* sp. as compared with the spider and the remaining myriapods might be an adaptation to the acceleration of neurogenesis relative to segment formation in these species. Similarly, differences in the number of neural precursors in a group and the initial arrangement of invagination sites seem to coincide with distinct morphologies of the neuroectoderm. The neuroectoderm of the diplopods and the geophilomorph centipede analysed seems to consist of many more cells than that of the spider and *Lithobius forficatus*. Correspondingly, up to 12 cells contribute to an individual invagination site in the diplopods and *Strigamia maritima*, while in the spider and *Lithobius forficatus* only five to nine cells were counted. Furthermore, in *Strigamia*, neural precursor groups are initially arranged in three rows. The invagination sites become rearranged to a pattern similar to the remaining myriapods and the spider during an expansion of the germ band along the longitudinal axis. This process does not take place during neural precursor formation in the spider and the remaining myriapod species analysed, and thus the arrangement of the invagination sites remains the same throughout neurogenesis.

To summarise, if we disregard the adjustments to the specific morphologies of embryogenesis in the individual species, the ground pattern of neurogenesis in arthropods seems to be the successive formation of about 30 neuroblasts/groups of neural precursors that are arranged in seven rows in each hemisegment. However, analyses of neurogenesis in outgroups to the euarthropods are necessary to confirm this assumption and to show whether the formation of groups of neural precursors, rather than the generation of individual neuroblasts, is the plesiomorphic state for this phylum.

ACKNOWLEDGEMENTS

I am grateful to Christian Berger for critical comments on the manuscript and preparing the schematic drawings in Figure 19.1. Thanks to Janina Seibert for the *Drosophila achaete* in situ hybridisation.

REFERENCES

Anderson, D. 1973. *Embryology and Phylogeny in Annelids and Arthropods*. Oxford: Pergamon Press.

Bate, M. 1976. Embryogenesis of an insect nervous system. I. A map of thoracic and abdominal neuroblasts in *Locusta migratoria*. *Journal of Embryology and Experimental Morphology* **35**, 107–123.

Bate, M. & Grunewald, E. B. 1981. Embryogenesis of an insect nervous system II. A class of neuron precursor cells and the origin of the intersegmental connectives. *Journal of Embryology and Experimental Morphology* **61**, 317–330.

Berger, C., Urban, J. & Technau, G. M. 2001. Stage-specific inductive signals in the *Drosophila* neuroectoderm control the temporal sequence of neuroblast specification. *Development* **128**, 3243–3251.

Chipman, A. D. & Stollewerk, A. 2006. Specification of neural precursor identity in the geophilomorph centipede *Strigamia maritima*. *Developmental Biology* **290**, 337–350.

Doe, C. Q. 1992. Molecular markers for identified neuroblasts and ganglion mother cells in the *Drosophila* central nervous system. *Development* **116**, 855–863.

Doe, C. Q., Chu-LaGraff, Q., Wright, D. M. & Scott, M. P. 1991. The *prospero* gene specifies cell fates in the *Drosophila* central nervous system. *Cell* **65**, 451–464.

Dohle, W. 1964. Die Embryonalentwicklung von *Glomeris marginata* (Villers) im Vergleich zur Entwicklung anderer Diplopoden. *Zoologische Jahrbücher, Abteilung für Anatomie* **81**, 241–310.

Dohle, W. & Scholtz, G. 1988. Clonal analysis of the crustacean segment: the discordance between genealogical and segmental borders. *Development* **104** (supplement), 147–160.

Dove, H. & Stollewerk, A. 2003. Comparative analysis of neurogenesis in the myriapod *Glomeris marginata* (Diplopoda) suggests more similarities to chelicerates than to insects. *Development* **130**, 2161–2171.

Friedrich, M. & Tautz, D. 1995. Ribosomal DNA phylogeny of the major extant arthropod classes and the evolution of myriapods. *Nature* **376**, 165–167.

Galant, R., Skeath, J. B., Paddock, S., Lewis, D. L. & Carroll, S. B. 1998. Expression pattern of a butterfly *achaete-scute* homolog reveals the homology of butterfly wing scales and insect sensory bristles. *Current Biology* **8**, 807–813.

Goodman, C. S. & Doe, C. Q. 1993. Embryonic development of the *Drosophila* central nervous system. In M. Bate & A. Martinez-Arias (eds.) *The Development of Drosophila melanogaster*. New York: Cold Spring Harbor Laboratory Press, pp. 1131–1206.

Hartenstein, V. & Campos-Ortega, J. A. 1984. Early neurogenesis in wildtype *Drosophila melanogster*. *Roux's Archive of Developmental Biology* **193**, 308–325.

Harzsch, S. 2006. Neurophylogeny: architecture of the nervous system and a fresh view on arthropod phylogeny. *Integrative and Comparative Biology* **46**, 162–194.

Harzsch, S., Müller, C. H. G. & Wolf, H. 2005. From variable to constant cell numbers: cellular characteristics of the arthropod nervous system argue against a sister-group relationship of Chelicerata and 'Myriapoda' but favour the Mandibulata concept. *Development Genes & Evolution* **215**, 53–68.

Heymons, R. 1901. Die Entwicklungsgeschichte der Scolopender. *Zoologica* **13**, 1–224.

Hwang, U. W., Friedrich, M., Tautz, D., Park, C. J. & Kim, W. 2001. Mitochondrial protein phylogeny joins myriapods with chelicerates. *Nature* **413**, 154–157.

Kadner, D. & Stollewerk, A. 2004. Neurogenesis in the chilopod *Lithobius forficatus* suggests more similarities to chelicerates than to insects. *Development Genes & Evolution* **214**, 367–379.

Knoll, H. J. 1974. Untersuchungen zur Entwicklungsgeschichte von *Scutigera coleoptrata* L. (Chilopoda). *Zoologische Jahrbücher, Abteilung für Anatomie* **92**, 47–132.

Kusche, K. & Burmester, T. 2001. Diplopod hemocyanin sequence and the phylogenetic position of the Myriapoda. *Molecular Biology and Evolution* **18**, 1566–1573.

Mallatt, J. M., Garey, J. R. & Shultz, J. W. 2004. Ecdysozoan phylogeny and Bayesian inference: first use of nearly complete 28S and 18S rRNA gene sequences to classify the arthropods and their kin. *Molecular Phylogenetics and Evolution* **31**, 178–191.

Mathew, A. P. 1956. Embryology of *Heterometrus scaber*. *Bulletin of the Center of Research of the Institute (University of Travancore)* **1**, 1–96.

Nardi, F., Spinsanti, G., Boore, J. L. *et al.* 2003. Hexapod origins: monophyletic or paraphyletic? *Science* **299**, 1887–1889.

Pioro, H. L. & Stollewerk, A. 2006. The expression pattern of genes involved in early neurogenesis suggests distinct and conserved functions in the diplopod *Glomeris marginata*. *Development Genes & Evolution* **216**, 417–430.

Pisani, D., Poling, L. L., Lyons-Weiler, M. & Hedges, S. B. 2004. The colonization of land by animals: molecular phylogeny and divergence times among arthropods. *BMC Biology* **2**, 1.

Pistillo, D., Skaer, N. & Simpson, P. 2002. *scute* expression in *Calliphora vicina* reveals an ancestral pattern of longitudinal stripes on the thorax of higher Diptera. *Development* **129**, 563–572.

Schachtner, J., Schmidt, M. & Homberg, U. 2005. Organization and evolutionary trends of primary olfactory centers in Tetraconata (Crustacea + Hexapoda). *Arthropod Structure & Development* **34**, 257–299.

Scholtz, G. 1992. Cell lineage studies in the crayfish *Cherax destructor* (Crustacea, Decapoda): germ band formation, segmentation and early neurogenesis. *Roux's Archive of Developmental Biology* **202**, 36–48.

Skaer, N., Pistillo, D. & Simpson, P. 2002. Transcriptional heterochrony of *scute* and changes in bristle pattern between two closely related species of blowfly. *Developmental Biology* **252**, 31–45.

Skeath, J. B. 1999. At the nexus between pattern formation and cell-type specification: the generation of individual neuroblast fates in the *Drosophila* embryonic central nervous system. *BioEssays* **21**, 922–931.

Stollewerk, A. 2002. Recruitment of cell groups through Delta/Notch signalling during spider neurogenesis. *Development* **129**, 5339–5348.

Stollewerk, A. & Chipman, A. D. 2006. Neurogenesis in chelicerates and myriapods and its importance for understanding arthropod relationships. *Integrative and Comparative Biology* **46**, 195–206.

Stollewerk, A. & Simpson, P. 2005. Evolution of early development of the nervous system: a comparison between arthropods. *BioEssays* **27**, 874–883.

Stollewerk, A., Tautz, D. & Weller, M. 2003. Neurogenesis in the spider: new insights from comparative analysis of morphological processes and gene expression patterns. *Arthropod Structure & Development* **32**, 5–16.

Stollewerk, A., Weller, M. & Tautz, D. 2001. Neurogenesis in the spider *Cupiennius salei*. *Development* **128**, 2673–2688.

Strausfeld, N. J., Strausfeld, C., Stowe, S., Rowell, D. & Loesel, R. 2006a. The organization and evolutionary implications of neuropils and their neurons in the brain of the onychophoran *Euperipatoides rowelli*. *Arthropod Structure & Development* **35**, 169–196.

Strausfeld, N. J., Strausfeld, C., Stowe, S., Rowell, D. & Loesel, R. 2006b. Arthropod phylogeny: onychophoran brain organization suggests an archaic relationship with a chelicerate stem lineage. *Proceedings of the Royal Society of London Biological Science* **273**, 1857–1866.

Tiegs, O. S. 1940. The embryology and affinities of the Symphyla based on the study of *Hanseniella agilis*. *Quarterly Journal of Microscopic Science* **82**, 1–225.

Tiegs, O. S. 1947. The development and affinities of the Pauropoda based on the study of *Pauropus sylvaticus*. *Quarterly Journal of Microscopic Science* **82**, 165–267 and 276–336.

Truman, J. W. & Ball, E. E. 1998. Patterns of embryonic neurogenesis in a primitive wingless insect, the silverfish, *Ctenolepisma longicaudata*: comparison with those seen in flying insects. *Development Genes & Evolution* **208**, 357–368.

Ungerer, P. 2006. *Aspekte der Neurogenese von* Orchestia cavimana (Amphipoda, Malacostraca, Crustacea). Unpublished Ph.D. thesis, Humboldt-University of Berlin.

Weller, M. & Tautz, D. 2003. *Prospero* and *Snail* expression during spider neurogenesis. *Development Genes & Evolution* **213**, 554–566.

Weygoldt, P. 1985. Ontogeny of the arachnid central nervous system. In F. G. Barth (ed.) *Neurobiology of Arachnids*. Berlin: Springer Verlag, pp. 20–37.

Wheeler, S. R., Carrico, M. L., Wilson, B. A., Brown, S. J. & Skeath, J. B. 2003. The expression and function of the *achaete-scute* genes in *Tribolium castaneum* reveals conservation and variation in neural pattern formation and cell fate specification. *Development* **130**, 4373–4381.

Wheeler, S. R., Carrico, M. L., Wilson, B. A. & Skeath, J. B. 2005. The *Tribolium* columnar genes reveal conservation and plasticity in neural precursor patterning along the embryonic dorsal-ventral axis. *Developmental Biology* **279**, 491–500.

Whitington, P. M. 2004. The development of the crustacean nervous system. In G. Scholtz (ed.) *Evolutionary Developmental Biology of Crustacea, Crustacean Issues 15*. Lisse, Netherlands: Balkema, pp. 135–167.

Whitington, P. M., Meier, T. & King, P. 1991. Segmentation, neurogenesis and formation of early axonal pathways in the centipede, *Ethmostigmus rubripes* (Brandt). *Roux's Archive of Developmental Biology* **199**, 349–363.

Winter, G. 1980. Beiträge zur Morphologie and Embryologie des vorderen Körperabschnitts (Cephalosoma) der Pantopoda Gerstaecker, 1863. I. Zur Entstehung des Zentralnervensystems. *Zeitschrift für systematische Zoologie und Evolutionsforschung* **18**, 27–61.

Wülbeck, C. & Simpson, P. 2000. Expression of *achaete-scute* homologues in discrete proneural clusters on the developing notum of the medfly *Ceratitis capitata*, suggests a common origin for the stereotyped bristle patterns of higher Diptera. *Development* **127**, 1411–1440.

Wülbeck, C. & Simpson, P. 2002. The expression of *pannier* and *achaete-scute* homologues in a mosquito suggests an ancient role of *pannier* as a selector gene in the regulation of the dorsal body pattern. *Development* **129**, 3861–3871.

Yoshikura, M. 1955. Embryological studies on the liphistiid spider, *Heptathela kimurai*, part II. *Kumamoto Journal of Science B* **2**, 1–86.

20

Arthropod appendages: a prime example for the evolution of morphological diversity and innovation

NIKOLA-MICHAEL PRPIC AND WIM G. M. DAMEN

The morphology of the appendages of the arthropods has been adapted to a large number of life styles that is virtually unparalleled in any other organ in the Metazoa. Different appendage types exist e.g. for walking, swimming, jumping, prey-capture, chewing, biting, mating, egg-laying, breathing in air, fresh water and salt water, and sensory perception (see Figure 20.1 for examples). Very specialised appendage types exist for specialised modes of life: for example, the spinnerets in spiders, brush legs for the distribution of pheromones (e.g. some moths) or stings for defence (e.g. bees and wasps). In many cases, appendages from a single segment or from several segments unite and form an entirely new structure capable of tapping into new resources, e.g. the labium of insects, formed by the fusion of an appendage pair, or the proboscis of ticks, mosquitoes and flies, all of which are composed of the appendages of at least two head segments.

A number of different appendage types can be present on a single individual. The number of different appendage types and their specific morphology depend on the species' life style, but in most cases at least three different types are present: appendages for sensory perception, feeding and locomotion (Figure 20.1).

The appendages of the arthropods thus have been a prime target of adaptive evolution. They are unparalleled in their sheer number of novel forms and functions. Therefore, they are an excellent model for the study of the principles of adaptive evolution and morphological change and innovation. The questions to be answered are: (1) What

Evolving Pathways: Key Themes in Evolutionary Developmental Biology, ed. Alessandro Minelli and Giuseppe Fusco. Published by Cambridge University Press.

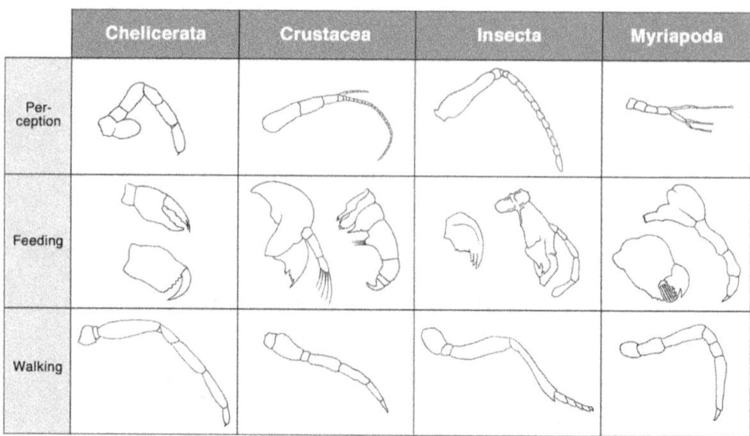

Figure 20.1 Diagram illustrating a small portion of the diversity of arthropod appendages. The upper row shows exemplary appendages for sensory perception, the centre row shows gnathal appendages and the lower row shows walking appendages from the arthropod classes indicated at the top. Note that the pedipalps in spiders (top row left) are used for perception, feeding and sperm transfer. The diagrams in the top row show simplified and generalised drawings of (from left to right): spider pedipalp, amphipod first antenna, bee antenna, pauropod antenna. The diagrams in the centre row show simplified and generalised drawings of (from left to right): scorpion chelicera (upper drawing) and spider chelicera (lower drawing), syncarid mandible, first maxilla of Remipedia, ectognathan mandible, ectognathan maxilla, millipede mandible, chilopod second maxilla. The bottom row shows simplified and generalised drawings of (from left to right): spider prosomal walking leg, amphipod pereiopod, dipteran thoracic leg, millipede trunk leg. Please note that the alignment of appendages in the diagram is based on similar functions and explicitly not on homology. Drawings after Westheide and Rieger (1996), Prpic and Damen (2004) and Prpic and Tautz (2003).

was the starting point for the evolution of this unique diversity? What is the 'ground state' of appendage that has served as the basis for further evolution? (2) What evolutionary changes have occurred to transform one appendage type into another? What are the underlying genetic mechanisms and how did they change during evolution to produce morphological innovations?

THE SEARCH FOR THE GROUND-STATE APPENDAGE

It is generally assumed that the diversity of extant appendages has evolved from a limited number of ancestral appendage forms or even from a single prototype appendage. The search for this 'ground-state' appendage has attracted considerable attention. Snodgrass (1935), for example, suggested that the leg segments (podomeres) of all extant arthropods are divided into a group of proximal podomeres and a group of distal podomeres. Snodgrass suggested that this common ground reflects an ancestral state of all arthropod appendages and that the ground-state appendage therefore consisted of two podomeres that he called coxopodite (the proximal one) and telopodite (the distal one) (Figure 20.2A). Interestingly, the early leg discs in *Drosophila* are subdivided into a proximal domain co-expressing the genes *extradenticle* (*exd*) and *homothorax* (*hth*), and a distal domain expressing the gene *Distal-less* (*Dll*) (e.g. Gonzalez-Crespo and Morata 1996, Gonzalez-Crespo *et al.* 1998, Wu and Cohen 1999). A different approach towards the ground-state appendage has been taken more recently. This approach is based on the finding that appendage morphology is also determined by

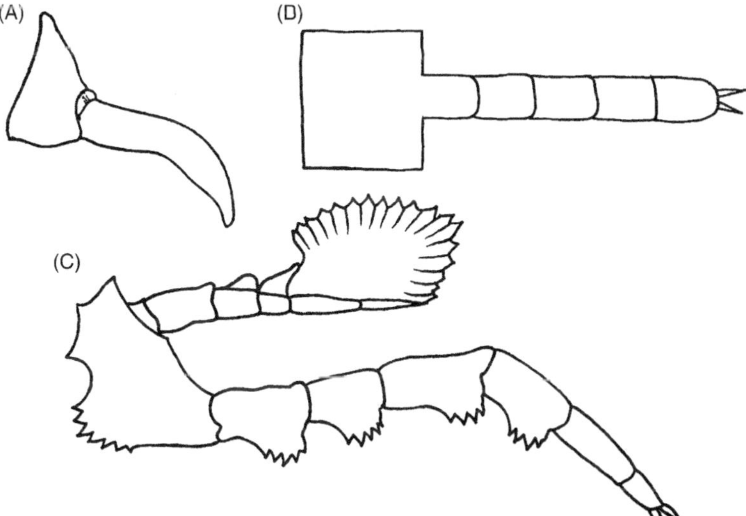

Figure 20.2 Different concepts of the ancestral or 'ground-state' appendage. A, The bipartite ancestral appendage according to Snodgrass (1935). B, The 'ground-state' appendage that results after the depletion of all known selector genes (after Casares and Mann 2001). C, A generalised trilobite leg (combined after Whittington and Almond 1987 and Westheide and Rieger 1996). According to the majority of authors, the ancestral appendage was relatively complex, similar to the appendages of the fossil arthropod group Trilobita (e.g. Walossek and Müller 1997, Boxshall 2004, Giorgianni and Patel 2005). See text for details.

selector genes, for example by the Hox genes (see below for details). Eliminating their influence, therefore, might reveal the ancestral, undiversified state. An experiment of that sort has been performed in *Drosophila*, where such genetic manipulations are possible. The appendage that develops in the absence of the influence of the known selector genes consists of a large proximal segment and a largely normal tarsus (Figure 20.2B) (Casares and Mann 2001). Experiments like this are intriguing, but selector genes like the Hox genes and their role in identity determination are phylogenetically much older than the arthropods. It is certainly wrong to assume that the ancestral arthropod did not have Hox genes or other selector genes. Thus the appendage obtained in *Drosophila* most probably does not correspond to any real stage of appendage evolution in the arthropods.

The question is not what the appendages look like without the influence of selector genes. Rather we have to ask what selector genes were present in the ancestral appendage and what functions they performed (see also Hughes and Kaufman 2002). Questions of this kind can only be answered by a comparative approach. Features common to all species are likely to be derived from a common ancestor. Such features, called symplesiomorphies, thus reveal details about the ground-state appendage. Apart from this indirect approach there is one alternative approach to learn about historical conditions: direct evidence in the form of fossils. There are a number of arthropod fossils that preserve the morphology of the appendages. Trilobites, for example, are among the earliest arthropod fossils, some dating from the Early Cambrian. They show very little appendage specialisation: their anterior-most appendage is a long antenna, but all following appendages are more or less identical (Whittington 1997). It is interesting to note that apart from the antenna the appendages of the trilobites unite functions that are relegated to different appendage types on separate body regions in extant arthropods: the trilobite post-antennal appendages were used for feeding, locomotion and breathing (Figure 20.2C). Thus, the trilobites do indeed appear to preserve a body plan from before the time of appendage specification and diversification, providing a direct image of the prototype arthropod appendage (see also Walossek and Müller 1997, Bitsch 2001, Boxshall 2004, Giorgianni and Patel 2005).

FROM GROUND STATE TO DIVERSITY: INSIGHTS FROM *DROSOPHILA*

Whatever the design of the prototype appendage, what possibilities are there to develop new appendage types from it? Principally, there are three different ways of change: (1) loss of an existing feature; (2) gain of a novel feature; or (3) modification of an existing feature.

There are many examples for each of these three modes or combinations of them. The mandible of insects is an example, where the entire distal portion of the appendage is lost. An example of a novel feature would be the brush legs of moths, as already mentioned in the introduction. Finally, modification of the number of leg joints in the tarsi of insects is an example of the third kind of change. These examples are all fine, but the interesting question is: how did it happen? What genes are involved in the changes? How did developmental pathways change during evolution? How can we explain new morphologies in terms of the underlying genetic and developmental mechanisms?

The data on leg development in the fruit fly *Drosophila melanogaster* is a starting point that can help address these questions. In *Drosophila*, the legs develop from so-called imaginal discs, which is unusual for arthropods. These discs start developing in the embryo as a small group of cells, and during the larval stages this group of cells folds into the body, forming the leg by ingrowth, rather than outgrowth as in most other arthropods (Cohen 1993). During metamorphosis this 'inward' leg turns outward, thus becoming a rather normal insect leg (Fristrom and Fristrom 1993). Despite this peculiar mode of development, the genetic mechanisms operating during *Drosophila* leg development can serve as a first guide to study the developmental mechanisms in other appendage types and other species. *Drosophila* leg development is governed by a hierarchic gene cascade (Figure 20.3) (Rauskolb and Irvine 1999). At the top level (Figure 20.3, top) two morphogens, Wingless (Wg) and Decapentaplegic (Dpp) generate a grid of morphogen concentrations and this information is read out by the next level in the cascade, the leg gap genes (Figure 20.3, centre) (e.g. Lecuit and Cohen 1997). These genes include *dachshund* (*dac*) and *Distalless* (*Dll*). The leg gap genes have two functions. First, they identify broad domains along the proximal–distal axis (e.g. Cohen and Jürgens 1989a,b, Mardon *et al.* 1994, Abu-Shaar and Mann 1998). Second, soon after the expression of the leg gap genes is initiated, their expression domains expand and thus partially overlap. These overlaps, together with overlaps with already expressed genes like *homothorax* (*hth*) and its co-factor *extradenticle* (*exd*), create areas of combinatorial gene expression that serve as a kind of 'address code' for the genes at the next level of the gene cascade (Rauskolb 2001). These genes, including mostly members of the *Notch* signalling pathway or its target genes, are activated in narrow rings along the leg by a specific combination

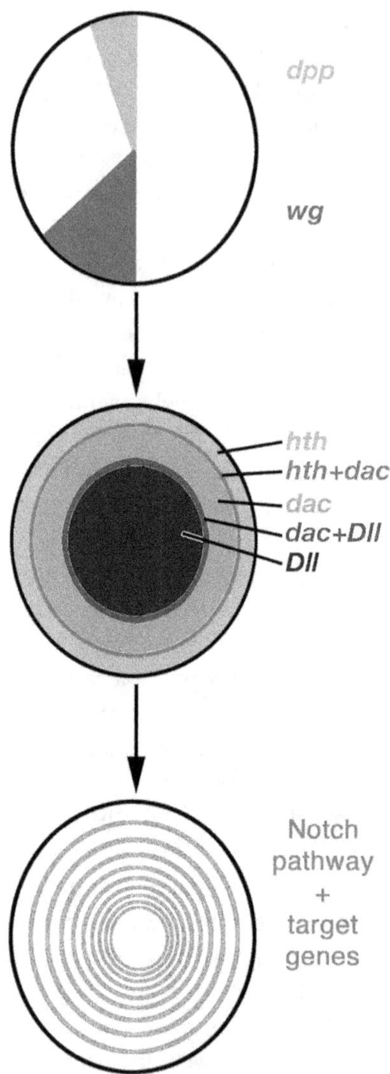

Figure 20.3 Genes and gene cascades involved in walking leg development. In *Drosophila* leg discs a hierarchic gene cascade guides proximal–distal axis development. At the top level are the two genes *dpp* and *wg* that are expressed in a dorsal and ventral sector, respectively. However, the Dpp and Wg proteins spread throughout the disc and activate the genes at the next level (e.g. *dac*, *Dll*). These genes are expressed in broad concentric but

of leg gap genes (Figure 20.3, bottom) (Rauskolb 2001). These rings define the locations where the boundaries between the segments are made (DeCelis *et al.* 1998, Bishop *et al.* 1999, Rauskolb and Irvine 1999). Of course the small number of leg gap genes are not enough to generate sufficient differential overlap zones to determine all nine segment borders (including the tarsal segments and the claw). Recently a number of additional genes have been identified that might be called 'tarsal gap genes'. These seem to supplement the already known leg gap genes in the tarsus to generate the additional segment borders there (genes such as *Bar* and *dlim1*; see Galindo *et al.* 2002 and references therein, reviewed in Kojima 2004). This entire process of leg patterning appears to be organised by the Hox gene *Antennapedia* (*Antp*), because, if misexpressed in the antennal segment, this gene is capable of generating a relatively normal leg, and if *Antp* function is lost in the legs, the tissue is transformed into antennae (Struhl 1982, Abbot and Kaufman 1986, Emerald and Cohen 2004).

How do the mechanisms in other appendage types in *Drosophila* differ from the one in the legs and how does this correlate with the differences in morphology between the appendage types? This question is rather difficult to answer, again because of the rather peculiar mode of appendage development in *Drosophila*. Not only do all appendages develop from imaginal discs, but also the mouthparts form the so-called proboscis, a structure peculiar to flies that is formed from parts of several highly modified appendages. Instead of *Antp*, it is the Hox genes *proboscipedia* (*pb*) and *Sex-combs-reduced* (*Scr*) that appear to trigger the crucial developmental steps necessary to generate the labial morphology in *Drosophila* (Abzhanov *et al.* 2001, Joulia *et al.* 2005, 2006). Abzhanov *et al.* (2001) were able to show that both Hox genes together repress several of the genes known to be important factors in leg development, such as *exd*, *hth*, *Dll* and *dac*. However, morphologically the proboscis is far from being simply a 'repressed leg' and it seems clear that additional unknown factors must also be involved in its development.

The development of the *Drosophila* antenna is independent from Hox gene function, because no Hox gene is expressed in the antennal

Fig. 20.3 (Cont.) overlapping domains, which establish a first crude subdivision of the developing leg and then activate, in a combinatorial fashion, the genes at the next level of the cascade, which comprise mainly members of the Notch signalling pathway or its target genes. These genes are expressed in concentric rings and define the location where the joints (i.e. the borders between the leg segments) are made. Simplified after Rauskolb and Irvine (1999).

segment. The role of an antennal selector gene is taken up by *hth*, which in the leg specifies proximal regions (Casares and Mann 1998). This substantial difference from the leg already hints at quite different genetic mechanisms operating in the developing antenna. Recent work has shown that in the antenna the subdivision into a proximal and a distal domain, which is so fundamental in the leg, does not exist (Dong *et al.* 2000, 2001); rather, *hth* and *Dll* overlap broadly. This co-expression of *hth* and *Dll* activates several antenna specific factors that have not been identified in any other appendage type, such as *cut*, *distal-antenna* (*dan*) or *spalt* (*sal*) (Chu *et al.* 2002, Dong *et al.* 2002, Emerald *et al.* 2003, Suzanne *et al.* 2003). Thus, unlike in the labial disc, in the antennal disc several appendage-specific factors have already been identified, but their exact correlation with the specific morphology of the antenna is still unclear.

FROM *DROSOPHILA MELANOGASTER* TO MILLIONS OF ARTHROPOD SPECIES

As already noted, the diversity of arthropod appendage morphology is immense. What about the genetic mechanisms generating this diversity? In the following, we will focus on the appendage types 'walking leg', 'mouth-part' and 'antenna', because for these types some details are known from *Drosophila* for comparison.

Walking legs in the form of unbranched, segmented appendages are present in representatives of all four arthropod classes. Comparative studies in a number of arthropod species have focused on the middle level of the leg genes cascade (e.g. Panganiban *et al.* 1994, Niwa *et al.* 1997, Abzhanov and Kaufman 2000, Jockusch *et al.* 2000, Prpic *et al.* 2001, 2003, Inoue *et al.* 2002, Prpic and Tautz 2003). These works revealed that the tripartite structure is conserved in all species. In addition, where functional analysis has been done, the role of these genes is also very similar to the *Drosophila* homologues (Beermann *et al.* 2001, Schoppmeier and Damen 2001, Angelini and Kaufman 2004). Details in the relative expression and expression dynamics of the genes, however, differ between the species. Since the areas of overlap of these genes in *Drosophila* determine the location where the genes at the lowest level are activated and the leg segment boundaries are made (Rauskolb 2001), the differences in the relative expression in other species might explain the differences in leg segment number.

Given the relatively high degree of conservation at the middle level of the cascade, it is surprising that not much seems to be conserved above that level. The *dpp* gene that in *Drosophila* is expressed along the dorsal side is expressed in the leg tip in most other species and later in one or several rings and dots, depending on the species (e.g. Sanchez-Salazar *et al.* 1996, Jockusch *et al.* 2000, Niwa *et al.* 2000, Prpic *et al.* 2003, Prpic 2004, Yamamoto *et al.* 2004). This suggests that there might be no cooperation between *dpp* and *wg* like in *Drosophila*. Indeed, recent studies were not able to confirm a similar role for *wg* in leg development in species other than *Drosophila* (Angelini and Kaufman 2005a,b). Also the role of *dpp* remains entirely unclear. Its role as a partner for *wg* has been questioned (Ober and Jockusch 2006), but the later expression in rings and dots has been linked to smaller differences in walking leg morphology (Niwa *et al.* 2000). Severe phenotypes of *Antp* RNAi in *Oncopeltus fasciatus* lead to the transformation of leg into antenna like in *Drosophila* (Angelini *et al.* 2005). Data from the beetle *Tribolium castaneum* also demonstrate a role of *Antp* in specifying leg identity (Beeman *et al.* 1989). Thus, *Antp* is obviously the selector gene for leg morphology in insects. However, other than in insects, the role of *Antp* in specifying walking leg morphology has been questioned. For example, it has been shown that in the crustacean *Daphnia magna*, *Antp* has the opposite function to *Drosophila Antp*: it represses *Dll* expression (Shiga *et al.* 2002). And in chelicerates *Antp* is not expressed in the walking legs at all (Figure 20.4D) (Damen *et al.* 1998, Telford and Thomas 1998, Abzhanov *et al.* 1999). The chelicerate walking legs are quite similar to the walking legs in crustaceans, myriapods and insects, but develop on segments that do not express *Antp* at all, except for the posterior portion of the fourth walking leg.

These data begin to illustrate an unexpected diversity of genetic mechanisms above a relatively conserved middle level with genes like *dac* and *Dll* (nothing is known about the lower level so far). Intriguingly, this introduces another level of diversity: the somewhat paradoxical phenomenon that there is diversity of genes and gene regulation on the top level, but relatively little differences in terms of morphological output, namely a walking leg.

Other appendage types such as mouthparts or antennae are not yet studied in such detail in other arthropod species. The studies on the development of the different mouthparts in crustaceans and insects seem to indicate a global role for the anterior Hox genes. Several crustacean species for example have additional mouthparts (so-called maxillipeds). Maxillipeds develop in segments where *Ubx* and *Abd-A* are not expressed, thus leaving space for the anterior Hox genes *pb*, *Deformed* (*Dfd*) and *Scr* (Averof and Patel 1997, Abzhanov and Kaufman 1999,

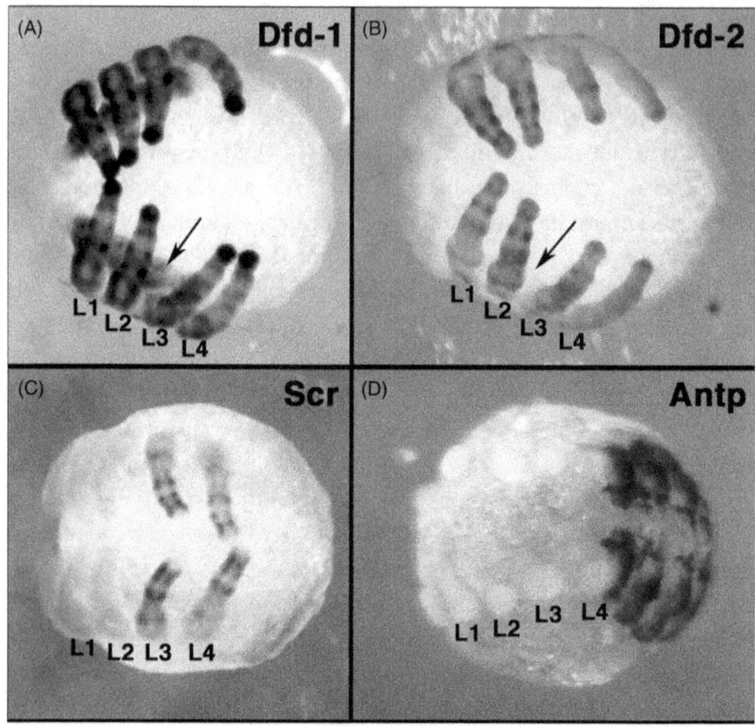

Figure 20.4 Expression of Hox genes in the spider *Cupiennius salei*. In *Drosophila*, walking leg morphology is determined by the Hox gene *Antp*, whereas the anterior Hox genes like *Dfd* and *Scr* specify mouthpart identity. This is not the case in chelicerates: first, the anterior Hox genes like *Dfd-1* (A), *Dfd-2* (B), and *Scr* (C) are expressed in all or some walking legs, but not in the mouthparts of *C. salei* (chelicera, pedipalp). Second, the *Antp* gene is not expressed in the walking legs at all (D), except for a small portion of the last walking leg (L4) (not visible in the figure). Note: *C. salei* has two *Dfd* paralogues (1, 2) that show differential expression in the neuroectoderm (arrows in A and B) and also some differences in detail in the walking legs, but the overall expression domain of both genes covers all four segments bearing walking legs. All embryos are oriented with anterior to the left. The embryos in A and B are late inversion stages, C is a mid inversion stage, and D is younger still (mid germband extension stage). Abbreviations: L1–L4, walking legs 1 to 4.

2004). Also in insects the anterior Hox genes are selector genes for mouthpart identity. Lack of the *pb* gene and expression of the *Dfd* gene have been connected with the specific stylet-shaped mouthparts in true bugs (Rogers *et al.* 2002). In addition, *Scr* is involved in labial identity and *Dfd* is involved in mandibular identity in *Oncopeltus fasciatus* (Hughes and Kaufman 2000). In *Tribolium castaneum pb*, *Dfd* and *Scr* appear to be involved in the specification of the pincer-shaped mouthparts (Beeman *et al.* 1989, Shippy *et al.* 2000, DeCamillis *et al.* 2001). The Hox targets in the mouthparts are unknown, although in *Tribolium pb* appears to repress *Dll* in the mouthparts (DeCamillis and ffrench-Constant 2003). It is intriguing, however, that again in chelicerates the anterior Hox genes *pb*, *Dfd* and *Scr* obviously do not specify mouthpart morphology, as the segments expressing these genes develop normal walking legs (Figure 20.4A–C) (Damen *et al.* 1998, Telford and Thomas 1998, Abzhanov *et al.* 1999).

With respect to the expression of other leg genes, such as *dac*, *Dll*, *hth* or *exd*, the patterns in maxilla and labium of (for example) *Tribolium castaneum* and *Schistocerca americana* are not dramatically different from the patterns in the leg (e.g. Prpic *et al.* 2001, Giorgianni and Patel 2004, Jockusch *et al.* 2004), indicating that there must be other factors as yet unidentified that are responsible for the specific morphology of the mouthparts. Significant differences exist, for example, in the maxillary stylet of the bug *Oncopeltus fasciatus* where several leg genes are broadly co-expressed (Angelini and Kaufman 2004, 2005a), and in the insect mandible where *Dll* expression is entirely missing (e.g. Popadic *et al.* 1996, 1998, Scholtz *et al.* 1998). Endites, finally, are proximal and ventral protrusions that are specific to mouthparts, but their specification is unclear. There seem to be mechanisms guiding the outgrowth of these structures that are different from the mechanisms in the main appendage axis (e.g. Giorgianni and Patel 2004, Jockusch *et al.* 2004). The *dac* gene is expressed in all endites of insect, myriapod and crustacean mouthparts that will adopt a biting or chewing tooth morphology later on (Abzhanov and Kaufman 2000, Prpic *et al.* 2001, Prpic and Tautz 2003), but is not expressed in endites that will be just weak shovel-like outgrowths like the gnathendite of spiders (Prpic and Damen 2004) or that will be brush-like like the pectinate lamella in millipede myriapods (Prpic and Tautz 2003). There might thus be a correlation of *dac* expression and tooth-shaped endite morphology.

Finally, we would like to mention a peculiar case of mouthpart: the chelicera of spiders. This is a stout appendage comprising a basal segment and a movable venom fang. The cheliceral body segment does not express any Hox genes and thus the chelicera develops without Hox input. The patterns of *hth*, *exd* and *Dll* largely overlap (Prpic and Damen 2004). This is quite different from the more typical gnathal appendages in other arthropod classes, but it is very reminiscent of

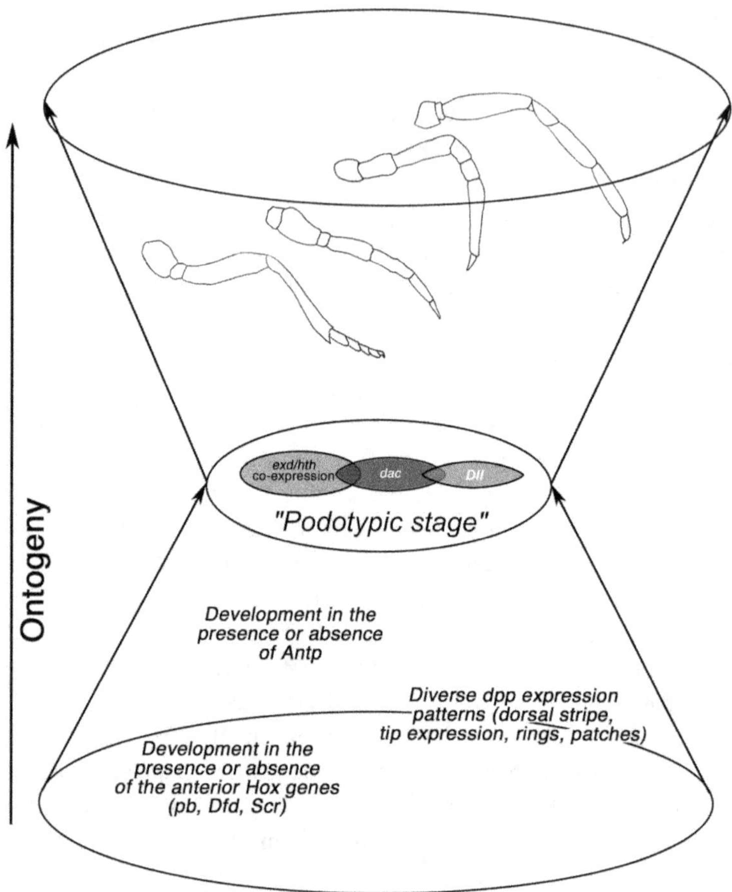

Figure 20.5 Is there a 'walking-leg-typic stage' at the junction between the diversity of early proximal–distal patterning mechanisms and the diversity of morphological form? It has been shown for *Drosophila* that the tripartite stage comprising a proximal domain with co-expression of *hth* and *exd*, a medial domain expressing *dac* and a distal domain expressing *Dll* is necessary for the development of the walking legs, but not of the antenna (Dong *et al.* 2000). This tripartite structure is present in the walking legs of other arthropods as well, but is absent from other appendage types like the chelicera in spiders (Prpic and Damen 2004) or the maxillary stylet in true bugs (Angelini and Kaufman 2004). It has been suggested that the tripartite structure is a necessary constraint to be passed through by all appendages that are to develop a walking-leg-like appearance (Prpic and Damen 2004). It has been shown that the diversity of genetic mechanisms before this tripartite stage is high, including the diversity of *dpp* expression patterns and the debated role of *wg*, the role of *Antp* and the role of the anterior Hox

the expression patterns in the antenna of *Drosophila*. In fact, the chelicera, although functionally a gnathal appendage, is homologous to the antenna (Damen *et al.* 1998, Telford and Thomas 1998, Mittmann and Scholtz 2003). The similarities in gene expression between the chelicera and the *Drosophila* antenna may therefore be based on a common origin of the appendage rather than on common function. However, an alternative and probably more likely explanation is that changing the leg-like sequence of the leg genes *hth/exd*, *dac*, *Dll* by broadly co-expressing some or all of these genes assists in the activation of different target genes that lead to novel morphologies such as the fang-like chelicera or the short and stubby antenna in *Drosophila* (Prpic and Damen 2004). This is further supported by the largely overlapping expression of *hth* and *dac* (and also *Dll*) in the maxillary stylet of *Oncopeltus fasciatus* (Angelini and Kaufman 2004, 2005a).

A KIND OF PHYLOTYPIC STAGE FOR THE DIFFERENT TYPES OF ARTHROPOD APPENDAGE?

The following sentence from Angelini and Kaufman (2005a) sums up our current ignorance with respect to the origin and evolution of one of the greatest diversifications of morphological form on Earth: 'We are still far from an explanation of biological diversity in which morphology may be unambiguously described by our knowledge of ontogenetic mechanisms.'

Research into the evolution of arthropod appendage development is still very much at its beginning. The broad comparative approach can identify features that are conserved among all species and thus reveal information about the ancestral appendage. This approach can also identify genes and gene regulations that are not conserved and thus may be responsible for morphological novelties. Unfortunately, the species studied so far cover only a tiny fraction of the diversity of the arthropods. What we also need are more detailed studies of gene function in selected species to achieve a more complete understanding of the relationship between gene expression and regulation and morphology.

Fig. 20.5 (Cont.) genes, like *pb*, *Dfd* and *Scr* (see text for details). Also the diversity of adult morphology of walking legs is high. The tripartite stage seems to channel the diversity of early patterning mechanisms into a path towards 'walking leg morphology' but still allows for a high amount of morphological diversity of adult structures. In this way it is similar to the phylotypic stage (Sander 1983, Raff 1996).

The current knowledge about the evolution of genetic mechanisms in appendage types like mouthparts or antennae is too fragmentary yet to allow for sound conclusions. But if the data from comparative studies of walking leg development are any indication, then it looks like there is not one level of appendage diversity but two: one level, as noted, is the diversity of adult walking leg morphology. But a second level is the diversity of developmental mechanisms above the level of genes like *dac* or *Dll* that, despite their significant differences, all converge again on creating a walking leg type of appendage (Figure 20.5). This, in a way, is reminiscent of the hour-glass model (Raff 1996) describing the diversity of early embryonic and adult body plans against the limited diversity of phenotypes at so-called phylotypic stage (Sander 1983). It seems that what is valid for animals as a whole, is also valid for single parts of them: one might construct an hour-glass model also for the walking legs of arthropods, with diversity in early development and adult morphology, but a 'podotypic' stage involving the genes *exd*, *hth*, *dac* and *Dll* connecting them in the middle (Figure 20.5).

ACKNOWLEDGEMENTS

We thank Alessandro Minelli and Giuseppe Fusco for the opportunity to contribute to this volume. The work of W.G.M.D. was supported in part by the DFG via SFB 572 of the University of Cologne and by the European Union via the Marie Curie Research and Training Network ZOONET (MRTN-CT-2004-005624). The work of N.M.P. was supported by the DFG (grant number PR 1109/1-1).

REFERENCES

Abbott, M. K. & Kaufman, T. C. 1986. The relationship between the functional complexity and the molecular organization of the *Antennapedia* locus of *Drosophila melanogaster*. *Genetics* **114**, 919–942.

Abu-Shaar, M. & Mann, R. S. 1998. Generation of multiple antagonistic domains along the proximodistal axis during *Drosophila* leg development. *Development* **125**, 3821–3830.

Abzhanov, A., Holtzman, S. & Kaufman, T. C. 2001. The *Drosophila* proboscis is specified by two Hox genes, *proboscipedia* and *Sex combs reduced*, via repression of leg and antennal appendage genes. *Development* **128**, 2803–2814.

Abzhanov, A. & Kaufman, T. C. 1999. Novel regulation of the homeotic gene *Scr* associated with a crustacean leg-to-maxilliped appendage transformation. *Development* **126**, 1121–1128.

Abzhanov, A. & Kaufman, T. C. 2000. Homologs of *Drosophila* appendage genes in the patterning of arthropod limbs. *Developmental Biology* **227**, 673–689.

Abzhanov, A. & Kaufman, T. C. 2004. Hox genes and tagmatization of the higher Crustacea (Malacostraca). In G. Scholtz (ed.) *Evolutionary Developmental Biology of Crustacea. (Crustacean Issues 15)*, Lisse: Balkema, pp. 43–74.

Abzhanov, A., Popadic, A. & Kaufman, T. C. 1999. Chelicerate Hox genes and the homology of arthropod segments. *Evolution & Development* **1**, 77–89.

Angelini, D. R. & Kaufman, T. C. 2004. Functional analyses in the hemipteran *Oncopeltus fasciatus* reveal conserved and derived aspects of appendage patterning in insects. *Developmental Biology* **271**, 306–321.

Angelini, D. R. & Kaufman, T. C. 2005a. Insect appendages and comparative ontogenetics. *Developmental Biology* **286**, 57–77.

Angelini, D. R. & Kaufman, T. C. 2005b. Functional analyses in the milkweed bug *Oncopeltus fasciatus* (Hemiptera) support a role for Wnt signaling in body segmentation but not appendage development. *Developmental Biology* **283**, 409–423.

Angelini, D. R., Liu, P. Z., Hughes, C. L. & Kaufman, T. C. 2005. Hox gene function and interaction in the milkweed bug *Oncopeltus fasciatus* (Hemiptera). *Developmental Biology* **287**, 440–455.

Averof, M. & Patel, N. H. 1997. Crustacean appendage evolution associated with changes in Hox gene expression. *Nature* **388**, 682–686.

Beeman, R. W., Stuart, J. J., Haas, M. S. & Denell, R. E. 1989. Genetic analysis of the homeotic gene complex (HOM-C) in the beetle *Tribolium castaneum*. *Developmental Biology* **133**, 196–209.

Beermann, A., Jay, D. G., Beeman, R. W. *et al.* 2001. The *Short antennae* gene of *Tribolium* is required for limb development and encodes the orthologue of the *Drosophila Distal-less* protein. *Development* **128**, 287–297.

Bishop, S. A., Klein, T., Martinez Arias, A. & Couso, J. P. 1999. Composite signaling from *Serrate* and *Delta* establishes leg segments in *Drosophila* through *Notch*. *Development* **126**, 2993–3003.

Bitsch, J. 2001. The hexapod appendage: basic structure, development and origin. *Annales de la Société Entomologique de France (Nouvelle Série)* **37**, 175–193.

Boxshall, G. A. 2004. The evolution of arthropod limbs. *Biological Reviews* **79**, 253–300.

Casares, F. & Mann, R. S. 1998. Control of antennal versus leg development in *Drosophila*. *Nature* **392**, 723–726.

Casares, F. & Mann, R. S. 2001. The ground state of the ventral appendage in *Drosophila*. *Science* **293**, 1477–1480.

Chu, J., Dong, P. D. & Panganiban, G. 2002. Limb type-specific regulation of *bric a brac* contributes to morphological diversity. *Development* **129**, 695–704.

Cohen, S. M. 1993. Imaginal disc development. In M. Bate & A. Martinez Arias (eds.) *The Development of Drosophila melanogaster. Volume II.* Cold Spring Harbor, New York: Cold Spring Harbor Laboratory Press, pp. 747–841.

Cohen, S. M. & Jürgens, G. 1989a. Proximal–distal pattern formation in *Drosophila*: graded requirement for *Distal-less* gene activity during limb development. *Roux' Archives of Developmental Biology* **198**, 157–169.

Cohen, S. M. & Jürgens, G. 1989b. Proximal-distal pattern formation in *Drosophila*: cell autonomous requirement for *Distal-less* gene activity in limb development. *EMBO Journal* **8**, 2045–2055.

Damen, W. G. M., Hausdorf, M., Seyfarth, E. A. & Tautz, D. 1998. A conserved mode of head segmentation in arthropods revealed by the expression pattern of Hox genes in a spider. *Proceedings of the National Academy of Sciences of the USA* **95**, 10665–10670.

DeCamillis, M. A. & ffrench-Constant, R. 2003. *Proboscipedia* represses distal signaling in the embryonic gnathal limb fields of *Tribolium castaneum*. *Development Genes & Evolution* **213**, 55–64.

DeCamillis, M. A., Lewis, D. L., Brown, S. J., Beeman, R. W. & Denell, R. E. 2001. Interactions of the *Tribolium Sex combs reduced* and *proboscipedia* orthologs in embryonic labial development. *Genetics* **159**, 1643–1648.

De Celis, J. F., Tyler, D. M., de Celis, J. & Bray, S. J. 1998. Notch signalling mediates segmentation of the *Drosophila* leg. *Development* **125**, 4617–4626.

Dong, P. D., Chu, J. & Panganiban, G. 2000. Coexpression of the homeobox genes *Distal-less* and *homothorax* determines *Drosophila* antennal identity. *Development* **127**, 209–216.

Dong, P. D., Chu, J. & Panganiban, G. 2001. Proximodistal domain specification and interactions in developing *Drosophila* appendages. *Development* **128**, 2365–2372.

Dong, P. D., Dicks, J. S. & Panganiban, G. 2002. *Distal-less* and *homothorax* regulate multiple targets to pattern the *Drosophila* antenna. *Development* **129**, 1967–1974.

Emerald, B. S. & Cohen, S. M. 2004. Spatial and temporal regulation of the homeotic selector gene *Antennapedia* is required for the establishment of leg identity in *Drosophila*. *Developmental Biology* **267**, 462–472.

Emerald, B. S., Curtis, J., Mlodzik, M. & Cohen, S. M. 2003. *distal antenna* and *distal antenna related* encode nuclear proteins containing pipsqueak motifs involved in antenna development in *Drosophila*. *Development* **130**, 1171–1180.

Fristrom, D. & Fristrom, J. W. 1993. The metamorphic development of the adult epidermis. In M. Bate & A. Martinez Arias (eds.) *The Development of Drosophila melanogaster, Vol. II*, Cold Spring Harbor, New York: Cold Spring Harbor Laboratory Press, pp. 843–897.

Galindo, M. I., Bishop, S. A., Greig, S. & Couso, J. P. 2002. Leg patterning driven by proximal-distal interactions and EGFR signaling. *Science* **297**, 256–259.

Giorgianni, M. W. & Patel, N. H. 2004. Patterning of the branched head appendages in *Schistocerca americana* and *Tribolium castaneum*. *Evolution & Development* **6**, 402–410.

Giorgianni, M. W. & Patel, N. H. 2005. Conquering land, air and water: the evolution and development of arthropod appendages. In D. E. G. Briggs (ed.) *Evolving Form and Function: Fossils and Development*. New Haven: Peabody Museum of Natural History, Yale University, pp. 159–188.

Gonzalez-Crespo, S., Abu-Shaar, M., Torres, M. *et al.* 1998. Antagonism between *extradenticle* function and Hedgehog signalling in the developing limb. *Nature* **394**, 196–200.

Gonzalez-Crespo, S. & Morata, G. 1996. Genetic evidence for the subdivision of the arthropod limb into coxopodite and telopodite. *Development* **122**, 3921–3928.

Hughes, C. L. & Kaufman, T. C. 2000. RNAi analysis of *Deformed*, *proboscipedia* and *Sex combs reduced* in the milkweed bug *Oncopeltus fasciatus*: novel roles for Hox genes in the hemipteran head. *Development* **127**, 3683–3694.

Hughes, C. L. & Kaufman, T. C. 2002. Hox genes and the evolution of the arthropod body plan. *Evolution & Development* **4**, 459–499.

Inoue, Y., Mito, T., Miyawaki, K. *et al.* 2002. Correlation of expression patterns of *homothorax*, *dachshund*, and *Distal-less* with the proximodistal segmentation of the cricket leg bud. *Mechanisms of Development* **113**, 141–148.

Jockusch, E. L., Nulsen, C., Newfeld, S. J. & Nagy, L. M. 2000. Leg development in flies versus grasshoppers: differences in *dpp* expression do not lead to differences in the expression of downstream components of the leg patterning pathway. *Development* **127**, 1617–1626.

Jockusch, E. L., Williams, T. A. & Nagy, L. M. 2004. The evolution of patterning of serially homologous appendages in insects. *Development Genes & Evolution* **214**, 324–338.

Joulia, L., Bourbon, H.-M. & Cribbs, D. L. 2005. Homeotic *proboscipedia* function modulates *hedgehog*-mediated organizer activity to pattern adult *Drosophila* mouthparts. *Developmental Biology* **278**, 496–510.

Joulia, L., Deutsch, J., Bourbon, H. M. & Cribbs, D. L. 2006. The specification of a highly derived arthropod appendage, the *Drosophila* labial palps, requires the joint action of selectors and signaling pathways. *Development Genes & Evolution* **216**, 431–442.

Kojima, T. 2004. The mechanism of *Drosophila* leg development along the proximo-distal axis. *Development Growth & Differentiation* **46**, 115–129.

Lecuit, T. & Cohen, S. M. 1997. Proximal-distal axis formation in the *Drosophila* leg. *Nature* **388**, 139–145.

Mardon, G., Solomon, N. M. & Rubin, G. M. 1994. *dachshund* encodes a nuclear protein required for normal eye and leg development in *Drosophila*. *Development* **120**, 3473–3486.

Mittmann, B. & Scholtz, G. 2003. Development of the nervous system in the 'head' of *Limulus polyphemus* (Chelicerata: Xiphosura): morphological evidence for a correspondence between the segments of the chelicerae and of the (first) antennae of Mandibulata. *Development Genes & Evolution* **213**, 9–17.

Niwa, N., Inoue, Y., Nozawa, A. *et al.* 2000. Correlation of diversity of leg morphology in *Gryllus bimaculatus* (cricket) with divergence in *dpp* expression pattern during leg development. *Development* **127**, 4373–4381.

Niwa, N., Saitoh, M., Ohuchi, H., Yoshioka, H. & Noji, S. 1997. Correlation between *Distal-less* expression patterns and structures of appendages in the development of the two-spotted cricket, *Gryllus bimaculatus*. *Zoological Science* **14**, 115–125.

Ober, K. A. & Jockusch, E. L. 2006. The roles of *wingless* and *decapentaplegic* in axis and appendage development in the red flour beetle, *Tribolium castaneum*. *Developmental Biology* **294**, 391–405.

Panganiban, G., Nagy, L. & Carroll, S. B. 1994. The role of the *Distal-less* gene in the development and evolution of insect limbs. *Current Biology* **4**, 671–675.

Popadic, A., Panganiban, G., Rusch, D., Shear, W. A. & Kaufman, T. C. 1998. Molecular evidence for the gnathobasic derivation of arthropod mandibles and for the appendicular origin of the labrum and other structures. *Development Genes & Evolution* **208**, 142–150.

Popadic, A., Rusch, D., Peterson, M., Rogers, B. T. & Kaufman, T. C. 1996. Origin of the arthropod mandible. *Nature* **380**, 395.

Prpic, N. M. 2004. Homologs of *wingless* and *decapentaplegic* display a complex and dynamic expression profile during appendage development in the millipede *Glomeris marginata* (Myriapoda: Diplopoda). *Frontiers in Zoology* **1**, 6.

Prpic, N. M. & Damen, W. G. M. 2004. Expression patterns of leg genes in the mouthparts of the spider *Cupiennius salei* (Chelicerata: Arachnida). *Development Genes & Evolution* **214**, 296–302.

Prpic, N. M., Janssen, R., Wigand, B., Klingler, M. & Damen, W. G. M. 2003. Gene expression in spider appendages reveals reversal of *exd/hth* spatial specificity, altered leg gap gene dynamics, and suggests divergent distal morphogen signaling. *Developmental Biology* **264**, 119–140.

Prpic, N. M. & Tautz, D. 2003. The expression of the proximodistal axis patterning genes *Distal-less* and *dachshund* in the appendages of *Glomeris marginata* (Myriapoda: Diplopoda) suggests a special role of these genes in patterning the head appendages. *Developmental Biology* **260**, 97–112.

Prpic, N. M., Wigand, B., Damen, W. G. M. & Klingler, M. 2001. Expression of *dachshund* in wild-type and *Distal-less* mutant *Tribolium* corroborates serial homologies in insect appendages. *Development Genes & Evolution* **211**, 467–477.

Raff, R. A. 1996. *The Shape of Life: Genes, Development, and the Evolution of Animal Form*. Chicago: University of Chicago Press.

Rauskolb, C. 2001. The establishment of segmentation in the *Drosophila* leg. *Development* **128**, 4511–4521.

Rauskolb, C. & Irvine, K. D. 1999. *Notch*-mediated segmentation and growth control of the *Drosophila* leg. *Developmental Biology* **210**, 339–350.

Rogers, B. T., Peterson, M. D. & Kaufman, T. C. 2002. The development and evolution of insect mouthparts as revealed by the expression patterns of gnathocephalic genes. *Evolution & Development* **4**, 96–110.

Sanchez-Salazar, J., Pletcher, M. T., Bennett, R. L. *et al.* 1996. The *Tribolium decapentaplegic* gene is similar in sequence, structure, and expression to the *Drosophila dpp* gene. *Development Genes & Evolution* **206**, 237–246.

Sander, K. 1983. The evolution of patterning mechanisms: gleanings from insect embryogenesis and spermatogenesis. In B. C. Goodwin, N. Holder & C. C. Wylie (eds.) *Development and Evolution: The Sixth Symposium of the British Society for Developmental Biology.* Cambridge: Cambridge University Press, pp. 137–159.

Scholtz, G., Mittmann, B. & Gerberding, M. 1998. The pattern of *Distal-less* expression in the mouthparts of crustaceans, myriapods and insects: new evidence for a gnathobasic mandible and the common origin of Mandibulata. *International Journal of Developmental Biology* **42**, 801–810.

Schoppmeier, M. & Damen, W. G. M. 2001. Double-stranded RNA interference in the spider *Cupiennius salei*: the role of *Distal-less* is evolutionarily conserved in arthropod appendage formation. *Development Genes & Evolution* **211**, 76–82.

Shiga, Y., Yasumoto, R., Yamagata, H. & Hayashi, S. 2002. Evolving role of Antennapedia protein in arthropod limb patterning. *Development* **129**, 3555–3561.

Shippy, T. D., Guo, J., Brown, S. J., Beeman, R. W. & Denell, R. E. 2000. Analysis of *maxillopedia* expression pattern and larval cuticular phenotype in wild-type and mutant *Tribolium*. *Genetics* **155**, 721–731.

Snodgrass, R. E. 1935. *Principles of Insect Morphology.* New York: McGraw-Hill.

Struhl, G. 1982. Genes controlling segmental specification in the *Drosophila* thorax. *Proceedings of the National Academy of Sciences of the USA* **79**, 7380–7384.

Suzanne, M., Estella, C., Calleja, M. & Sanchez-Herrero, E. 2003. The *hernandez* and *fernandez* genes of *Drosophila* specify eye and antenna. *Developmental Biology* **260**, 465–483.

Telford, M. J. & Thomas, R. H. 1998. Expression of homeobox genes shows chelicerate arthropods retain their deutocerebral segment. *Proceedings of the National Academy of Sciences of the USA* **95**, 10671–10675.

Walossek, D. & Müller, K. J. 1997. Cambrian 'Orsten'-type arthropods and the phylogeny of Crustacea. In R. A. Fortey & R. H. Thomas (eds.) *Arthropod Relationships.* London: Chapman and Hall, pp. 139–153.

Westheide, W. & Rieger, R. M. 1996. *Spezielle Zoologie, Band 1, Einzeller und Wirbellose Tiere.* Stuttgart: Gustav Fischer Verlag.

Whittington, H. B. 1997. The trilobite body. In R. L. Kaesler (ed.) *Treatise on Invertebrate Palaeontology. Part O. Arthropoda 1. Trilobita, revised. Vol. 1: Introduction, Order Agnostida, Order Redlichiida.* Boulder, CO: The Geological Society of America and Lawrence, KS: The University of Kansas, pp. 87–135.

Whittington, H. B. & Almond, J. E. 1987. Appendages and habits of the Upper Ordovician trilobite *Triarthrus eatoni*. *Philosphical Transactions of the Royal Society of London B* **317**, 1–46.

Wu, J. & Cohen, S. M. 1999. Proximodistal axis formation in the *Drosophila* leg: subdivision into proximal and distal domains by Homothorax and Distal-less. *Development* **126**, 109–117.

Yamamoto, D. S., Sumitani, M., Tojo, K., Lee, J. M. & Hatakeyama, M. 2004. Cloning of a *decapentaplegic* orthologue from the sawfly, *Athalia rosae* (Hymenoptera), and its expression in the embryonic appendages. *Development Genes & Evolution* **214**, 128–133.

21

Ontogeny of the spiralian brain

CLAUS NIELSEN

Spiral cleavage is a characteristic feature of several protostomian taxa, sometimes united as Spiralia (Dohle 1996), but its presence in a number of these groups has been debated. This could be the result of too vague definitions, so I will emphasise here the presence of both a spiral pattern with shifting direction of the spindles in the early cleavages and a cell lineage including prototroch cells (trochoblasts) differentiating from cells along the border between first and second micromere quartet. This automatically excludes the non-ciliated groups, but their cleavage types could be discussed in the light of the conclusions reached here.

The cleavage pattern defines two regions of the larvae: the episphere, consisting of cells from the first micromere quartet, including the primary and accessory trochoblasts, and the hyposphere 'below' the prototroch (Figures 21.1 and 21.2). The origin of different parts of the central nervous systems from these two regions has been documented sporadically in a number of older papers on embryology of various species, and some more recent studies provide information obtained by modern methods including cell labelling. The literature on cell lineage up to about 2004 has been summarised earlier (Nielsen 2004, 2005a). Here I will try to update the information and to incorporate new information obtained from studies of Hox genes (see also Nielsen 2005b), with special emphasis on the origin of the nervous system.

CLEAVAGE PATTERNS AND CELL LINEAGE

Annelids and molluscs are the 'core spiralians' and the homology of their cleavage patterns and cell lineages seems unquestioned (Henry

Evolving Pathways: Key Themes in Evolutionary Developmental Biology, ed. Alessandro Minelli and Giuseppe Fusco. Published by Cambridge University Press.

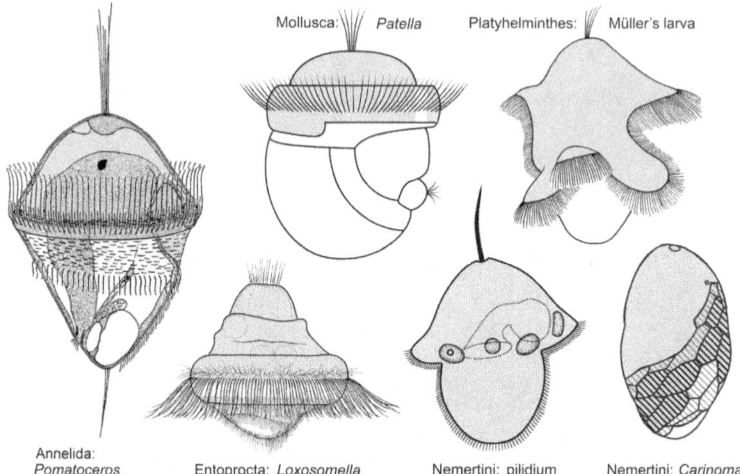

Figure 21.1 Examples of spiralian larvae. The episphere, i.e. the cells from the first micromere quartet of the spiralian cleavage, is light grey. *Carinoma* has no prototroch, but large cells of the lineages of primary trochoblasts (heavy hatching) and accessory and secondary trochoblasts (light hatching) are cleavage-arrested and are resorbed at metamorphosis like trochoblasts of other spiralians (see text).

and Martindale 1999, Nielsen 2004). The general pattern can be illustrated by the cell lineage of the annelid *Podarke* (Figure 21.2), which shows the origin of the primary, secondary and accessory prototroch cells. Many smaller or larger deviations from this pattern have been described, especially in the development of accessory and secondary prototroch cells (review in Nielsen 2004). Many species with yolk-rich eggs show more important deviations, the most conspicuous example being the cephalopods, which have very large eggs and a discoidal cleavage without any trace of the spiral pattern. However, it seems generally accepted that this represents highly specialised conditions. Species with planktotrophic larvae, and many species with lecithotrophic development, have a conspicuous prototroch consisting of compound cilia on multiciliated cells (the only known exception being the annelid *Owenia*, which has single cilia on monociliate cells; see Emlet and Strathmann 1994). At metamorphosis, the trochoblasts degenerate and are either resorbed or cast off.

Sipunculans show a spiral cleavage pattern, and the serosa of the *Sipunculus* larva appears to develop as a fold from the periphery of the episphere (Hatschek 1883), but the cell lineage has not been described.

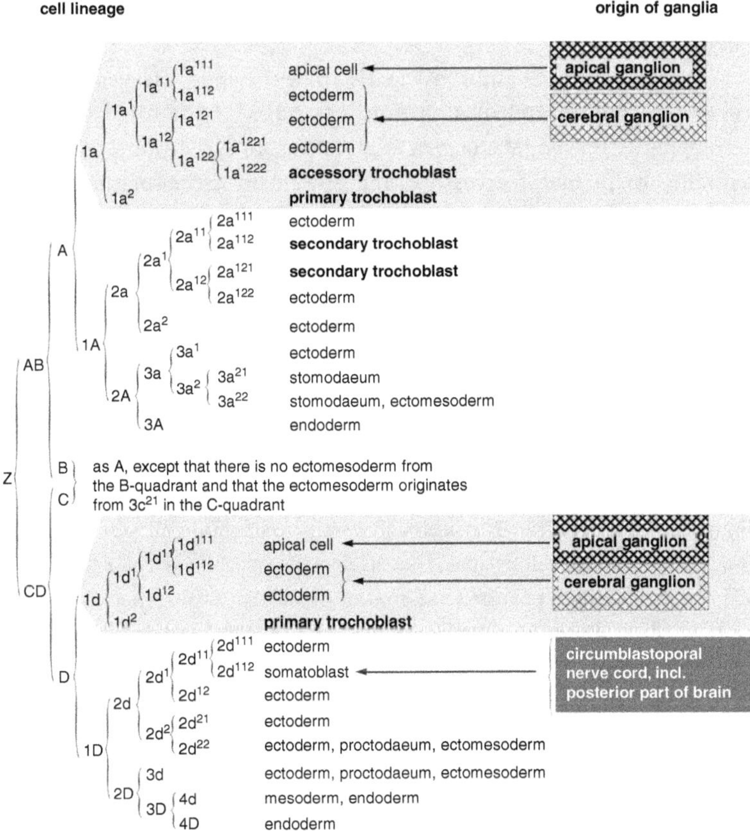

Figure 21.2 Cell lineage of an annelid as represented by *Podarke* (based on Treadwell 1901). Blastomeres of the episphere are in the light grey areas. The grey tones and the two types of cross-hatching correspond to similar areas in Figures 21.1, 21.3 and 21.4. The origin of different parts of the central nervous system is indicated (see text). It might be expected that descendants of the cells 2a–c contribute to the ventral nervous system, but this has not been investigated.

Very large primary and accessory trochoblasts have been identified in *Golfingia* and *Phascolopsis*, but a prototroch with compound cilia never develops. The prototroch cells degenerate at metamorphosis (Gerould 1906).

Entoprocts have spiral cleavage with primary and accessory trochoblasts (Marcus 1939, Malakhov 1990). The prototroch is retracted at metamorphosis, and degenerating prototroch cells with slowly beating compound cilia in a large vacuole have been observed in the primary body cavity of newly settled juveniles of several species (Nielsen 1971).

Early nemertine embryos clearly show the spiral cleavage pattern (review in Nielsen 2005a). Cell lineage studies of the almost 'direct' developing *Carinoma* (Maslakova *et al.* 2004a, b) show that cells corresponding to primary, accessory and secondary trochoblasts become very large and look different from the other epithelial cells of the epi- and hyposphere, although they do not develop a prototroch (Figure 21.1). These cells degenerate at a 'metamorphosis', probably by apoptosis, exactly like prototroch cells of annelids and molluscs. The episphere is much larger than the hyposphere at this stage, but the following development has not been studied. A cell lineage study of *Cerebratulus lacteus*, which has a pilidium larva (Henry and Martindale 1998), shows that the band of longer cilia along the edges of the larva is formed by cells along the border between the first and second micromere quartet (Figure 21.1) of a typical prototroch (with additional cells from the 3c and 3d cells). The ciliary band is discarded together with other parts of both epi- and hyposphere at metamorphosis. Other species with a more or less direct development bridge the gap between these two types of development. A number of species develop an outer layer of large cells with or without cilia, and these cells are shed at metamorphosis, exposing the adult ectoderm (Iwata 1958, 1960, Hickman 1963). Thus, the episphere and prototroch appear to have a cell lineage and a developmental fate like those of the other spiralians and they are probably homologous.

Platyhelminths (catenulids and rhabditophorans) show the spiral cleavage pattern in early stages, whereas the later stages show an enormous variation (review in Nielsen 2005a). Some polyclads have the characteristic ciliated, probably planktotrophic larvae called Müller's or Goette's larvae, with a circumferential, lobed band of longer cilia (Figure 21.1). The cell lineage of this band is unfortunately not known, but a comparison of the position of the borderline between the first two micromere quartets and the position of the band indicates that it has the same position as that of typical prototrochs (Nielsen 2005a) and may thus be an ancestral spiralian character. The ciliary band degenerates at metamorphosis (Ruppert 1978), just as other prototrochs.

Thus, it can safely be concluded that annelids, molluscs, sipunculans, entoprocts, nemertines and platyhelminths fulfil the narrower definition of spiral cleavage mentioned above. The ontogeny of the nervous systems in these phyla will be discussed below.

A spiral cleavage pattern has been reported from a gnathostomulid (Riedl 1969), but no details were given. Spiralian-like characters of rotifers and ectoprocts have been discussed elsewhere (Nielsen 2005a). These groups will not be treated here.

The report of a spiralian pattern in the development of a phoronid by Rattenbury (1954) is now generally interpreted as a misunderstanding (Zimmer 1991).

ONTOGENY OF THE NERVOUS SYSTEM

It has long been recognised that the various elements of the central nervous systems of spiralians develop in a sequence, and 'larval' and 'adult' parts of the CNS have been recognised for example in molluscs (review in Croll and Dickinson 2004). However, there has usually been no emphasis on the cell lineage of the various ganglia (see Nielsen 2005a).

Apical ganglion

An apical organ with a tuft of long cilia is described in most of the ciliated spiralian larvae. However, as shown below, it is necessary to distinguish between an apical ganglion and cerebral ganglia which may develop in close apposition to the apical ganglion or more laterally.

The apical ganglion is usually stated to be derived from the rosette cells, $1a^{111}–1d^{111}$. The cells bearing the apical tuft are descendents of these cells, but actual cell lineage studies are lacking, and several species lack ciliated cells in one or more quadrants. Other species have extensive ciliated areas around the apical pole, but their cell lineage and association with a ganglion are uncertain.

Annelid trochophores generally have an apical ciliary tuft, but the cell lineage of the cells of the apical ganglion has not been documented. The apical tuft is always lost long before, e.g. in *Polygordius* (Woltereck 1902), or at metamorphosis, but the fate of the apical ganglion itself seems to be unknown. There is much variation in the origin and number of the various cell types of the apical ganglion. *Spirobranchus* larvae (Lacalli 1981, 1984) have an apical ganglion comprising 16 cells, one of which has a tuft of long cilia. Larvae of *Phyllodoce* (Lacalli 1981) have an apical organ with five ciliated cells, three of which form a long apical tuft, and three to four associated cells.

An apical tuft situated at the most apical cells occurs in almost all ciliated molluscan larvae. However, there is much variation between the species. A number of 'apical organs' have been studied, but it is not always clear whether the cephalic ganglia have been distinguished (discussion in Page and Parries 2000). Conklin (1897) observed four small apical ganglionic cells in early veligers of the gastropod *Crepidula* and observed their disappearance in later, still intracapsular stages (see also Dickinson *et al.* 1999). Some species are reported to develop apical cilia on cells from all four quadrants, but some species have cilia only on 1c- and 1d-cells, and some have only one ciliated cell. The apical ganglion of *Ilyanassa* consists of about 25 neurons, five of which are serotonergic; it degenerates at metamorphosis through apoptosis (Dickinson and Croll 2003, Leise *et al.* 2004, Gifondorwa and Leise 2006). Descriptions of ultrastructure and immunocytochemistry of a number of species agree on this general picture (e.g. Marois and Carew 1997, on *Aplysia*; Page 2002, on *Tectura*; and Hinman *et al.* 2003, on *Haliotis*). The other classes have been studied less intensively, but information in general agreement with that given for the gastropods is available for solenogasters (Okusu 2002), polyplacophorans (e.g. Friedrich *et al.* 2002, Voronezhskaya *et al.* 2002, Henry *et al.* 2004), scaphopods (e.g. van Dongen and Geilenkirchen 1974) and bivalves (e.g. Tardy and Dongard 1993) (reviews in Croll and Dickinson 2004, Nielsen 2004).

Trochophores of the sipunculans *Golfingia* and *Phascolopsis* have an apical ganglion with four ciliated cells with a very long ciliary tuft (Gerould 1906). The *Phascolion* larva has a small apical ganglion with an apical tuft and two to three FMRFamide-positive cells, but no serotonergic cells. The prototroch nerve ring found in most trochophores is lacking (Wanninger *et al.* 2005), in accordance with the interpretation of the large ring of compound cilia not as a prototroch but as an accessory post-trochal band (Nielsen 2005b).

The very large and complicated apical organ of the entoprocts is believed to be a specialised apical ganglion, as indicated by the cell lineage study of early stages of the development of *Pedicellina* (Marcus 1939); it disappears at metamorphosis.

Nemertines show a large variation in larval types, but almost all are ciliated and have an apical organ with an apical tuft. In the pilidium larva of *Cerebratulus lacteus*, the cells with the tuft develop from all four quadrants (Henry and Martindale 1998). The almost 'direct' developing *Carinoma* has a small apical organ originating from all four quadrants (Maslakova *et al.* 2004b). The apical organ appears to be non-neuronal (Lacalli and West 1985, Hay-Schmidt 1990). Most of the episphere with the apical organ is shed at metamorphosis in the pilidium larvae (Hickman 1963). In *Micrura*, the apical organ is shed with the larval ectoderm (Iwata 1958), and this may be the case for other species too, but direct statements are lacking (Iwata 1960).

The origin of platyhelminth nervous systems is poorly known. Lacalli (1982, 1983) made detailed transmission electron microscope studies of newly hatched Müller's larvae of the polyclad *Pseudoceros* and reported the presence of a well-defined apical ganglion with four to five monociliate apical cells surrounded by narrow gland cells; it was well demarcated from the cerebral ganglia (brain). In *Hoploplana*, the apical tuft develops from the 1a and 1c cells (Boyer *et al.* 1998). The apical tuft disappears at metamorphosis (Ruppert 1978).

Thus it appears that the apical ganglion differentiates from the cells of the apical rosette in all species investigated. The ganglion degenerates before or just after metamorphosis in all the species where its fate has been followed.

Cerebral ganglia

The paired cerebral ganglia with a commissure below the apical ganglion develop from cells of the first micromere quartet, close to or even apposed to the apical ganglion (apical organ, see above) or more laterally at the episphere. The cerebral ganglia are retained through metamorphosis as the anterior part of the adult brain.

There are many reports of development of cerebral ganglia in annelids. They agree that the ganglia develop from paired thickenings of the epithelium of the episphere lateral to the apical ganglion, but a more precise cell lineage has only been indicated for *Arenicola*, where Child (1900) suggested that the cells $1d^{112112}$ and $1c^{112112}$ should give rise to the ganglia. The best-studied annelid is *Platynereis*, where Ackermann *et al.* (2005) reported that the cerebral ganglia develop from all four cells of the first micromere quartet, but with the major part originating from the 1c- and 1d-cells. Some neurites were reported to extend along the circumoesophageal connectives to the ventral longitudinal nerves from the 1a–1c cells, and many neurites extend to the ventral side from the 1d cell. The direct developing leech *Helobdella* has a very small first micromere quartet and develops no apical ganglion, but the cells give rise to the prostomium and part of the foregut plus the cerebral (supra-oesophageal) ganglion, which is retained in the adult brain (Nardelli-Haeflinger and Shankland 1993, Huang *et al.* 2002).

The development of cerebral ganglia from thickened epithelial areas of the episphere has been reported from several gastropods, e.g. *Crepidula* (Conklin 1897) and *Lymnaea* (Verdonk and van den Biggelaar

1983), and similarly from solenogasters (Okusu 2002), polyplacophorans (Henry *et al.* 2004) and bivalves (Meisenheimer 1901, Cragg and Crisp 1991) (review in Nielsen 2004).

In the sipunculans *Golfingia*, *Phascolopsis* and *Phascolion*, cells on the ventral side of the apical cells form a thickening which develops into the cerebral (supra-oesophageal) ganglia (Gerould 1906, Åkesson 1958, 1961). Wanninger *et al.* (2005) reported a few FMRFamidergic and serotonergic cells in the cerebral ganglia of early larvae of *Phascolion*.

Entoproct larvae have a paired (most loxosomatids) or single (colonial forms) frontal ganglion, which is usually associated with ciliated cells and sometimes with eyes together forming the frontal organ. The cell lineage has not been studied, but since nerves connect this organ both to the apical organ and to a prototroch nerve (Nielsen 1971) it is probably homologous to the cerebral ganglia of other spiralians. It is lost at metamorphosis.

Nemertine development is indirect through the characteristic planktotrophic pilidium or more or less 'direct' through a ciliated or unciliated, spindle-shaped to spherical larva. The indirect development is now quite well known through a number of classical studies and the modern study of *Cerebratulus lacteus* using the fluorescent tracer technique (Henry and Martindale 1998). The ectoderm of the anterior part of the juvenile worm inside the 'amniotic cavity' is formed from the two cephalic discs, which develop from the micromeres 1a and 1b. The cephalic ganglia develop from the ectoderm of the cephalic discs (Salensky 1912, Henry and Martindale 1998), i.e. from the episphere as in the other spiralians (Figure 21.3). The 'direct' developing species (and various intermediate types) are not so well known, and the cell lineage of the cerebral ganglia is unknown. However, development of cerebral ganglia from groups of cells lateral to the apical ganglion has been reported from a number of groups, e.g. *Procephalothrix*, *Tubulanus* and *Emplectonema* (Iwata 1960) and *Carinoma* (Maslakova *et al.* 2004a). It should be noted that the 'dorsal' and 'ventral' brain ganglia are preoral, encircling the proboscis, and have been reported to develop through differentiation from the paired cephalic ganglia (Friedrich 1979). Hay-Schmidt (1990) stained nerve cells in late pilidium larvae with a young juvenile inside. Concentrations of serotonergic processes were observed representing dorsal and ventral cerebral ganglia together surrounding the developing proboscis.

As mentioned above, the origin of the platyhelminth nervous system is poorly known. Surface (1907) reported that the cerebral ganglia of the polyclad *Hoploplana* should differentiate from descendants

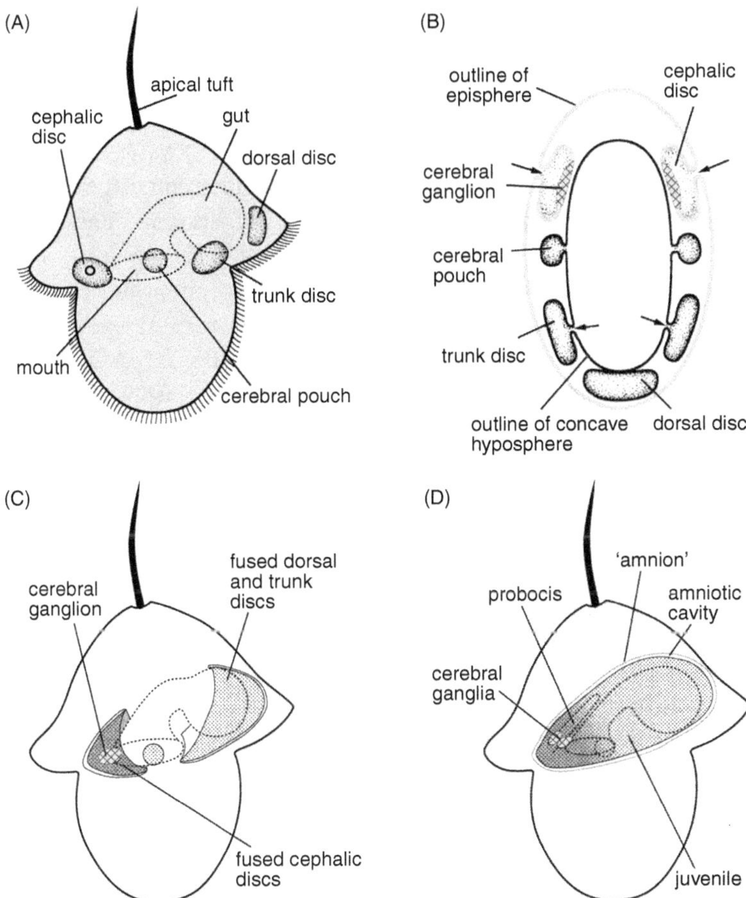

Figure 21.3 Development of the juvenile inside a pilidium larva. A, C and D, larvae seen from the left side; B is a schematic horizontal section passing through the openings of the ectodermal pouches with the imaginal discs. The ectoderm of the juvenile develops from the inner walls of the cephalic, trunk and dorsal discs whereas the cerebral pouches give rise only to the cerebral organs. The light grey colour indicates the whole episphere in A and B, and the cephalic imaginal discs in C and D. The brain-anlagen at the cephalic imaginal discs are cross-hatched. Based on Salensky (1912) and Henry and Martindale (1998).

of the cells $1a^{11221}$–$1d^{11221}$. This was questioned by Kato (1940) who studied a number of polyclads, both species with direct development and species with Müller's larvae, but his alternative interpretation is very difficult to follow. Lacalli (1982, 1983) studied newly hatched

Müller's larvae of *Pseudoceros* and found a well-defined cerebral ganglion with two paired nerve cords extending posteriorly; the short, anterolateral pair abutted the peripheral nerve along the frontal ciliated lobe and the longer posterolateral pair reached the same nerve at the lateral lobes. Boyer *et al.* (1998) re-studied the cell lineage of *Hoploplana* using tracer techniques and found that the eyes are differentiated from the 1a and 1c cells, but the brain was not observed. Scattered nerve cells originating from the first micromere quartet were observed in a pattern generally resembling the larval nervous system described by Lacalli (1982, 1983) for *Pseudoceros*; which in turn resembles the nervous system in the episphere of annelid larvae (Lacalli 1984).

Thus, the cerebral ganglia develop from the episphere and are retained as part of the adult brain in all well-studied species, except in the entoprocts, where the frontal ganglion is lost at metamorphosis.

Ventral nerves

Most of the spiralians have a pair of ventral nerves connected to the brain through perioral connectives. The embryological origin of these nerves is in most cases unknown, but a number of studies on annelid embryology show origin from the somatoblast, i.e. the 2d cell.

Cell lineage studies of a number of polychaetes, e.g. *Arenicola* (Child 1900), *Chaetopterus* (Henry and Martindale 1987) and *Platynereis* (Dorresteijn 1990, Ackermann *et al.* 2005), show that, at least in species with unequal cleavage, the descendants of the 2d cell cover almost the entire hyposphere, finally fusing ventrally along the fusing blastopore lips. A similar pattern was observed in the equally dividing *Polygordius* by Woltereck (1904). The paired ventral nerves differentiate from the ectoderm along the fusion line, and Dorresteijn (1998) demonstrated that the cholinergic cells of the ventral ganglia in 3-setiger larvae of *Platynereis* are descendants of the cells $2d^{1121}$ and $2d^{1122}$. The two posterior 'pioneer cells' descend from $2d^2$ and $2d^{12}$. The elegant cell labelling study of *Platynereis* by Ackermann *et al.* (2005: their Fig. 31) showed that cells of the 2d lineage extend pre-orally along the posterior side of the cerebral ganglia. With the fusion of the ventral nerves posterior to the mouth, a composite perioral brain is formed (see Figure 21.4). The leeches, such as *Helobdella*, are direct developers and their cleavage pattern is somewhat modified. The 1D cell divides into two large cells, the M cell, which gives rise to the mesoderm, and the NOPQ cell, which gives off three micromeres and then divides bilaterally giving rise to the two sides of the whole segmented ectoderm. Each of

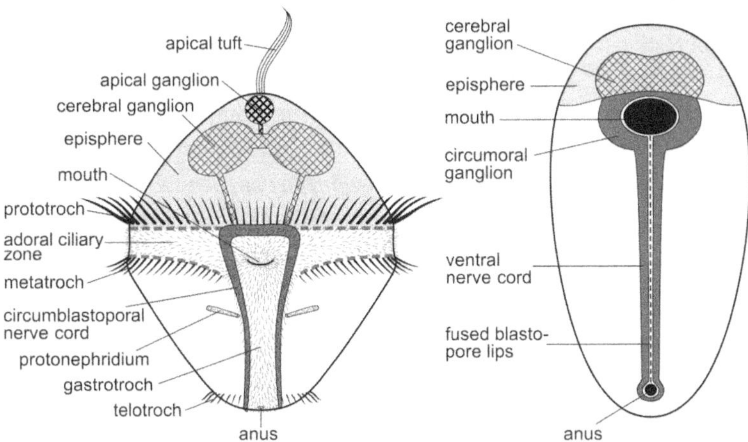

Figure 21.4 Larval and adult central nervous systems of a hypothetical ancestral spiralian with a planktotrophic trochophora larva. The grey shades and the two types of cross-hatching correspond to the similar areas in Figure 21.2.

the large NOPQ cells divides into a transverse row of four teloblasts, and by their anterior proliferation the two germinal bands elongate and spread laterally, so that they finally fuse along the ventral midline. Anteriorly, the two germ bands move around the small episphere with the mouth and brain (Nardelli-Haeflinger and Shankland 1993, Shankland and Savage 1997). Neuroblasts differentiate from all teloblasts, but the main parts of the ventral ganglia develop from cells of the median row of cells, called N (Weisblat *et al.* 1984). The anterior part of the germinal plate fills the ventral 'gap' in the horseshoe-shaped cerebral (supra-oesophageal) ganglion, together forming the circumoral brain.

The ventral nervous system of most molluscan groups is a complicated array of specialised ganglia, which in many cases fuse to form a complicated perioral loop of ganglia. Only the polyplacophorans have an unganglionated system; here Kowalevsky (1883) reported that the lateral and ventral nerves differentiate from the postoral ectoderm, but the anterior parts were not studied. The bivalve *Mytilus* develops a double row of lateroventral ganglia, pleural, parietal, visceral and pedal, from the ectoderm of the hyposphere (Raineri 1995), and a pre-oral commissure between the pleural ganglia gives the ventral nervous system a shape resembling that of the hypothetical ancestor shown in Figure 21.4. The development of the gastropod nerves is more difficult to follow, and most studies do not state the origin of the ganglia, but the general pattern can in most cases be interpreted as modified from that of *Mytilus* (several studies, e.g. Dickinson and Croll 2003, on *Ilyanassa*, and Hinman *et al.* 2003, on *Haliotis*). 'Pioneer cells' resembling those described from *Platynereis* have been found, for example in the gastropods *Crepidula* and *Ilyanassa* (Dickinson *et al.* 1999, Dickinson and Croll 2003).

In the sipunculans *Sipunculus*, *Golfingia*, *Phascolopsis* and *Phascolion*, the development of a median nerve cord from the ventral ectoderm was reported by Hatschek (1883), Gerould (1906) and Åkesson (1958), but the origin of the connections to the cerebral ganglia was not observed. In tear-drop stages of *Phascolion*, Wanninger *et al.* (2005) observed FMRFamide-positive cells forming a paired rudiment of the ventral nerve cords some distance behind the cerebral ganglia. During the following development, these two cell groups become more solid and grow anteriorly to reach and fuse with the cerebral ganglia and posteriorly to become paired components of the ventral nerve cord. Serotonergic cells form a delicate loop around the oesophagus from an area just behind the few serotonergic cells of the cerebral ganglia to a ventral area with a number of nuclei. In later stages, the posterior parts of this loop extend posteriorly as a paired, and more posteriorly unpaired, mid-ventral nerve cord.

The sessile entoprocts develop a dumbbell-shaped ventral ganglion after metamorphosis (Nielsen 1971), but its cell lineage has not been studied.

The origin of the ventral part of the central nervous system of nemertines is not well known and the old accounts are partially contradictory. In *Geonemertes*, Hickman (1963) reported that a pair of longitudinal nerves extending from the brain to the tail originates from thickenings of the adult ectoderm below the larval ectoderm, i.e. from the hyposphere. A late pilidium larva with a young juvenile inside stained for serotonin showed a nerve loop around the mouth confluent with the ventral cephalic nerve concentrations (ganglia and their commissure), posteriorly extending ventrolaterally almost to the posterior tip of the juvenile (Hay-Schmidt 1990). The shape and position of this bundle of nerve processes indicate that it represents the ventral nerve system.

Platyhelminths generally have a pair of longitudinal main nerves, but their origin is unknown (Reuter and Halton 2001).

HOX GENE EXPRESSION

There are surprisingly few reports of Hox gene expression in spiralians. To my knowledge, only a few annelids and molluscs have been studied.

The polychaete *Chaetopterus* has been studied by Irvine and Martindale (2000) and Peterson *et al.* (2000b). In the early larval stage, *Hox1* expression was seen in an area described as the foregut–midgut boundary, which corresponds to the blastopore lips around the anterior part of the blastopore, i.e. in the zone where the ventral nerves differentiate. In the middle larval stages, *Hox1* and *Hox2* expression was seen in the anterior parts of the ventral nerves, which are widely separated in this highly aberrant polychaete. All *Hox1-5* genes are expressed in the ventral nerves and the pygidium in the middle and late stages, and low concentrations of transcripts are found already at the early gastrula stage. No Hox gene expression was seen in the prostomium. In the leech *Helobdella*, Kourakis *et al.* (1997) found expression of *Hox1* (*Lox7*) throughout the segmented ventral nervous system, but not in the prostomium.

Hox genes have been identified in a number of gastropods and bivalves, and the bivalve *Pecten maximus* shows all the genes of the 'typical lophotrochozoan Hox cluster' (Canapa *et al.* 2005). However, only the studies of Giusti *et al.* (2000) and Hinman *et al.* (2003) on *Haliotis* deal with the gene expression. No Hox gene expression was found in pre-trochophore stages and in the trochophores, the Hox gene expression was restricted to the posttrochal ectoderm, i.e. the hyposphere. *Hox1* was expressed only in a ring along the shell gland in the trochophore, *Hox2* (and *Hox3* with low intensity) only in the pedal ganglion of the pre-torsional veliger, and *Hox3–5* in an antero-posterior temporal sequence in the pleural, oesophageal and branchial ganglia. *Hox4* further became expressed along the mantle edge of the competent veliger, indicating that some of the Hox genes also become involved in the organising of later developmental stages.

Nevertheless, these few studies indicate that Hox gene expression is limited to the hyposphere of the spiralian larvae, and this should make it possible to distinguish the cerebral component (without Hox gene expression) from the ventral component (with Hox expression) in the circumoral brains of dual origin.

CONCLUSIONS

The above discussions support the conclusions of Peterson *et al.* (2000a) and Peterson and Eernisse (2001) that the spiralians (Annelida, Mollusca, Sipuncula, Entoprocta, Nemertini and Platyhelminthes s.str.) form a monophyletic group having an ancestor with a trochophora larva. They were uncertain about the inclusion of the 'Lophophorata', but it should be emphasised once again that phoronids and brachiopods show none of the characters discussed here (except the larval apical ganglion which appears to be a eumetazoan synapomorphy). The trochophore is highly modified in many groups, for example in protobranch bivalves, in nemertines, both in types with a pilidium larva and with a 'direct' development, and in certain annelids, e.g. *Polygordius* (see reviews in Nielsen 2004, 2005a). However, the occurrence of (for example) pericalymma larvae of various types within genera (e.g. the polychaete *Polygordius* and the bivalve *Lyrodus*, see Nielsen 2001) where other members of the genus have unmodified trochophores clearly indicates that the trochophore can easily be modified in various directions.

Three components of the central nervous system (Figure 21.4) can be recognised with more or less certainty in all spiralian phyla.

The apical ganglion is found in almost all species having a ciliated larva, but its nature as a nervous centre is not well documented. It disappears before or at metamorphosis in all eumetazoans (perhaps with the exception of the ctenophores).

The origin of the cerebral ganglia from the episphere, separate from the apical ganglion is well documented in all phyla, and it is retained as an important component in the adult brain in all groups but the entoprocts. Hox gene expression has not been observed in the cerebral ganglia, and is probably lacking totally in the episphere.

The origin of the ventral nerves is less well documented, but the studies of the polychaete *Platynereis* demonstrate that the ventral nervous structures surround the mouth, extending posteriorly along the fused blastopore lips to the anus. The nervous system of the sipunculan *Phascolion* shows a morphological development which fits the predicted development. Its juvenile brain is indeed composed of the cerebral ganglia and anterior elements of the perioral nerve ring, i.e. the pre-oral part of the periblastoporal nerve ring, but the cell lineage is unfortunately unknown. Also, the gastropod *Haliotis* shows development of post-trochal ganglia with Hox gene expression.

The general picture which has emerged through this study will, it is hoped, inspire new studies of the development of nervous systems in spiralians (and other bilaterians), which combine observations on cell lineage, origin of ganglia and nerve cords, and gene expression.

ACKNOWLEDGEMENTS

My best thanks to Birgitte Rubæk (Copenhagen) and Giuseppe Fusco (Padua) for help with the illustrations.

REFERENCES

Ackermann, C., Dorresteijn, A. & Fischer, A. 2005. Clonal domains in post-larval *Platynereis dumerilii* (Annelida: Polychaeta). *Journal of Morphology* **266**, 258–280.

Åkesson, B. 1958. A study of the nervous system of the Sipunculoideae. *Undersökninger över Öresund* **38**, 1–249.

Åkesson, B. 1961. The development of *Golfingia elongata* Keferstein (Sipunculidea) with some remarks on the development of neurosecretory cells in sipunculids. *Arkiv för Zoologi*, 2. ser. **13**, 511–531.

Boyer, B. C., Henry, J. J. & Martindale, M. Q. 1998. The cell lineage of a polyclad turbellarian embryo reveals close similarity to coelomate spiralians. *Developmental Biology* **204**, 111–123.

Canapa, A., Biscotti, M. A., Olmo, E. & Barucca, M. 2005. Isolation of Hox and ParaHox genes in the bivalve *Pecten maximus*. *Gene* **348**, 83–88.

Child, C. M. 1900. The early development of *Arenicola* and *Sternaspis*. *Archiv für Entwicklungsmechanik der Organismen* **9**, 587–723, plates 21–25.

Conklin, E. G. 1897. The embryology of *Crepidula*. *Journal of Morphology* **13**, 1–226.

Cragg, S. M. & Crisp, D. J. 1991. The biology of scallop larvae. In S. E. Shumway (ed.) *Biology, Ecology and Aquaculture of Scallops*. Amsterdam: Elsevier, pp. 75–91.

Croll, R. P. & Dickinson, A. J. G. 2004. Form and function of the larval nervous system in molluscs. *Invertebrate Reproduction and Development* **46**, 173–187.

Dickinson, A. J. G. & Croll, R. P. 2003. Development of the larval nervous system of the gastropod *Ilyanassa obsoleta*. *Journal of Comparative Neurology* **466**, 197–218.

Dickinson, A. J. G., Nason, J. & Croll, R. P. 1999. Histochemical localization of FMRFamide, serotonin and catecholamines in embryonic *Crepidula fornicata* (Gastropoda, Prosobranchia). *Zoomorphology* **119**, 49–62.

Dohle, W. 1996. Spiralia. In W. Westheide (ed.) *Spezielle Zoologie, Teil 1: Einzeller und Wirbellose Tiere*, Stuttgart: Gustav Fischer, pp. 205–209.

Dorresteijn, A. W. C. 1990. Quantitative analysis of cellular differentiation during early embryogenesis of *Platynereis dumerilii*. *Roux's Archive of Developmental Biology* **199**, 14–30.

Dorresteijn, A. W. C. 1998. How do spiralian embryos accomplish cell diversity? *Zoology*, **100**, 307–319.

Emlet, R. B. & Strathmann, R. R. 1994. Functional occurrences of simple cilia in the mitraria of oweniids (an anomalous polychaete) and comparisons with other larvae. In W. H. Wilson, S. A. Stricker & G. L. Shinn (eds.) *Reproduction and Development of Marine Invertebrates*. Baltimore: Johns Hopkins University Press, pp. 143–157.

Friedrich, H. 1979. Nemertini. In F. Seidel (ed.) *Morphogenese der Tiere, Deskriptive Morphogenese*, 3. Lieferung. Jena: Gustav Fischer, pp. 1–136.

Friedrich, S., Wanninger, A., Brückner, M. & Haszprunar, G. 2002. Neurogenesis in the mossy chiton, *Mopalia muscosa* (Gould) (Polyplacophora): Evidence against molluscan metamerism. *Journal of Morphology* **253**, 109–117

Gerould, J. H. 1906. The development of *Phascolosoma*. *Zoologische Jahrbücher, Anatomie* **23**, 77–162.

Gifondorwa, D. J. & Leise, E. M. 2006. Programmed cell death in the apical ganglion during larval metamorphosis of the marine mollusc *Ilyanassa obsoleta*. *Biological Bulletin* **210**, 109–120.

Giusti, A. F., Hinman, V. F., Degnan, S. M., Degnan, B. M. & Morse, D. E. 2000. Expression of a *Scr/Hox5* gene in the larval central nervous system of the gastropod *Haliotis*, a non-segmented spiralian lophotrochozoan. *Evolution & Development* **2**, 294–302.

Hatschek, B. 1883. Über Entwicklung von *Sipunculus nudus*. *Arbeiten aus den Zoologischen Instituten der Universität Wien und der Zoologischen Station in Triest* **5**, 61–140.

Hay-Schmidt, A. 1990. Catecholamine-containing, serotonin-like and neuropeptide FMRFamide-like immunoreactive cells and processes in the nervous system of the pilidium larva (Nemertini). *Zoomorphology*, **109**, 321–244.

Henry, J. J. & Martindale, M. Q. 1987. The organizing role of the D quadrant as revealed through the phenomenon of twinning in the polychate *Chætopterus variopedatus*. *Roux's Archives of Developmental Biology* **196**, 499–510.

Henry, J. J. & Martindale, M. Q. 1998. Conservation of the spiralian developmental program: cell lineage of the nemertean, *Cerebratulus lacteus*. *Developmental Biology* **201**, 253–269.

Henry, J. J. & Martindale, M. Q. 1999. Conservation and innovation in spiralian development. *Hydrobiologia* **402**, 255–265.

Henry, J. Q., Okusu, A. & Martindale, M. Q. 2004. The cell lineage of the polyplacophoran *Chaetopleura apiculata*: variation in the spiralian program and implications for molluscan evolution. *Developmental Biology* **272**, 145–160.

Hickman, V. V. 1963. The occurrence in Tasmania of the land nemertine, *Geonemertes australiensis* Dendy, with some account of its distribution, habits, variations and development. *Papers and Proceedings of the Royal Society of Tasmania* **97**, 63–75.

Hinman, V. F., O'Brien, E. K., Richards, G. S. & Degnan, B. M. 2003. Expression of anterior *Hox* genes during larval development of the gastropod *Haliotis asinina*. *Evolution & Development* **5**, 508–521.

Huang, F. Z., Kang, D., Ramirez-Weber, F.-A., Bissen, S. T. & Weisblat, D. A. 2002. Micromere lineages in the glossiphoniid leech *Helobdella*. *Development* **129**, 719–732.

Irvine, S. Q. & Martindale, M. Q. 2000. Expression patterns of anterior Hox genes in the polychaete *Chaetopterus*: correlation with morphological boundaries. *Development* **217**, 333–351.

Iwata, F. 1958. On the development of the nemertean *Micrura akkeshiensis*. *Embryologia* **4**, 103–131.

Iwata, F. 1960. Studies on the comparative embryology of nemerteans with special reference to their interrelationships. *Publications from the Akkeshi Marine Biological Station* **10**, 1–51.

Kato, K. 1940. On the development of some Japanese polyclads. *Japanese Journal of Zoology* **8**, 537–573.

Kourakis, M. J., Master, V. A., Lokhorst, D. K. *et al.* 1997. Conserved anterior boundaries of Hox gene expression in the central nervous system of the leech *Helobdella*. *Developmental Biology* **190**, 284–300.

Kowalevsky, M. A. 1883. Embryogénie du *Chiton polii* (Philippi). *Annales du Musée d'Histoire naturelle de Marseille, Zoologie* **1**(5), 1–46, 8 plates.

Lacalli, T. C. 1981. Structure and development of the apical organ in trochophores of *Spirobranchus polycerus*, *Phyllodoce maculata* and *Phyllodoce mucosa* (Polychaeta). *Proceedings of the Royal Society of London B* **212**, 381–402.

Lacalli, T. C. 1982. The nervous system and ciliary band of Müller's larva. *Proceedings of the Royal Society of London B* **217**, 37–58.

Lacalli, T. C. 1983. The brain and central nervous system of Müller's larva. *Canadian Journal of Zoology* **61**, 39–51.

Lacalli, T. C. 1984. Structure and organization of the nervous system in the trochophore larva of *Spirobranchus*. *Philosophical Transactions of the Royal Society B* **306**, 79–135.

Lacalli, T. C. & West, J. E. 1985. The nervous system of a pilidium larva: evidence from electron microscope reconstructions. *Canadian Journal of Zoology* **63**, 1909–1916.

Leise, E. M., Kempf, S. C., Durham, N. R. & Gifondorwa, D. J. 2004. Induction of metamorphosis in the marine gastropod *Ilyanassa obsoleta*: 5HT, NO and programmed cell death. *Acta Biologica Hungarica* **55**, 293–300.

Malakhov, V. V. 1990. Description of the development of *Ascopodaria discreta* (Coloniales, Barentsiidae) and discussion of the Kamptozoa status in the animal kingdom. *Zoologicheskii Zhurnal* **69**, 20–30.

Marcus, E. 1939. Briozoarios marinhos brasileiros III. *Boletim da Faculdade de Filosofia, Ciências e Letras, Universidade de Sao Paulo, Zoologia* **3**, 111–354.

Marois, R. & Carew, T. J. 1997. Fine structure of the apical ganglion and its serotonergic cells in the larva of *Aplysia californica*. *Biological Bulletin* **192**, 388–398.

Maslakova, S. A., Martindale, M. Q. & Norenburg, J. L. 2004a. Vestigial prototroch in a basal nemertean, *Carinoma tremaphoros* (Nemertea; Palaeonemertea). *Evolution & Development* **6**, 219–226.

Maslakova, S. A., Martindale, M. Q. & Norenburg, J. L. 2004b. Fundamental properties of the spiralian development program are displayed by the basal nemertean *Carinoma tremaphoros* (Palaeonemertea, Nemertea). *Developmental Biology* **267**, 342–360.

Meisenheimer, J. 1901. Entwicklungsgeschichte von *Dreissensia polymorpha* Pall. *Zeitschrift für Wissenschaftliche Zoologie* **69**, 1–137, plates 1–13.

Nardelli-Haeflinger, D. & Shankland, M. 1993. *Lox10*, a member of the *NK-2* homeobox gene class, is expressed in a segmental pattern in the endoderm and in the cephalic nervous system of the leech *Helobdella*. *Development* **118**, 877–892.

Nielsen, C. 1971. Entoproct life-cycles and the entoproct/ectoproct relationship. *Ophelia* **9**, 209–341.

Nielsen, C. 2001. Animal Evolution: Interrelationships of the Living Phyla. 2nd edn. Oxford: Oxford University Press.

Nielsen, C. 2004. Trochophora larvae: cell-lineages, ciliary bands and body regions. 1. Annelida and Mollusca. *Journal of Experimental Zoology B (Molecular and Developmental Evolution)* **302**, 35–68.

Nielsen, C. 2005a. Trochophora larvae: cell-lineages, ciliary bands and body regions. 2. Other groups and general discussion. *Journal of Experimental Zoology B (Molecular and Developmental Evolution)* **304**, 401–447.

Nielsen, C. 2005b. Larval and adult brains. *Evolution & Development* **7**, 483–489.

Okusu, A. 2002. Embryogenesis and development of *Epimenia babai* (Mollusca Neomeniomorpha). *Biological Bulletin* **203**, 87–103.

Page, L. R. 2002. Apical sensory organ in larvae of the patellogastropod *Tectura scutum*. *Biological Bulletin* **202**, 6–22.

Page, L. R. & Parries, S. C. 2000. Comparative study of the apical ganglion in planktotrophic caenogastropod larvae: ultrastructure and immunoreactivity to serotonin. *Journal of Comparative Neurology* **418**, 383–401.

Peterson, K. J., Cameron, R. A. & Davidson, E. H. 2000a. Bilaterian origins: significance of new experimental observations. *Developmental Biology* **219**, 1–17.

Peterson, K. J. & Eernisse, D. J. 2001. Animal phylogeny and the ancestry of bilaterians: inferences from morphology and 18S rDNA sequences. *Evolution & Development* **3**, 170–205.

Peterson, K. J., Irvine, S. Q., Cameron, R. A. & Davidson, E. H. 2000b. Quantitative assessment of *Hox* complex expression in the direct development of the polychaete annelid *Chaetopterus* sp. *Proceedings of the National Academy of Sciences of the USA* **97**, 4487–4492.

Raineri, M. 1995. Is a mollusc an evolved bent metatrochophore? A histochemical investigation of neurogenesis in *Mytilus* (Mollusca: Bivalvia). *Journal of the Marine Biological Association of the United Kingdom* **75**, 571–592.

Rattenbury, J. C. 1954. The embryology of *Phoronopsis viridis*. *Journal of Morphology* **95**, 289–349.

Reuter, M. & Halton, D. W. 2001. Comparative neurobiology of Platyhelminthes. In D. T. J. Littlewood & R. A. Bray (eds.) *Interrelationships of the Platyhelminthes* (Systematics Association Special Volume 60). London: Taylor and Francis, pp. 239–249.

Riedl, R. J. 1969. Gnathostomulida from America. *Science* **163**, 445–462.

Ruppert, E. E. 1978. A review of metamorphosis of turbellarian larvae. In F. S. Chia (ed.) *Settlement and Metamorphosis of Marine Invertebrate Larvae*. New York: Elsevier, pp. 65–81.

Salensky, W. 1912. Über die Metamorphose der Nemertinen. I. Entwicklungsgeschichte der Nemertine im Inneren des Pilidiums. *Mémoires de l'Académie impériale des Sciences de St.-Pétersbourg, 8. série, Classe physico-mathématique* **30** (10), 1–74.

Shankland, M. & Savage, R. M. 1997. Annelids, the segmented worms. In S. F. Gilbert & A. M. Raunio (eds.) *Embryology. Constructing the Organism*. Sunderland: Sinauer Associates, pp. 219–235.

Surface, F. M. 1907. The early development of a polyclad, *Planocera inquilina* Wh. *Proceedings of the Academy of Natural Sciences Philadelphia* **59**, 514–559, plates 35–40.

Tardy, J. & Dongard, S. 1993. Le complexe apical de la véligère du *Ruditapes philippinarum* (Adams et Reeve, 1850) Mollusque Bivalve Vénéridé. *Comptes Rendus de l'Académie des Sciences, Sciences de la Vie* **316**, 177–184.

Treadwell, A. L. 1901. Cytogeny of *Podarke obscura* Verrill. *Journal of Morphology* **17**, 199–486, plates 36–40.

van Dongen, C. A. M. & Geilenkirchen, W. L. M. 1974. The development of *Dentalium* with special reference to the significance of the polar lobe. I–III. Division chronology and development of the cell pattern in *Dentalium dentale* (Scaphopoda). *Proceedings of the Koninklijke Nederlandse Akademie van Wetenschappen, Series C* **77**, 57–100.

Verdonk, N. H. & van den Biggelaar, J. A. M. 1983. Early development and the formation of the germ layers. In K. M. Wilbur (ed.) *The Mollusca*, Vol. 3, New York: Academic Press, pp. 91–122.

Voronezshskaya, E. E., Tyurin, S. A. & Nezlin, L. P. 2002. Neuronal development in larval chiton *Ischnochiton hakodakensis* (Mollusca: Polyplacophora). *Journal of Comparative Neurology* **444**, 25–38.

Wanninger, A., Koop, D., Bromham, L., Noonan, E. & Degnan, B. M. 2005. Nervous and muscle system development in *Phascolion strombus* (Sipuncula). *Development, Genes & Evolution* **215**, 509–518.

Weisblat, D. A., Kim, S. Y. & Stent, G. S. 1984. Embryonic origins of cells in the leech *Helobdella triserialis*. *Developmental Biology* **104**, 65–85.

Woltereck, R. 1902. Trochophora-Studien I. Histologie der Larve und die Entstehung des Annelids bei den *Polygordius*-Arten der Nordsee. *Zoologica (Stuttgart)* **13**(34), 1–71, 11 plates.

Woltereck, R. 1904. Beiträge zur praktischen Analyse der *Polygordius*-Entwicklung nach dem 'Nordsee-' und dem 'Mittelmeer-Typus'. *Archiv für Entwicklungsmechanik der Organismen* **18**, 377–403.

Zimmer, R. L. 1991. Phoronida. In A. C. Giese (ed.) *Reproduction of Marine Invertebrates*, Vol. 6. Pacific Grove, CA: Boxwood Press, pp. 1–45.

Index